TOPICS FOR A STATISTICAL DESCRIPTION OF RADAR CROSS SECTION

WILEY SERIES IN REMOTE SENSING

Jin Au Kong, Editor

Tsang, Kong, Shin • THEORY OF MICROWAVE REMOTE SENSING
Hord • REMOTE SENSING: METHODS AND APPLICATIONS
Elachi • INTRODUCTION TO THE PHYSICS AND TECHNIQUES OF REMOTE SENSING
Szekielda • SATELLITE MONITORING OF THE EARTH
Maffett • TOPICS FOR A STATISTICAL DESCRIPTION OF RADAR CROSS SECTION
Asrar • THEORY AND APPLICATIONS OF OPTICAL REMOTE SENSING

TOPICS FOR A STATISTICAL DESCRIPTION OF RADAR CROSS SECTION

Andrew Lewis Maffett

WILEY

A WILEY-INTERSCIENCE PUBLICATION

JOHN WILEY & SONS

New York • Chichester • Brisbane • Toronto • Singapore

Library of Congress Cataloging in Publication Data:

Maffett, Andrew Lewis.
  Topics for a statistical description of radar cross section.

  (Wiley series in remote sensing)
  "A Wiley-Interscience publication."
  Bibliography: p.
  Includes index.
  1. Radar—Statistical methods. 2. Radar—Mathematical models. I. Title. II. Series
TK6580.M34 1989 621.3848 88-20698
ISBN 0-471-61357-6

Printed in the United States of America

10 9 8 7 6 5 4 3 2 1

To Pauline

CONTENTS

PREFACE

This book treats the subject of radar cross section (RCS), with special emphasis on the statistical aspects and applications thereof. To that purpose, it is organized into four parts, of which the third part forms the central core:

I. Fundamentals including notation, definitions, and formulas,
II. Scattering and diffraction of electromagnetic waves; approximations from asymptotic and numerical methods,
III. Practical RCS analyses and applications using statistical methods, and
IV. The role of constitutive parameters in RCS.

The material is arranged so that each part consists of five chapters, each of which is designed to occupy a two-hour lecture period. Thus, the 20 chapters can be used for a standard 40-hour course of lectures. The material has, in fact, been used twice in this manner for groups of working engineers; so it has benefited considerably from constructive comment and criticism offered by the listeners.

Chapters 1 through 5 review and establish the notation, the basic definitions and theorems of electromagnetic theory. Chapters 6 through 10 present and discuss some results of asymptotic approximation theories as well as numerical and measurement methods. In Chapters 11 through 15 these RCS estimation theories are combined with statistical ideas to show that the RCS of a complex target is β-distributed (Chapter 12). This salient result is verified with measured data in Chapter 13 and is then applied, in Chapter 15, to the subject of detection theory.

All of Chapters 1 through 15 are develped under the assumptions that (1) the medium of propagation is free space and (2) the scattering bodies are perfect conductors (i.e., metallic). Chapters 16 through 20, in contrast, look at the nature of results that can be established when assumptions 1 and 2 are lifted.

Each chapter has its own equation and figure numbers beginning with unity.

This work is aimed at RCS estimation for complex targets, based largely on statistical methods. Siegel, beginning in the 1950s, promoted the use of first order statistics of RCS (mean and variance), described by Crispin and Maffett (1965), to this purpose. In the 1980s, Follin used higher order statistics of RCS to show that the RCS of a complex target can be described, in the neighborhood of some fixed but arbitrary aspect, by a β-distribution; this idea was further developed by Follin and Maffett (1981), and by Follin, Paddison, and Maffett (1984) and is expanded with additional detail and examples (especially as regards detection theory and the comparison of RCS data) in Chapters 11 through 15 of this work.

ANDREW LEWIS MAFFETT

Dexter, Michigan
November 1988

ACKNOWLEDGMENTS

I wish to express my sincere gratitude to David E. Stein, who read my manuscript with tireless care. He performed the yeoman task of checking my equations and made a host of timely and useful suggestions. Improvements in accuracy and consistency of the text are due to him; any and all remaining errors, however, are my own.

Many of my graphs were made with a graphics code, GRAPH, formulated by Herbert W. Klimach and David C. Ager, who kindly made it available to me for experimentation. I thank both of them for their helpful tutelage. In addition, I acknowledge the coding expertise of the former, who programmed the RCS code, without which the RCS statistics study would not have been possible.

Further, I wish to express my pleasure and appreciation for the privilege of working on the RCS statistics development with Fletcher C. Paddison and James W. Follin, Jr. at The Johns Hopkins University Applied Physics Laboratory in the early 1980s.

Last, but by no means least, I am grateful to Robert A. Shuchman for his encouragement to commence this project; to many others of my colleagues and friends who listened to me with patience and offered judicious and valuable suggestions; and to the Wiley editorial staff for their unfailingly enthusiastic and helpful efforts on my behalf.

TOPICS FOR A STATISTICAL DESCRIPTION OF RADAR CROSS SECTION

HISTORICAL BACKGROUND AND PERSPECTIVE

The development of any field of human endeavor is, by and large, the historical, or more precisely, the biographical record of contributors to that field. RCS development is no different from any other field. And yet it *is* different, in that many of the fundamental tenets of the subject were established long before (in some cases, centuries before) the advent of radar and the concept of RCS.

For these reasons we cite, beginning with Empedocles and Euclid, some of the outstanding intellects in, together with their contributions to, optics and, more generally, electromagnetics, who laid the foundations for the concept of RCS. The cast has some interesting characters: for example, Fermat, who is remembered for his Last Theorem (number theory); Hooke, who is known for his theory of springs; and Bunsen, whose gas burner is familiar to every student of elementary chemistry.

There have been some interesting twists of fashion regarding an acceptable description of the propagation and scattering of electromagnetic energy. The great Newton promoted the corpuscular theory of light based on particle propagation through an aether. His position appeared to be untenable when Fresnel showed that diffraction phenomena could be explained by wave theory. But particle theory regained its validity when Einstein published his results on photon propagation. Today, both wave and particle theories are required to explain all of the observed phenomena of electromagnetic energy propagation and scattering.

The inventions and discoveries of electromagnetic theory go side by side with noteworthy events occurring elsewhere in the world. For example, Laplace was an official in Napoleon's government as well as in the government of Louis XVI. Ampère and Faraday corresponded about their work in magnetic and electric induction in spite of the hostile attitudes of their countries (the latter being at the height of his career when gold was discovered in California). Newton was Master of the Mint at the time Captain Kidd was stashing his treasure. And so on through a number of such events that are included to show the state of the world when developments of interest here were made.

The reader may wish to choose his own list of entries from modern time. I have selected individuals whose contributions have done much to advance the understanding of the theory, application, and measurement of RCS.

Preradar People Whose Ideas Contributed to RCS Concepts

EMPEDOCLES (490–430 B.C.) EUCLID (c. 300 B.C.)	First *systematic* writings on OPTICS
DESCARTES (1596–1650)	Light a pressure transmitted through a perfectly elastic medium (aether)
SNELL (1591–1626)	Law of *refraction* (experimentally)
FERMAT (1601–1665)	Principle of *least times* (nature always acts by the shortest course.)
BOYLE (1627–1691) HOOKE (1635–1703) GRIMALDI (1618–1663)	Phenomenon of *interference* (Newton's rings); *diffraction* (light in shadowed region)
NEWTON (1642–1727)	*Refrangibility* (white light into colors by a prism); *emission* (or *corpuscular*) theory
RÖMER (1644–1710)	Speed of light (from eclipses of Jupiter's satellites)
HUYGENS (1629–1695)	*Wave* theory; *polarization* (from refraction studies)
FRANKLIN (1706–1790)	Practical studies of electrical phenomena
EULER (1707–1783)	Wave theory supporter (transversality)
COULOMB (1736–1806)	Attraction of charges
LAGRANGE (1736–1813)	Variational calculus and analytical mechanics
YOUNG (1773–1829)	*Interference* (first steps toward acceptance of wave theory)
VOLTA (1745–1827)	Electric circuits
LAPLACE (1749–1827)	Emission theory supporter
LEGENDRE (1752–1833)	Legendre functions (early text on integral calculus)
AMPÈRE (1775–1836)	Law of *magnetic induction*
BESSEL (1784–1846)	Bessel functions
FRESNEL (1788–1827)	Treatment of *diffraction* by *wave theory*

ARAGO (1786–1853)	Showed, together with Fresnel, that two rays polarized at right angles never interfere (which could not be reconciled with the existence of longitudinal waves)
OHM (1787–1854)	Resistance
NAVIER (1785–1836) CAUCHY (1789–1857)	Elastic media (aether)
FARADAY (1791–1867)	Law of *electric induction*
HENRY (1797–1878)	Inductance
RIEMANN (1826–1866)	Complex variables
POISSON (1781–1840)	Prediction of bright spot behind disk; verified experimentally by Arago
GREEN, G. (1793–1841)	Mathematical theory of electromagnetism (1828); chose phrase "potential function" for solution of Laplace's differential equation
JOULE (1818–1889) KELVIN (Wm. Thomson) (1824–1908)	Connections with heat
RAYLEIGH (John Wm. Strutt) (1842–1919)	*Low-frequency theory*
DIRICHLET (1805–1859) NEUMANN (1832–1925)	*Boundary value* formulations
MAXWELL (1831–1879)	Showed possibility of EM waves propagated with a velocity that could be calculated from electrical measurements
KOHLRAUSCH (1809–1858) WEBER (1804–1891)	Carried out these measurements and found the velocity to be that of *light.* Maxwell then conjectured that *light waves* are *electromagnetic waves*
HERTZ (1857–1894)	Verified Maxwell's conjecture that light waves are electromagnetic waves
FRAUNHOFER (1787–1826)	Dark lines in solar spectrum
BUNSEN (1811–1899)	Found these lines to be absorption lines
HELMHOLTZ (1821–1894)	Wave theory
KIRCHHOFF (1824–1887)	Circuit theory laws
MINKOWSKI (1864–1909)	Mathematical foundation for Einstein's relativity
CLEBSCH (1833–1872)	Solved boundary value problem of a wave in an elastic medium impinging on a sphere (1861: pre-Maxwell, 1864)
LORENZ (1829–1891)	Reflection and refraction of light by a transparent sphere

HEAVISIDE (1850–1925)	Operator theory
MICHELSON (1852–1931)	Aether; speed of light
BROMWICH (1875–1929)	Scattering and diffraction of a plane wave by a sphere
DOPPLER (1803–1853)	Motion of sources

The Appearance of Radar

HULSMEYER 1904 patent	Obstacle detector and navigational aid for ships
MARCONI (c. 1922)	Urged use of short waves for radio detection
BREIT & TUVE (c. 1925)	Measured height of ionosphere using pulse ranging
WATSON–WATT 1934/5:	Plan for detection of aircraft by pulse method
1936:	Five stations (about 25 miles apart) set up on the Thames
1938:	Signal Corps Labs designed and built position-finding equipment for control of AA guns and searchlights
1939:	NRL installed radar on USS New York
1940:	British developed multicavity magnetron as a practical source of pulsed microwave energy
1940:	*RAdio Detection And Ranging* = *RADAR* code word adopted by U.S. Navy
DuBRIDGE	MIT Radiation Lab., 1940–1945. Established by the Microwave Subcommittee of the National Defense Research Committee
1941:	Aircraft interception model delivered to the Army Air Corps; a sea search radar was delivered to the Navy and installed on the U.S.S. Semmes

By 1945, under the direction of the Radiation Laboratory, $27 billion worth of radar equipment had been delivered to the U.S. Army and Navy.

The RCS Concept

PURCELL, E. M. (1912–) KERR, D. E.	Radar systems

GOLDSTEIN, Herbert	The concept of RCS (c. 1942 or 1943)
WATSON, G. N.	Contour integral representations
LUNEBERG, R. K. KLINE, Morris (1908–)	Low-frequency methods
FOCK, V. A. (1898–)	Creeping waves
LEONTOVICH, M. A.	Surface (impedance) boundary condition
SILVER, S. (1915–1974)	Antenna theory
SIEGEL, K. M. (1923–1976)	Development of practical RCS methods
KELLER, J. B. (1923–)	Geometrical theory of diffraction
UFIMTSEV, P. Y.	Physical theory of diffraction
OSHIRO, F. K. (1923–)	Numerical methods
HARRINGTON, R. F. (1925–)	Method of Moments
STRATTON, J. A. (1901–)	Treatise on electromagnetic theory
PETERS, L., Jr. (1923–)	Equivalent current method; traveling waves

Historical Perspective†

1558	Queen Elizabeth I succeeds to British throne
1564	Shakespeare born
1588	Spanish Armada
1620	Pilgrims land at Plymouth Rock
1643	Louis XIV succeeds to French throne
1658	Cromwell dies
1685	Bach born
1698	Captain Kidd buries treasure
1722	Peter the Great attacks Persia
1732	Haydn born
1756	Mozart born
1770	Beethoven born
1776	Declaration of Independence
1789	Washington elected first president; French revolution
1799	Napoleon rules France
1803	Louisiana purchase ($12M)
1818	Marx born
1845	President Tyler annexes Texas
1846	War with Mexico (over California, Polk president) 1850: California becomes state
1848	Gold discovered (Sacramento valley)

† Samuel Eliot Morison, 1965, *The Oxford History of the American People,* Oxford University Press, New York.

1864	Civil War ends
1867	Alaska purchase (Seward's folly; A. Johnson president)
1876	Bell invents telephone (1885: 135,000 subscribers)
1879	Edison invents light bulb
1883	T. Roosevelt invests in Bad Lands ranch; buffalo nearly extinct
1884	Railroads reach west coast (Northern, Central, Union Pacific)
1888	Electric streetcars
1898	Spanish–American war
1903	Wright brothers flight (T. Roosevelt president)
1914	World War I
1924	Lenin dies; Stalin succeeds
1927	Lindbergh's flight (Paris 33 hours 39 minutes)
1929	Stock market crash (followed by Great Depression)
1933	National Industrial Recovery Act (The New Deal—F. Roosevelt president)
1939	Hitler attacks Poland; World War II begins
1941	Pearl Harbour; the Battle of Britain
1944	Invasion of Europe
1945	Hiroshima; Yalta
1946	United Nations
1948	Marshall Plan
1950	Korean War
1952	Stalin dies
1954	Salk vaccine
1957	Sputnik
1959	Castro
1962	Cuban missile crisis
1963	Vietnam
1969	Armstrong lands on moon

PART I

FUNDAMENTALS

In Part I we review the fundamental facts necessary to the study of RADAR CROSS SECTION (RCS): some general mathematical matters and an introduction to electromagnetic fields in Chapter 1; the constitutive composition of scattering bodies in Chapter 2; the boundary conditions that must be satisfied by an electromagnetic field at the surface of a scatterer in Chapter 3; Green's functions, Huygens's principle, and their relation with equivalent sources in Chapter 4; and, finally, in Chapter 5, a definition of the subject of this text: RCS.

There are many good and more detailed discussions of these basics to be found in other works. We mention a few favorites: For mathematical definitions, theorems, and formulas: Magnus and Oberhettinger, 1954; Morse and Feshbach, I, II, 1953; Fulks, 1969; and Churchill and Brown, 1984. For electromagnetic theory and RCS: Stratton, 1941; Bowman, Senior, and Uslenghi, 1969; Crispin and Siegel, 1968; Kong, 1975; and Knott, Shaeffer, and Tuley, 1986.

1

DEFINITION OF FIELD QUANTITIES AND MAXWELL'S EQUATIONS

We assume that the reader is familiar with the sets of real and complex numbers and the operations that may be defined over them. We refer to these sets as scalars.

Vectors can be defined, then, as arrays of scalars. For example, let $\hat{x}$, $\hat{y}$, $\hat{z}$ represent mutually orthogonal rectangular unit (denoted by a caret) vectors. An arbitrary vector $\mathbf{v}$ can be written as a linear combination of these unit vectors:

$$\mathbf{v} = v_x\hat{x} + v_y\hat{y} + v_z\hat{z}, \tag{1-1}$$

where $v_{x,y,z}$ represent components of $\mathbf{v}$ in the $\hat{x}$, $\hat{y}$, $\hat{z}$-directions, respectively. Employing matrix multiplication, Eq. 1-1 may be rewritten as a product of arrays:

$$\mathbf{v} = (v_x,\ v_y,\ v_z)\begin{pmatrix} \hat{x} \\ \hat{y} \\ \hat{z} \end{pmatrix}. \tag{1-2}$$

Notice that it would be just as convenient to write the row of scalars v_x, v_y, v_z as a column and the column of unit vectors $\hat{x}$, $\hat{y}$, $\hat{z}$ as a row; to do so we need only exchange the order of multiplication. We say that the row (v_x, v_y, v_z), or the column $\begin{pmatrix} v_x \\ v_y \\ v_z \end{pmatrix}$, forms a representation for $\mathbf{v}$.

We review quickly the operations of scalars on vectors and vectors on vectors. Let $\mathbf{v}$ and $\mathbf{w}$ be two vectors and a and b scalars (these may be either real or complex and the representations of the vectors may also have real or complex entries). Then define

Scalar multiplication and vector addition:

$$a\mathbf{v} + b\mathbf{w} = (av_x + bw_x,\ av_y + bw_y,\ av_z + bw_z); \tag{1-3}$$

Dot (or scalar) multiplication:

$$\mathbf{v} \cdot \mathbf{w} = (v_x,\ v_y,\ v_z)(w_x,\ w_y,\ w_z)' \tag{1-4}$$

where the prime denotes the transpose of a matrix:

$$= v_x w_x + v_y w_y + v_z w_z;$$

Vector (or cross-) multiplication:

$$\mathbf{v} \times \mathbf{w} = \det \begin{bmatrix} \hat{x} & \hat{y} & \hat{z} \\ v_x & v_y & v_z \\ w_x & w_y & w_z \end{bmatrix}, \tag{1-5}$$

where det denotes the determinant operation on a matrix,

$$= \hat{x}(v_y w_z - v_z w_y) + \hat{y}(v_z w_x - v_x w_z) + \hat{z}(v_x w_y - v_y w_x)$$

Note that the first product yields a vector, the second a scalar, and the third a vector. The vector product can be defined in terms of matrix (or second rank tensor) operations and we shall illustrate this representation below when we review differential operators.

It is useful to have a polar as well as a rectangular system. We define

$$\hat{r} = \hat{x} \sin\theta \cos\phi + \hat{y} \sin\theta \sin\phi + \hat{z} \cos\theta,$$
$$\hat{\theta} = \hat{x} \cos\theta \cos\phi + \hat{y} \cos\theta \sin\phi - \hat{z} \sin\theta,$$
$$\hat{\phi} = \hat{r} \times \hat{\theta} = -\hat{x} \sin\phi + \hat{y} \cos\phi;$$

or, as a matrix operation,

$$\begin{pmatrix} \hat{r} \\ \hat{\theta} \\ \hat{\phi} \end{pmatrix} = \begin{pmatrix} \sin\theta\cos\phi & \sin\theta\sin\phi & \cos\theta \\ \cos\theta\cos\phi & \cos\theta\sin\phi & -\sin\theta \\ -\sin\phi & \cos\phi & 0 \end{pmatrix} \begin{pmatrix} \hat{x} \\ \hat{y} \\ \hat{z} \end{pmatrix} = T \begin{pmatrix} \hat{x} \\ \hat{y} \\ \hat{z} \end{pmatrix}. \tag{1-6}$$

Notice first that $T' = T^{-1}$ so that

$$\begin{pmatrix} \hat{x} \\ \hat{y} \\ \hat{z} \end{pmatrix} = \begin{pmatrix} \sin\theta\cos\phi & \cos\theta\cos\phi & -\sin\phi \\ \sin\theta\sin\phi & \cos\theta\sin\phi & \cos\phi \\ \cos\theta & -\sin\theta & 0 \end{pmatrix} \begin{pmatrix} \hat{r} \\ \hat{\theta} \\ \hat{\phi} \end{pmatrix} = T' \begin{pmatrix} \hat{r} \\ \hat{\theta} \\ \hat{\phi} \end{pmatrix}. \tag{1-7}$$

We call this feature of T orthogonality; indeed, orthonormality, since $TT' = I$, the identity matrix. The nine matrix entries of Eq. 1-6 may be identified with the direction cosines of the unit vector $\hat{r}$. Then notice also that $\hat{\theta} = (d/d\theta)\hat{r}$.

The Pythagorean theorem gives the square of the length of $\mathbf{v}$:

$$\|\mathbf{v}\|^2 = v_x^2 + v_y^2 + v_z^2 = \mathbf{v} \cdot \mathbf{v}. \tag{1-8}$$

We may of course write $\mathbf{v}$ in the polar system as well as the rectangular

$$\mathbf{v} = (v_x, v_y, v_z)\begin{pmatrix}\hat{x}\\ \hat{y}\\ \hat{z}\end{pmatrix} = (v_r, v_\theta, v_\phi)\begin{pmatrix}\hat{r}\\ \hat{\theta}\\ \hat{\phi}\end{pmatrix}, \tag{1-9}$$

whence

$$(v_x, v_y, v_z) = (v_r, v_\theta, v_\phi)T \tag{1-10}$$

and from the orthogonality of T,

$$\|\mathbf{v}\|^2 = v_r^2 + v_\theta^2 + v_\phi^2. \tag{1-11}$$

We remarked above that the vector components can be complex numbers. This situation arises in the so-called phasor notation. Let $\mathbf{v}(t) = \hat{x}v_x \cos(\omega t + \phi_x) + \hat{y}v_y \cos(\omega t + \phi_y) + \hat{z}v_z(\omega t + \phi_z)$, where t may stand for time and ω for (circular) frequency; the amplitudes $v_{x,y,z}$ and phases $\phi_{x,y,z}$ are arbitrary real numbers. Making use of facts about complex numbers, we write $\mathbf{v}(t)$ as

$$\begin{aligned}\mathbf{v}(t) &= \mathrm{Re}\{[\hat{x}v_x \exp(j\phi_x) + \hat{y}v_y \exp(j\phi_y) + \hat{z}v_z \exp(j\phi_z)] \exp(j\omega t)\}\\ &= \mathrm{Re}\{\mathbf{v} \exp(j\omega t)\},\end{aligned} \tag{1-12}$$

where Re means "real part of," $j = \sqrt{-1}$, and $\exp(A)$ stands for the exponential e^A.

Time averages can be defined for each component of $\mathbf{v}(t)$ as

$$\begin{aligned}\langle v_x(t)\rangle &= \left(\frac{1}{T}\right)\int_0^T v_x(t)\,dt = \left(\frac{1}{T}\right)\mathrm{Re}\left\{v_x \int_0^T e^{j\omega t + \phi_x}\,dt\right\}\\ &= 0 \quad \text{if} \quad T = \frac{2\pi}{\omega}\end{aligned} \tag{1-13}$$

(or, if $T = 1/f$, where $f = \omega/2\pi$ is frequency in Hz). Then

$$\langle v_x^2(t)\rangle = \left(\frac{v_x^2}{T}\right)\int_0^T \cos^2(\omega t + \phi_x)\,dt = \frac{1}{2}v_x^2. \tag{1-14}$$

Examples. Let $\mathbf{v}(t) = \mathrm{Re}\{\mathbf{v}e^{j\omega t}\} = \mathbf{v}_R \cos(\omega t) - \mathbf{v}_I \sin(\omega t)$, where $\mathbf{v} = \mathbf{v}_R + j\mathbf{v}_I$; and let $\mathbf{w}(t)$ be defined in a similar manner.

Example 1

$$\begin{aligned}\mathbf{v}(t) \times \mathbf{w}(t) = \mathbf{v}_R \times \mathbf{w}_R \cos^2 \omega t + \mathbf{v}_I \times \mathbf{w}_I \sin^2 \omega t\\ -\frac{1}{2}(\mathbf{v}_R \times \mathbf{w}_I + \mathbf{v}_I \times \mathbf{w}_R)\sin(2\omega t)\end{aligned} \tag{1-15}$$

$$\langle \mathbf{v}(t) \times \mathbf{w}(t)\rangle = \frac{1}{2}(\mathbf{v}_R \times \mathbf{w}_R + \mathbf{v}_I \times \mathbf{w}_I). \tag{1-16}$$

Example 2

$$\mathbf{v} \times \mathbf{w}^* = (\mathbf{v}_R + j\mathbf{v}_I) \times (\mathbf{w}_R - j\mathbf{w}_I) \quad \text{(note } no \text{ time dependence)}$$
$$= [\mathbf{v}_R \times \mathbf{w}_R + \mathbf{v}_I \times \mathbf{w}_I] + j[\mathbf{v}_I \times \mathbf{w}_R - \mathbf{v}_R \times \mathbf{w}_I]. \qquad (1\text{-}17)$$

So we also have then, combining Eqs. 1-16 and 1-17,

$$\langle \mathbf{v}(t) \times \mathbf{w}(t) \rangle = \frac{1}{2} \operatorname{Re}(\mathbf{v} \times \mathbf{w}^*). \qquad (1\text{-}18)$$

1-1 DIFFERENTIAL OPERATORS

The differential vector operator del (∇) can be defined as

$$= \hat{x}(\partial/\partial x) + \hat{y}(\partial/\partial y) + \hat{z}(\partial/\partial z).$$

The application of the ∇ operator gives three expressions that are much used in scattering theory. They are

Gradient: the operator ∇ maps a scalar function to a vector function:

$$\nabla f = \mathbf{grad}\, f,$$
$$\text{or,} = \hat{x}(\partial f/\partial x) + \hat{y}(\partial f/\partial y) + \hat{z}(\partial f/\partial z);$$

Divergence: the operator $\nabla \cdot$ maps a vector function to a scalar:

$$\operatorname{div} \mathbf{v}, \text{ or } \nabla \cdot \mathbf{v} = \partial v_x/\partial x + \partial v_y/\partial y + \partial v_z/\partial z; \qquad (1\text{-}19)$$

Curl: the operator **curl** or $\nabla \times$ maps a vector function to a vector function:

$$\mathbf{curl}\, \mathbf{v} \text{ or } \nabla \times \mathbf{v} = \hat{x}(\partial v_z/\partial y - \partial v_y/\partial z) + \hat{y}(\partial v_x/\partial z - \partial v_z/\partial x)$$
$$+ \hat{z}(\partial v_y/\partial x - \partial v_x/\partial y). \qquad (1\text{-}20)$$

Recall that these nice arrangements work only in rectangular coordinates; see Magnus and Oberhettinger, 1954 for suitable expressions for other coordinate systems.

The theorems of Green (also called the divergence theorem by many authors; see Fulks, 1969) and Stokes provide relationships among the above expressions. We state each:

Green's Theorem. Let V be a volume in three-dimensional space enclosed by a smooth surface S; and let $\mathbf{v}(\mathbf{P})$ be a vector function that is continuously differentiable at every point $\mathbf{P}$ within and on S; then

$$\iiint_V \operatorname{div} \mathbf{v}\, dV = \iint_S \mathbf{v} \cdot \hat{n}\, dS,$$

where $\hat{n}$ is the exterior normal to S.

Corollary

$$\iiint_V \mathbf{curl}\, \mathbf{v}\, dV = \iint_S \hat{n} \times \mathbf{v}\, dS.$$

Stokes's Theorem. Let S be a smooth convex surface bounded by a smooth curve C; and let $\mathbf{v}(\mathbf{P})$ be a function with continuous second derivatives defined over S and on C; then

$$\int_C \mathbf{v}(\mathbf{P})\, \mathbf{dP} = \iint_S (\mathbf{curl}\, \mathbf{v}(\mathbf{P})) \cdot \hat{n}\, dS,$$

where $\hat{n}$ is the unit normal of S directed in a positive right-handed sense with respect to the integration direction on C.

Stokes's theorem can be used to obtain a convenient expression for the **curl** operation. Consider $(\mathbf{curl}\, \mathbf{v}) \cdot \hat{n}$ and use Stokes's theorem, assuming $\mathbf{v}$ analytic:

$$\int_C \mathbf{v} \cdot \mathbf{dP} = \iint_S (\mathbf{curl}\, \mathbf{v}) \cdot \hat{n}\, dS.$$

Shrink C to a point Q:

$$(\mathbf{curl}\, \mathbf{v}) \cdot \hat{n}\, dS = \lim_{C \to 0} \int_C \mathbf{v} \cdot \mathbf{dP},$$

$$= \lim_{C \to 0} \int_C (v_x\, dx + v_y\, dy + v_z\, dz).$$

Let C be a square with sides a, b, c, and d (beginning with a parallel to the y-axis, continuing counterclockwise with b parallel to the z-axis, and so on) in the yz-plane; evaluate the right-hand side:

$$(\mathbf{curl}\, \mathbf{v}) \cdot \hat{x} = \frac{\{[v_z]_b - [v_z]_d\}\, \Delta z + \{[v_y]_a - [v_y]_c\}\, \Delta y}{\Delta y\, \Delta z}$$

$$\rightarrow \partial v_z/\partial y - \partial v_y/\partial z.$$

(Note that the partial derivative of v with respect to z requires the subtraction $[v_y]_c - [v_y]_a$.) Treating the other two components similarly, we have the expression for **curl v** in the form of a determinant:

$$\mathbf{curl}\, \mathbf{v} = \det \begin{pmatrix} \hat{x} & \hat{y} & \hat{z} \\ \partial/\partial x & \partial/\partial y & \partial/\partial z \\ v_x & v_y & v_z \end{pmatrix}. \tag{1-21}$$

The **curl** operation $\nabla\times$ can also be given a matrix (or tensor) representation:

$$\nabla\times = \begin{pmatrix} 0 & -\partial/\partial z & \partial/\partial y \\ \partial/\partial z & 0 & -\partial/\partial x \\ -\partial/\partial y & \partial/\partial x & 0 \end{pmatrix}. \tag{1-22}$$

This operator is then applied, using standard matrix multiplication, to a column representation for a vector $\mathbf{v}$:

$$\mathbf{v} = \begin{pmatrix} v_x \\ v_y \\ v_z \end{pmatrix}.$$

Thus,

$$\nabla \times \mathbf{v} = \begin{pmatrix} 0 & -\partial/\partial z & \partial/\partial y \\ \partial/\partial z & 0 & -\partial/\partial x \\ -\partial/\partial y & \partial/\partial x & 0 \end{pmatrix} \begin{pmatrix} v_x \\ v_y \\ v_z \end{pmatrix} = \begin{pmatrix} \partial v_z/\partial y - \partial v_y/\partial z \\ \partial v_x/\partial z - \partial v_z/\partial x \\ \partial v_y/\partial x - \partial v_x/\partial y \end{pmatrix};$$

or

$$= \hat{x}(\partial v_z/\partial y - \partial v_y/\partial z) + \hat{y}(\partial v_x/\partial z - \partial v_z/\partial x) + \hat{z}(\partial v_y/\partial x - \partial v_x/\partial y) \qquad (1\text{-}23)$$

in our standard notation using unit vectors $\hat{x}, \hat{y}, \hat{z}$. **curl** is a singular, generally noncommutative operator.

We can also deal with the gradient operator ∇ in a similar way: ∇ is a vector operator:

$$\nabla \equiv \begin{pmatrix} \partial/\partial x \\ \partial/\partial y \\ \partial/\partial z \end{pmatrix}. \qquad (1\text{-}24)$$

It is of course generally applied to a scalar function. However, it can be applied to a vector, producing a tensor of rank 2 (i.e., a 3×3 matrix), by left-multiplying the transpose of the vector. For example, it can be applied to itself:

$$\nabla\nabla \equiv \begin{pmatrix} \partial/\partial x \\ \partial/\partial y \\ \partial/\partial z \end{pmatrix} (\partial/\partial x, \partial/\partial y, \partial/\partial z). \qquad (1\text{-}25)$$

The result may be applied to either a scalar or a vector. Equation 1-25 is expanded and used (applied to a scalar) in Section 4 below. In this connection an interesting identity is developed. Additional formulas for vector operators may be found in Stratton (1948).

1-2 FLUX DENSITIES AND FIELDS

Coulomb's experimental law, which gives the force between two electric charges in an infinite medium, includes the following information:

1. Force is proportional to the product of charge magnitudes and

inversely proportional to the square of the distance between them; it is dependent on the medium and acts along a line joining the charges.

2. Like charges repel, opposites attract.

Letting **f** be force, q_1 and q_2 signed charges separated by a distance r, ϵ the permittivity of the medium, and $\hat{a}_r$ a unit vector pointing from one charge directly away from the other, the law of Coulomb may be written:

$$\mathbf{f} = h\left(\frac{q_1 q_2}{\epsilon r^2}\right)\hat{a}_r; \tag{1-26}$$

h is a proportionality constant dependent on the units chosen; if these are mks (meters, kilograms, seconds; introduced by Giorgi in 1901), then $h = 1/4\pi$. [For example, if q_1 and q_2 have opposite signs, then $q_1 q_2$ is negative; so **f** has direction opposite to $\hat{a}_r$ (i.e., from one charge toward the other. Hence item 2 above).]

The electric field intensity, or electric field vector **E**, is defined as the force **f** acting on an infinitesimal charge Δq (small enough not to disturb the original charge distribution of the system), that is,

$$\mathbf{E} = \frac{\mathbf{f}}{\Delta q} = \left(\frac{q}{4\pi\epsilon r^2}\right)\hat{a}_r \tag{1-27}$$

is the electric field arising from a point charge q in a homogeneous dielectric medium (of permittivity ϵ); $\hat{a}_r$ points from positive toward negative charges. **E** has dimensions (coulombs meter)/farads/(meter)^2 = volts/meter.

For isotropic materials,

$$\mathbf{D} = \epsilon\mathbf{E} = \left(\frac{q}{4\pi r^2}\right)\hat{a}_r \tag{1-28}$$

and is called the electric flux density.

The magnetic field **H** is defined in terms of magnetic flux density **B** as $\mu\mathbf{H} = \mathbf{B}$ for linear isotropic media. **B** can be described in terms of the force **df** produced on a small current element of length **dP** carrying current J as

$$\mathbf{df} = J\,\mathbf{dP} \times \mathbf{B}. \tag{1-29}$$

(Refer to Panofsky and Phillips, 1956, p. 112 ff, for a detailed discussion of the magnetic flux density **B**.) Chapter 2 provides a more thorough examination of the nature of the constitutive parameters ϵ and μ.

1-3 MAXWELL'S EQUATIONS

Set $\mathbf{v} = \mathbf{E}$ in Stokes's theorem. Then

$$\int_C \mathbf{E} \cdot \mathbf{dP} = \iint_S (\nabla \times \mathbf{E}) \cdot \hat{n}\, dS.$$

But Faraday's law states:

$$\int_C \mathbf{E} \cdot \mathbf{dP} = -(\partial/\partial t) \iint_S \mathbf{B} \cdot \hat{n}\, dS.$$

Therefore,

$$\text{(F): } \nabla \times \mathbf{E} = -\partial \mathbf{B}/\partial t \tag{1-30}$$

which is Maxwell's equation for Faraday's law.

Ampère's law can be stated as: the line integral of a static magnetic field **H** about a closed path C is equal to the normal component of current **J** flowing across the surface enclosed by that path. Thus:

$$\int_C \mathbf{H} \cdot \mathbf{dP} = \iint_S \mathbf{J} \cdot \hat{n}\, dS.$$

Maxwell observed that this statement is not complete for time-varying fields and postulated the additional term $\partial \mathbf{D}/\partial t$. The resulting Maxwell–Ampère law can then be written as

$$\int_C \mathbf{H} \cdot \mathbf{dP} = \iint_S (\partial \mathbf{D}/\partial t) \cdot \hat{n}\, dS + \iint_S \mathbf{J} \cdot \hat{n}\, dS.$$

Applying Stokes's theorem to the left-hand side of this equation and assuming that the time differentiation can be exchanged with the surface integration, we find

$$\text{(A): } \nabla \times \mathbf{H} = (\partial/\partial t)\mathbf{D} + \mathbf{J}. \tag{1-31}$$

Now take the divergence of Eq. 1-30. Since $\nabla \cdot \nabla \times \mathbf{v} = 0$, $\nabla \cdot (\partial \mathbf{B}/\partial t) = 0$. And since **B** is analytic, we may exchange $\nabla \cdot$ and $\partial/\partial t$ so that

$$\text{(GB): } \nabla \cdot \mathbf{B} = 0 \tag{1-32}$$

which is Maxwell's equation for Gauss's law. **B** is said to be solenoidal. In an analogous fashion we can connect Maxwell's equation:

$$\text{(GD): } \nabla \cdot \mathbf{D} = \rho \tag{1-33}$$

for Gauss's electric flux law with Ampère's law (Eq. 1-31).

Using the identity $\nabla \cdot \nabla \times \mathbf{v} = 0$, apply the operator $\nabla \cdot$ to both sides of Ampère's law, so that with an interchange of space and time derivatives,

$$\nabla \cdot \mathbf{J} + (\partial/\partial t)\nabla \cdot \mathbf{D} = 0;$$

or, using Eq. 1-33,

$$\nabla \cdot \mathbf{J} + \partial\rho/\partial t = 0,$$

or

$$\text{(CJ): } \nabla \cdot \mathbf{J} = -\partial\rho/\partial t \tag{1-34}$$

which is the law of conservation of electric charge, (i.e., $\nabla \cdot \mathbf{J}$ is the flow of electric current out of a differential volume, and $-\partial\rho/\partial t$ is the rate of decrease of electric charge in the volume); so Eq. 1-34 implies that electric charge is *conserved.*

Note: If Gaussian rather than Giorgi mks units are used, then Maxwell's equations are written in the following form:

$$\text{(F): } \nabla \times \mathbf{E} + (1/c)(\partial \mathbf{B}/\partial t) = 0; \tag{1-35}$$

$$\text{(A): } \nabla \times \mathbf{H} - (1/c)(\partial \mathbf{D}/\partial t) = (4\pi/c)\mathbf{J}; \tag{1-36}$$

$$\text{(GB): } \nabla \cdot \mathbf{B} = 0; \quad \text{and} \tag{1-37}$$

$$\text{(GD): } \nabla \cdot \mathbf{D} = 4\pi\rho. \tag{1-38}$$

Here c is the speed of light. ($c = 299{,}792{,}458$ m/s in a vacuum.)

It is sometimes very useful for purposes of symmetry (such as in numerical work) to introduce into the Maxwell–Faraday equation a vector term that has no physical reality but may be called magnetic current $\mathbf{K}$; it is of course analogous to the electric current $\mathbf{J}$. Faraday's equation then reads

$$\text{(F): } \nabla \times \mathbf{E} + \partial \mathbf{B}/\partial t = -\mathbf{K}. \tag{1-39}$$

To maintain this fiction, the Gauss law for divergence of magnetic flux must then be modified by introducing a magnetic charge m (also fictitious):

$$\text{(GB): } \nabla \cdot \mathbf{B} = m. \tag{1-40}$$

Finally, then, we may also invent a conservation law for the magnetic charge. Apply $\nabla \cdot$ to both sides of Faraday's law:

$$\text{(CM): } \nabla \cdot \mathbf{K} = -\partial m/\partial t. \tag{1-41}$$

Here, $\nabla \cdot \mathbf{K}$ represents the flow of magnetic current from a differential volume, while $-\partial m/\partial t$ gives the rate of decrease of magnetic charge in that volume. Therefore, magnetic charge (if it exists) is conserved.

In summary then, Maxwell's set of symmetric equations now reads:

$$\text{(F): } \nabla \times \mathbf{E} + \partial \mathbf{B}/\partial t = -\mathbf{K}; \tag{1-42}$$

$$\text{(A): } \nabla \times \mathbf{H} - \partial \mathbf{D}/\partial t = \mathbf{J}; \tag{1-43}$$

$$\text{(GD): } \nabla \cdot \mathbf{D} = \rho; \tag{1-44}$$

$$\text{(GB): } \nabla \cdot \mathbf{B} = m. \tag{1-45}$$

The standard Maxwell set is obtained by simply setting $\mathbf{K} = 0$ and $m = 0$. An application for these representations for Maxwell's equations is given in Section 9-7.

1-4 TIME-HARMONIC FIELDS

All field and current quantities discussed so far are functions of both position (x, y, z) and time t: $\mathbf{v} = \mathbf{v}(x, y, z, t)$. If the time dependence is a function of a *single* frequency,† we assume that the variation is cosinusoidal, i.e.,

$$\begin{aligned}\mathbf{v}(x, y, z, t) &= \mathbf{v}_1(x, y, z)\cos(\omega t + \phi_1) \\ &= \operatorname{Re}[\mathbf{v}_1(x, y, z)e^{j\omega t}e^{j\phi_1}]. \end{aligned} \tag{1-46}$$

Set $\mathbf{v}(x, y, z) = \mathbf{v}_1(x, y, z)e^{j\phi_1}$. This complex function, together with ω, gives all of the information necessary to describe $\mathbf{v}(x, y, z, t)$. Henceforth, we take all vector functions $\mathbf{v}(x, y, z)$ to have the *complex* form above. Note that for the single-frequency case the operator $\partial/\partial t$ can be replaced by the operator $j\omega$.

Maxwell's equations now become (with both $\mathbf{K}$ and m set to zero)

$$\begin{aligned}&(\text{F}): \nabla \times \mathbf{E} = -j\omega\mathbf{B};\\ &(\text{A}): \nabla \times \mathbf{H} = j\omega\mathbf{D} + \mathbf{J};\\ &(\text{GB}): \nabla \cdot \mathbf{B} = 0;\\ &(\text{GD}): \nabla \cdot \mathbf{D} = \rho.\end{aligned}$$

Furthermore, the law of charge conservation becomes

$$(\text{CE}): \nabla \cdot \mathbf{J} = -j\omega\rho,$$

where the left-hand side represents total time-harmonic current flowing out of a unit volume and the right-hand side the amount of time-harmonic charge in that volume.

1-5 PLANE WAVES

Suppose electromagnetic fields are generated by $\mathbf{J}$ and ρ in a localized region. Then outside of this region $\mathbf{J} = 0$ and $\rho = 0$. So Maxwell's equations become

$$\begin{aligned}&(\text{F}): \nabla \times \mathbf{E} = -j\omega\mu_0\mathbf{H};\\ &(\text{A}): \nabla \times \mathbf{H} = j\omega\epsilon_0\mathbf{E};\\ &(\text{GB}): \nabla \cdot \mathbf{H} = 0;\\ &(\text{GD}): \nabla \cdot \mathbf{E} = 0;\end{aligned}$$

where we have used $\mathbf{D} = \epsilon_0\mathbf{E}$, $\mathbf{B} = \mu_0\mathbf{H}$, about which more later. Take the **curl** of (F) above and substitute into (A). Using the vector identity $\nabla \times (\nabla \times \mathbf{E}) = \nabla(\nabla \cdot \mathbf{E}) - \nabla^2\mathbf{E}$, we obtain the *wave equation:*

$$(\text{W}): \nabla^2\mathbf{E} + \omega^2\mu_0\epsilon_0\mathbf{E} = 0. \tag{1-47}$$

† It should be remarked here that the time dependence $e^{-i\omega t}$ is also in common usage, with i and j having the common meaning of $\sqrt{-1}$. To switch from the usage above, simply replace j by $-i$.

A simple solution of (W) is $\mathbf{E} = \hat{x}E_0e^{-jkz}$ which yields when substituted into (W):

$$(-k^2 + \omega^2\mu_0\epsilon_0)E_0 = 0.$$

But $E_0 \neq 0$. So we have the *dispersion relation*

$$k^2 = \omega^2\mu_0\epsilon_0. \tag{1-48}$$

Thus,

$$\mathbf{E}(z, t) = \text{Re}[\mathbf{E}(z)e^{j\omega t}] = \hat{x}E_0\cos(\omega t - kz). \tag{1-49}$$

$E_x|_{z=0} = E_0\cos\omega t$ which is periodic with period T such that $\omega T = 2\pi$. $f = 1/T = \omega/2\pi$; $\omega = 2\pi f$ is the angular frequency of the wave. At $z_0 = \lambda$ such that $k\lambda = 2\pi$, E_x repeats itself. So $\lambda = 2\pi/k$ is called *wavelength*. The *number* of wavelengths contained in the spatial distance 2π is $k = 2\pi/\lambda$ and so k is called *wave number*.

At what velocity must we move to keep up with the wave? We must have $\cos(\omega t - kz) = \text{constant}$; or, $\omega t - kz = \text{constant}$. For fixed, but arbitrary, values of t we have planes on which the phase is constant; hence *plane waves*. Since $dz/dt = v = \omega/k$, we have, using $k^2 = \omega^2\mu_0\epsilon_0$, $\omega/k = 1/\sqrt{\mu_0\epsilon_0} = c$, from the dispersion relation. This number is the velocity of light c in free space.

The magnetic field is, for this case, from (F) and (A) above,

$$\mathbf{H} = \hat{y}\sqrt{\epsilon_0/\mu_0}\,E_0e^{-jkz}; \tag{1-50}$$

or, with the time dependence included,

$$\mathbf{H} = \hat{y}\sqrt{\frac{\epsilon_0}{\mu_0}}\,E_0\cos(\omega t - kz). \tag{1-51}$$

$$\eta = 1\Big/\sqrt{\frac{\epsilon_0}{\mu_0}} \tag{1-52}$$

is the *intrinsic impedance* of free space.

More generally, for homogeneous media, plane wave solutions can be written in the form

$$e^{-j(\mathbf{k}\cdot\mathbf{r} - \omega t)}, \tag{1-53}$$

where $\mathbf{k}$ is the wave, or propagation, vector and $\mathbf{r}$ is a position vector to a point of interest. Maxwell's equations become, upon substitution of the above expression,

$$\text{(F): } \mathbf{k}\times\mathbf{E} = \omega\mathbf{B};$$

$$\text{(A): } \mathbf{k}\times\mathbf{H} = -\omega\mathbf{D};$$

$$\text{(GB): } \mathbf{k}\cdot\mathbf{B} = 0;$$

$$\text{(GB): } \mathbf{k}\cdot\mathbf{D} = 0.$$

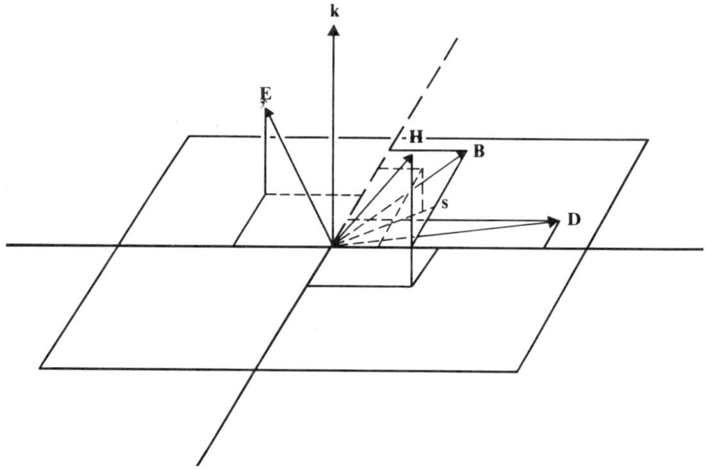

Figure 1-1 Relation between vector sets **E**, **H**, **k**, and **D**, **B**, **k** for a plane wave propagating in the direction, $\hat{k}$.

Geometrically, then, **k** is perpendicular to **B** and **D**; **B** is perpendicular to **k** and **E**; and **D** is perpendicular to **k** and **H** (see Fig. 1-1). If $\mathbf{D} = \epsilon \mathbf{E}$ and $\mathbf{B} = \mu \mathbf{H}$, where ϵ and μ are scalars, then the wave equation becomes

$$\mathbf{k} \times (\mathbf{k} \times \mathbf{E}) = -\omega^2 \mu \mathbf{D} = -\omega^2 \epsilon \mu \mathbf{E}. \tag{1-54}$$

Using $\mathbf{D} = \epsilon \mathbf{E}$ and $\mathbf{k} \cdot \mathbf{D} = 0$ produces

$$(k^2 - \omega^2 \epsilon \mu)\mathbf{E} = 0.$$

But $\mathbf{E} \neq 0$ and so we have the dispersion relation (again):

$$k^2 - \omega^2 \epsilon \mu = 0, \tag{1-48}$$

k being the magnitude of **k**. **k** has the dimension of inverse distance, $1/\sqrt{\epsilon\mu}$ the dimension of velocity. $N = c\sqrt{\epsilon\mu} = ck/\omega$, which is dimensionless, is the *refractive index* of the medium and $\omega/k = 1/\sqrt{\epsilon\mu}$ is the velocity of the wave in the medium.

1-6 POLARIZATION

The polarization of a plane wave is defined in terms of the behavior of the tip of the E-vector as a function of time.

Consider

$$\mathbf{E} = \hat{x} a_x e^{-j(kz - \omega t - \delta_x)} + \hat{y} a_y e^{-j(kz - \omega t - \delta_y)} \tag{1-55}$$

that is, with a_x and a_y real,

$$E_x = a_x \cos(\omega t - kz + \delta_x); \qquad E_y = a_y \cos(\omega t - kz + \delta_y) \tag{1-56}$$

in real space-time. Elimination of $\omega t - kz$ from Eqs. 1-56 yields

$$\left(\frac{E_x}{a_x}\right)^2 + \left(\frac{E_y}{a_y}\right)^2 - \left(\frac{2E_xE_y}{a_xa_y}\right)\cos\delta = \sin^2\delta, \qquad \delta = \delta_y - \delta_x, \tag{1-57}$$

which is the general equation for an ellipse. In other words, the projection of the trace of the tip of the E-vector on the xy-plane forms an ellipse (or some degenerate form of an ellipse depending on the value of the phase difference δ).

Equation 1-57 can be viewed in its canonical form (i.e., with the cross-term eliminated) by introducing the linear transformation (or rotation through an angle ψ) (see Fig. 1-2):

$$\begin{aligned} E_\xi &= E_x \cos\psi + E_y \sin\psi \\ E_\eta &= -E_x \sin\psi + E_y \cos\psi. \end{aligned} \tag{1-58}$$

The rotated ellipse has the form

$$\left(\frac{E_\xi}{a_\xi}\right)^2 + \left(\frac{E_\eta}{a_\eta}\right)^2 = 1 \tag{1-59}$$

and, as would be expected from application of the linear transformation

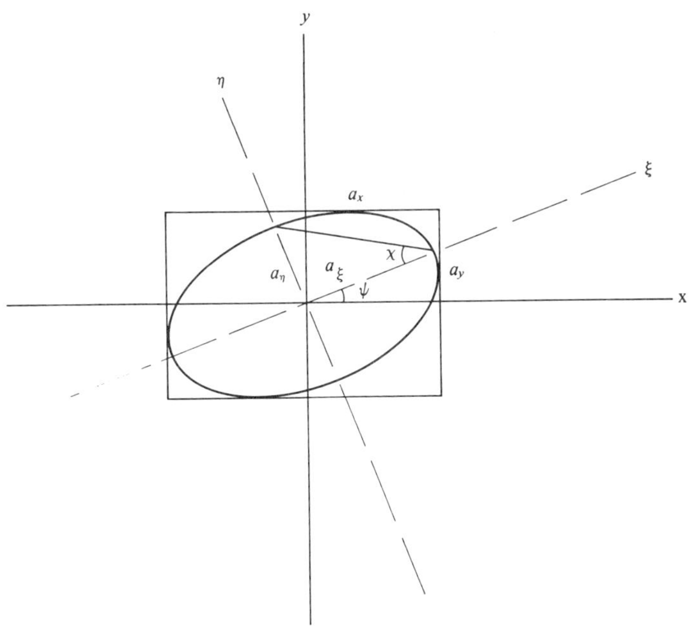

Figure 1-2 Polarization ellipse, showing rotation angle ψ and ellipticity angle χ.

(Eq. 1-58), Eq. 1-59 can be expressed parametrically as

$$\begin{aligned} E_\xi &= a_\xi \cos(\omega t - kz + \delta) \\ E_\eta &= a_\eta \sin(\omega t - kz + \delta). \end{aligned} \tag{1-60}$$

The rotation angle ψ can be written in terms of the original coefficients as

$$\tan 2\psi = \frac{2a_x a_y \cos\delta}{a_x^2 - a_y^2}; \tag{1-61}$$

or, if we put

$$\tan\alpha = \frac{a_y}{a_x},$$

then

$$\tan 2\psi = \tan 2\alpha \cos\delta. \tag{1-62}$$

Clearing the fractions in both Eqs. 1-59 and 1-57, we find, since constant terms remain unchanged under the linear transformation (Eq. 1-58),

$$a_\xi^2 a_\eta^2 = a_x^2 a_y^2 \sin^2\delta. \tag{1-63}$$

Further, set

$$\tan\chi = \frac{a_\eta}{a_\xi}; \tag{1-64}$$

then, using Eq. 1-62, together with the facts that

$$\sin 2\chi = \frac{2a_\xi a_\eta}{a_\xi^2 + a_\eta^2}$$

and

$$a_x^2 + a_y^2 = a_\xi^2 + a_\eta^2,$$

we find

$$\sin 2\chi = \sin 2\alpha \sin\delta. \tag{1-65}$$

The angle χ is called the ellipticity angle. (It is related to ellipse eccentricity e by $e = \sqrt{1 - \tan^2\chi}$.) (Fig. 1-2.)

We can display the entire gamut of polarization conditions as locations on the surface of a sphere, called a Poincaré sphere (Fig. 1-3), with a polar coordinate system defined in terms of the angles 2χ and 2ψ, which we take to be latitude and longitude, respectively. Choose the radius S_0 of the sphere such that

$$S_0^2 = a_x^2 + a_y^2 = a_\xi^2 + a_\eta^2. \tag{1-66}$$

In terms of rectangular coordinates S_1, S_2, S_3, a point on the sphere is

$$\begin{aligned} S_1 &= S_0 \cos 2\chi \cos 2\psi \\ S_2 &= S_0 \cos 2\chi \sin 2\psi \\ S_3 &= S_0 \sin 2\chi \end{aligned}$$

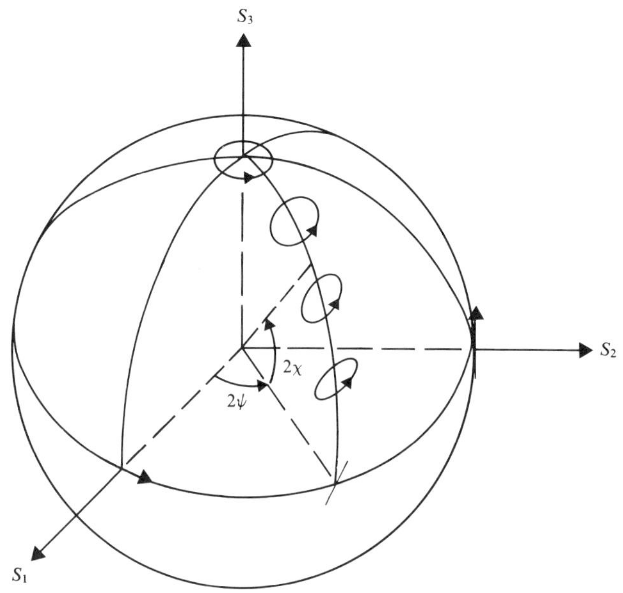

Figure 1-3 Poincaré sphere with Stokes parameters as axes.

and we have

$$S_0^2 = S_1^2 + S_2^2 + S_3^2.$$

From Eq. 1-66 above we also have

$$S_1^2 = a_x^2 - a_y^2; \qquad S_2^2 = 2a_x a_y \cos\delta = S_1^2 \tan 2\psi.$$

The parameters S_i, $i = 0, 1, 2, 3$, are called Stokes parameters.

An elliptical polarization with ellipticity χ and orientation ψ can be uniquely mapped to a point on the Poincaré sphere having latitude 2χ and longitude 2ψ. Notice that the ranges $-45° \le \chi \le 45°$ and $-90° \le \psi \le 90°$ suffice to represent all physical polarizations. The sense (or handedness) of the elliptical polarization is described by the sign of χ, with positive sign denoting left-handedness and negative sign right-handedness. (According to IEEE standard definitions (IEEE Trans. on Antenna and Propagation, 1983), the polarization of a wave receding from an observer is called right-handed if the electric vector appears to be rotating clockwise and left-handed if the electric vector appears to be rotating counterclockwise.) Thus, the upper hemisphere (positive values of χ) displays left-handed polarizations while the lower hemisphere (negative values of χ) displays right-handed polarizations. Positive rotations of the polarization ellipse (see Fig. 1-2) correspond to positive values of ψ and could be described as "east longitudes"; negative rotations of the polarization ellipse correspond to negative values of ψ and could be described as "west longitudes" (i.e., $\psi = -35°$ would be 35° west longitude). With such nomenclature the great circle formed by the intersection of the S_1S_3-plane with the Poincaré sphere must be designated as the "Greenwich meridian."

Figure 1-3 shows several examples of polarization ellipses: horizontal polarization at Greenwich meridian on the equator, vertical polarization at 90° east longitude on the equator, left circular at the north pole, right circular at the south pole, as well as several ellipses of differing fatness (eccentricity) in the northern hemisphere at 45° east longitude.

The topic of polarization, together with its various representations, is at present very popular and its study has been termed variously by the names of polarimetry and polarimetrics. (These words seem to have appeared in the language within the past 5 to 10 years.) The application of greatest current interest is in the area of remote sensing where scattered waves are usually partially polarized. But the real impetus for polarization study lies in the area of experimental remote sensing measurements as a result of recent advances in the capabilities of multipolarization radars. The interested reader may enjoy such representative expositions on remote sensing polarimetrics as those by the following authors: Van Zyl et al. (1987), Zebker et al. (1987), and Ulaby et al. (1987), as well as a report with a somewhat wider viewpoint by Mieras et al. (1978), in which Boerner gives an extensive historical treatment of the subject.

2

CONSTITUTIVE RELATIONS

We saw above that the Maxwell equations describing Gauss's laws are *not* independent of the two describing Faraday's and Ampère's laws. So we have only two vector, or six scalar, equations that are independent from which to determine four vector, or 12 scalar, quantities. We need six more scalar equations, and these are furnished by the constitutive relations. These relations provide information about the environment in which the electromagnetic fields occur.

A simple medium is characterized by a

permittivity ϵ

and by a

permeability μ

such that

$$\mathbf{D} = \epsilon \mathbf{E} \tag{2-1}$$

and

$$\mathbf{B} = \mu \mathbf{H}. \tag{2-2}$$

For free space:

$$\mu = \mu_0 = 4\pi \times 10^{-7} \frac{H}{m}$$

$$\epsilon = \epsilon_0 = 8.85 \times 10^{-12} \frac{F}{m}.$$

If either ϵ or μ is a function of frequency, the medium is called dispersive; such a medium is water. The water molecule tends to line up with the E-field. Quantitatively, this line-up ability is described by *susceptibility* χ_e. **D** is then composed of a free-space part ϵ_0 and a material part characterized by a dipole moment density $\mathbf{P}$ such that

$$\mathbf{P} = \epsilon_0 \chi_e \mathbf{E}. \tag{2-3}$$

The dipole moment density $\mathbf{P}$ induced by $\mathbf{E}$ may be thought of as composed of many individual dipoles $P = Nqd$, where N is the number and d is the distance between the $-q$ and $+q$ charges. Then

$$\mathbf{D} = \epsilon_0 \mathbf{E} + \mathbf{P} = \epsilon_0(1 + \chi_e)\mathbf{E}$$

so that

$$\epsilon = \epsilon_0(1 + \chi_e). \tag{2-4}$$

At low frequencies the water molecules can align with the slowly varying field easily and so χ_e is large, for example $\epsilon \approx 80\epsilon_0$. At optical frequencies, $\epsilon \approx 2\epsilon_0$.

In diamagnetic materials, magnetic fields produced by orbital and spinning electrons cancel out completely. But under an external magnetic field, the balance is distorted. So a magnetic moment density is induced in the medium opposite to the applied external field:

$$\mathbf{B}_{\text{total}} = \mu_0(\mathbf{H}_{\text{external}} + \mathbf{M}).$$

The magnitude of $\mathbf{M}$ is proportional to the externally applied $\mathbf{H}$: $\mathbf{M} = \chi_m \mathbf{H}_{\text{external}}$. So

$$\mathbf{B}_{\text{total}} = \mu_0(1 + \chi_m)\mathbf{H}_{\text{external}}.$$

χ_m is called magnetic susceptibility (and is negative for diamagnetic materials). Some typical values are:

Copper	-0.94×10^{-5}
Water	-0.88×10^{-5}
Air	3.60×10^{-7}
Al	2.10×10^{-5}
Liquid O	3.50×10^{-3}

In their most general form, constitutive relations can be written† (cf. Kong, 1975.)

$$c\mathbf{D} = \mathfrak{P} \cdot \mathbf{E} + \mathfrak{L} \cdot c\mathbf{B},$$
$$\mathbf{H} = \mathfrak{M} \cdot \mathbf{E} + \mathfrak{Q} \cdot c\mathbf{B},$$

where $\mathfrak{P}$, $\mathfrak{Q}$, $\mathfrak{L}$, and $\mathfrak{M}$ are 3×3 matrices whose entries are called constitutive parameters, and $c = 3 \times 10^8$ m/s (or, see Section 1-3).

† This interesting form is chosen to facilitate the demonstration that the relations are invariant under a Lorentz transformation.

Alternatively,

$$\begin{pmatrix} c\mathbf{D} \\ \mathbf{H} \end{pmatrix} = \mathfrak{C} \cdot \begin{pmatrix} \mathbf{E} \\ c\mathbf{B} \end{pmatrix}, \qquad \mathfrak{C} = \begin{pmatrix} \mathfrak{P} & \mathfrak{L} \\ \mathfrak{M} & \mathfrak{Q} \end{pmatrix}, \tag{2-5}$$

where the latter is a 6×6 matrix with the dimensions of admittance. Media may be classified according to the functional dependence of $\mathfrak{C}$:

1. Inhomogeneous: function of space coordinates
2. Nonstationary: function of time
3. Time-dispersive: function of time derivatives
4. Spatial-dispersive: function of spatial derivatives
5. Nonlinear: function of the electromagnetic field

$\mathfrak{C}$ may also be coupled to other physical disciplines (see, for example, Kong, 1975). We have, after some manipulation of Eq. 2-5,

$$\begin{pmatrix} \mathbf{D} \\ \mathbf{B} \end{pmatrix} = \mathfrak{C}_{EH} \begin{pmatrix} \mathbf{E} \\ \mathbf{H} \end{pmatrix}, \qquad \mathfrak{C}_{EH} = \begin{pmatrix} \boldsymbol{\epsilon} & \boldsymbol{\xi} \\ \boldsymbol{\zeta} & \boldsymbol{\mu} \end{pmatrix} = \begin{pmatrix} \mathfrak{P} - \mathfrak{L} \cdot \mathfrak{Q}^{-1} \cdot \mathfrak{M} & \mathfrak{L} \cdot \mathfrak{Q}^{-1} \\ -\mathfrak{Q}^{-1} \cdot \mathfrak{M} & \mathfrak{Q}^{-1} \end{pmatrix}$$

or, solving for **E** and **H**,

$$\begin{pmatrix} \mathbf{E} \\ \mathbf{H} \end{pmatrix} = \mathfrak{C}_{DB} \begin{pmatrix} \mathbf{D} \\ \mathbf{B} \end{pmatrix}, \qquad \mathfrak{C}_{DB} = \begin{pmatrix} \boldsymbol{\kappa} & \boldsymbol{\chi} \\ \boldsymbol{\gamma} & \boldsymbol{\nu} \end{pmatrix} = \begin{pmatrix} \mathfrak{P}^{-1} & -\mathfrak{P}^{-1} \cdot \mathfrak{L} \\ \mathfrak{M} \cdot \mathfrak{P}^{-1} & \mathfrak{Q} - \mathfrak{M} \cdot \mathfrak{P}^{-1} \cdot \mathfrak{L} \end{pmatrix}. \tag{2-6}$$

If $\mathfrak{L}$ and $\mathfrak{M}$ are not identically zero, the medium is bianisotropic. If $\mathfrak{L} = \mathfrak{M} = 0$, the medium is anisotropic. We consider this case further in Part IV. Here we have

$$\mathbf{D} = \boldsymbol{\epsilon} \cdot \mathbf{E} \tag{2-7}$$

and

$$\mathbf{B} = \boldsymbol{\mu} \cdot \mathbf{H}. \tag{2-8}$$

2-1 WAVES IN DISSIPATIVE MEDIA

By Ohm's law, the conduction current $\mathbf{J}_c$ is

$$\mathbf{J}_c = \sigma \mathbf{E},$$

where σ represents conductivity (in siemens or mhos/m). The **J** in Maxwell's equation for Ampère's law can encompass two kinds of current: source $\mathbf{J}_0$ and conduction $\mathbf{J}_c$; and $\mathbf{J}_0$ itself can be composed of a conduction current and a convection current. So we can rewrite Maxwell's equation for Ampère's law as

$$\text{(A)}\ \nabla \times \mathbf{H} = j\omega\left(\epsilon - \frac{j\sigma}{\omega}\right)\mathbf{E} + \mathbf{J}_0. \tag{2-9}$$

Now if we replace $\epsilon - j\sigma/\omega$ by ϵ, we can write this equation for a source-free medium as

$$\nabla \times \mathbf{H} = j\omega\epsilon\mathbf{E},$$

where ϵ is complex for a conducting medium. The wave equation is

$$\text{(W)}\ (\nabla^2 + \omega^2\epsilon\mu)\mathbf{E} = 0, \tag{2-10}$$

and plane wave solutions still work, viz.,

$$\mathbf{E} = \hat{x}E_0 e^{-jkz} \tag{2-11}$$

$$\mathbf{H} = \hat{y}\left(\frac{E_0}{\eta}\right)e^{-jkz} \tag{2-12}$$

with $k = \omega\sqrt{\epsilon\mu}$ and $\eta = \sqrt{\dfrac{\mu}{\epsilon}}$, k and η complex.

Write $k = k_R - jk_I$ and $\eta = \eta e^{j\phi}$. Then

$$\mathbf{E} = \hat{x}E_x = \hat{x}E_0 e^{-k_I z} e^{-jk_R z},$$

$$\mathbf{H} = \hat{y}\left(\frac{E_0}{|\eta|}\right)e^{-k_I z}e^{-jk_R z}e^{-j\phi}.$$

Including the time variation, we can write

$$E_x(z, t) = \text{Re}(E_x e^{j\omega t}) = E_0 e^{-k_I z}\cos(\omega t - k_R z),$$

which is a plane wave traveling in the $+z$-direction with velocity $v = \omega/k_R$ and amplitude which decays like $e^{-k_I z}$.

Suppose the wave at $z = d$ has decayed to $1/e$ of its value at $z = 0$; then $k_I d = 1$ and $d = 1/k_I$. For conducting media

$$k = \omega\sqrt{\epsilon\mu}\sqrt{1 - \frac{j\sigma}{\omega\epsilon}} = k_R - jk_I; \tag{2-13}$$

$\sigma/\omega\epsilon$ is called the loss tangent of a conducting medium. If $\sigma/\omega\epsilon \ll 1$ (slightly conducting), then

$$k \approx \omega\sqrt{\epsilon\mu}\left(1 - \frac{j\sigma}{2\omega\epsilon}\right). \tag{2-14}$$

Here $d = (2/\sigma)\sqrt{\epsilon/\mu}$ since $k_I = \frac{1}{2}\sigma\sqrt{\mu/\epsilon}$.

If $\sigma/\omega\epsilon \gg 1$ (highly conducting), then

$$k \approx \omega\sqrt{\epsilon\mu}\sqrt{-\frac{j\sigma}{\omega\epsilon}} = (1 - j)\sqrt{\frac{\omega\mu\sigma}{2}}. \tag{2-15}$$

So $d = \sqrt{2/\omega\mu\sigma}$, which is usually called skin depth δ (very small). Very little conduction current therefore flows inside the conductor, that is, the current flows essentially on the conductor surface. This phenomenon is known as the *skin effect*. (One often sees $\delta = 1/\sqrt{\pi\mu\sigma f}$, where $\omega = 2\pi f$.)

2-2 PERFECT CONDUCTORS

From Ohm's law $\mathbf{J} = \sigma\mathbf{E}$ again; if $\mathbf{E} = 0$, then $\sigma \to \infty$ in order to have a finite current. So the skin depth $\delta = 0$ here. Copper, aluminum, gold, and silver are usually regarded as perfect conductors. For example, for copper $\sigma = 5.7 \times 10^7$ mhos/m at room temperature.

A useful quantity called resistivity, or surface resistivity R_s, can be defined by looking at the current at and near the surface of a plane conductor. On the basis of the Fig. 2-1, one may find

$$E_z = E_0 e^{-x(1+j)/\delta}, \qquad H_y = H_0 e^{-x(1+j)/\delta}, \qquad i_z = i_0 e^{-x(1+j)/\delta}, \tag{2-16}$$

where H_0 and i_0 are magnitudes of the magnetic field and the current density at the surface $x = 0$. The total current, found by integrating i_z from the surface to an infinite depth, is, for a unit width,

$$J_z = \int_0^\infty i_z \, dx = i_0 \int_0^\infty e^{-x(1+j)/\delta} \, dx = \frac{i_0 \delta}{1+j}. \tag{2-17}$$

Since the electric field at the surface is

$$E_{z0} = \frac{i_0}{\sigma},$$

the internal impedance, for a unit length and a unit width, is

$$Z_s = \frac{E_{z0}}{J_z} = \frac{1+j}{\sigma\delta},$$

which we set $= R_s + j\omega L_i$. Then

$$R_s = \frac{1}{\sigma\delta} \quad (\text{or } \sqrt{\pi\mu f/\sigma}) \tag{2-18}$$

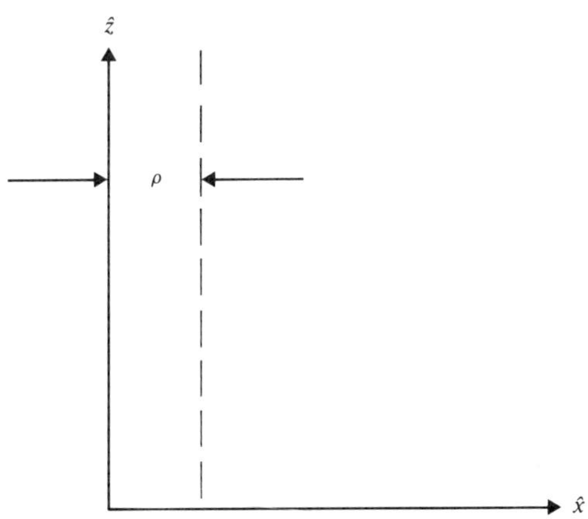

Figure 2-1 Current near the surface of a plane conductor.

($= \omega L_i$ as well, i.e., the resistance and internal reactance of a plane conductor are equal at any frequency).

An alternative interpretation of δ is then that the skin effect resistance of a semiinfinite plane conductor is exactly the same as the dc resistance of a plane conductor of depth δ; or the resistance of this conductor with exponential decrease in current density is exactly the same as if the current were uniformly distributed over a depth δ.

R_s, the resistance of a plane conductor for a *unit length* and *unit width,* is called *surface resistivity*. The dimension of R_s is ohms or, as it is usually called, *ohms per square*. R_s may be a useful parameter in the analysis of other than plane surfaces. (For a finite area of a plane conductor, resistance is found by multiplying R_s by length and dividing by width, since width elements are essentially in parallel.)

3

BOUNDARY CONDITIONS AND SURFACE CURRENTS

At a boundary Σ separating two media, conditions on the tangential and normal components of the electric and magnetic fields can be obtained from Maxwell's equations. The method of derivation is quite standard; so we discuss briefly the nature of the conditions and refer the reader to Born and Wolf (1959, p. 4), or Stratton (1941, p. 34), or Kong (1975, p. 5) for details.

The differential forms of Maxwell's equations are not applicable here since, at Σ, the vector field quantities may become discontinuous, while the charge and current densities require conversion into appropriate surface expressions. However, we can choose convenient integral forms which are themselves suggestive of the boundary conditions. Let V be a volume enclosed by a surface S with unit outward normal $\hat{n}'$. With the help of the volume and surface integral expressions given by Green's (Divergence) theorem and corollary (Section 1-1) we can put Maxwell's equations into the form:

$$\text{(F)} \qquad \iint dS\,(\hat{n}' \times \mathbf{E}) + \iiint dV\, \partial \mathbf{B}/\partial t = 0, \tag{3-1}$$

$$\text{(A)} \qquad \iint dS\,(\hat{n}' \times \mathbf{H}) - \iiint dV\, \partial \mathbf{D}/\partial t = \iiint dV\, \mathbf{J}, \tag{3-2}$$

$$\text{(GB)} \qquad \iint dS\,(\hat{n}' \cdot \mathbf{B}) = 0, \tag{3-3}$$

$$\text{(GD)} \qquad \iint dS\,(\hat{n}' \cdot \mathbf{D}) = \iiint dV\, \rho. \tag{3-4}$$

Notice that the surface integrals of the first two equations have integrands involving the tangential components of the electric and magnetic fields, while the last two equations have integrands involving the normal components of the electric and magnetic flux densities. The classical derivations use small areas or volumes that cross or penetrate Σ, for which the equations hold, and on which the role of the integrand expressions can be emphasized and easily followed. The only serious difficulties are encountered with the current density **J** and the charge density ρ. In the first case, the boundary must support a surface current $\mathbf{J}_S$ which is related to the current density **J** in such a way that the former remains finite even though the latter may become infinite. In the second case the charge density ρ must be reducible to a surface charge layer expressible as a surface charge density ρ_S. As with the surface current, the surface charge density must also remain finite even though the charge density were to become infinite. These results can be brought about when the small areas and volumes are shrunk to a limiting point on Σ, with a resulting coincidence of their normal $\hat{n}'$ with the normal $\hat{n}$ of Σ.

The boundary conditions, stated in terms of tangential and normal components of the vector field quantities, may then be summarized as follows:

> The tangential components of the electric field are continuous at Σ but the tangential components of the magnetic field undergo a jump discontinuity equal to the surface current $\mathbf{J}_S$ at Σ. The following two equations express these facts, with subscripts 1 and 2 referring to the two media which are separated by the boundary Σ:
>
> $$\hat{n} \times (\mathbf{E}_1 - \mathbf{E}_2) = 0 \tag{3-5}$$
>
> from the Faraday equation (3-1); and
>
> $$\hat{n} \times (\mathbf{H}_1 - \mathbf{H}_2) = \mathbf{J}_S \tag{3-6}$$
>
> from the Ampère equation (3-2).
>
> The normal components of the magnetix flux density are continuous at Σ but the normal components of the electric flux density suffer a jump discontinuity equal to the surface charge density ρ_S at Σ. The following two equations express these facts:
>
> $$\hat{n} \cdot (\mathbf{B}_1 - \mathbf{B}_2) = 0 \tag{3-7}$$
>
> from the Gauss equation (3-3); and
>
> $$\hat{n} \cdot (\mathbf{D}_1 - \mathbf{D}_2) = \rho_S \tag{3-8}$$
>
> from the Gauss equation (3-4).

3-1 POYNTING'S THEOREM

Since **E** is in volts/m and **H** is in amperes/m, a product will be in watts/m^2. For a time-harmonic wave, the power density per unit area is defined as the complex Poynting vector:

$$\mathbf{S} = \mathbf{E} \times \mathbf{H}^* \; W/m^2. \tag{3-9}$$

Form

$$\mathbf{H}^* \cdot \nabla \times \mathbf{E} = -j\omega\mu\mathbf{H} \cdot \mathbf{H}^*$$

and

$$\mathbf{E} \cdot \nabla \times \mathbf{H}^* = -j\omega\varepsilon\mathbf{E} \cdot \mathbf{E}^* + \mathbf{J}^* \cdot \mathbf{E}$$

from Maxwell's curl equations. Subtract the second from the first and use the identity

$$\nabla \cdot (\mathbf{E} \times \mathbf{H}^*) = \mathbf{H}^* \cdot \nabla \times \mathbf{E} - \mathbf{E} \cdot \nabla \times \mathbf{H}^*$$

to get

$$\begin{aligned}\nabla \cdot \mathbf{S} &= -j\omega\mu\mathbf{H} \cdot \mathbf{H}^* + j\omega\varepsilon\mathbf{E} \cdot \mathbf{E}^* - \mathbf{J}^* \cdot \mathbf{E} \\ &= -j\omega\mu\, |\mathbf{H}|^2 + j\omega\varepsilon\, |\mathbf{E}|^2 - \mathbf{J}^* \cdot \mathbf{E}.\end{aligned}$$

The last term represents power supplied by the current source **J**.

The *time-average Poynting vector* $\langle \mathbf{S} \rangle$ is the average of **S** over a period $T = 2\pi/\omega$:

$$\begin{aligned}\langle \mathbf{S} \rangle &= \frac{1}{2\pi} \int_0^{2\pi} d(\omega t)\mathbf{E}(x, y, z, t) \times \mathbf{H}(x, y, z, t) \\ &= \frac{1}{2} \operatorname{Re}\{\mathbf{E} \times \mathbf{H}^*\}. \qquad \text{(cf. Eq. 1-18)}\end{aligned} \tag{3-10}$$

3-2 REFLECTION AND TRANSMISSION AT A PLANE BOUNDARY

Let a wave be linearly polarized and propagate (in medium 1; see above) in other than the z-direction, say

$$\mathbf{E} = \hat{y}E_0 \exp(-jxk_x - jzk_z). \tag{3-11}$$

A vector wave can be written as $\mathbf{k} = \hat{x}k_x + \hat{z}k_z$ and a position vector as $\mathbf{r} = x\hat{x} + z\hat{z}$ so that Eq. 3-11 can be rewritten as

$$\mathbf{E} = \hat{y}E_0 \exp(-j\mathbf{k} \cdot \mathbf{r}). \tag{3-12}$$

$\mathbf{k} \cdot \mathbf{r} =$ constant defines the phase front of the wave, which is a plane (Fig. 3-1). From Faraday's law,

$$\begin{aligned}\mathbf{H} &= \frac{j}{\omega\mu_1} \nabla \times \mathbf{E} = (-k_z\hat{x} + k_x\hat{z})\left(\frac{E_0}{\omega\mu_1}\right) \exp(-j\mathbf{k} \cdot \mathbf{r}) \\ &= \left(\frac{E_0}{k_0\eta_0\mu_1}\right)(-k_z\hat{x} + k_x\hat{z}) \exp(-j\mathbf{k} \cdot \mathbf{r}),\end{aligned} \tag{3-13}$$

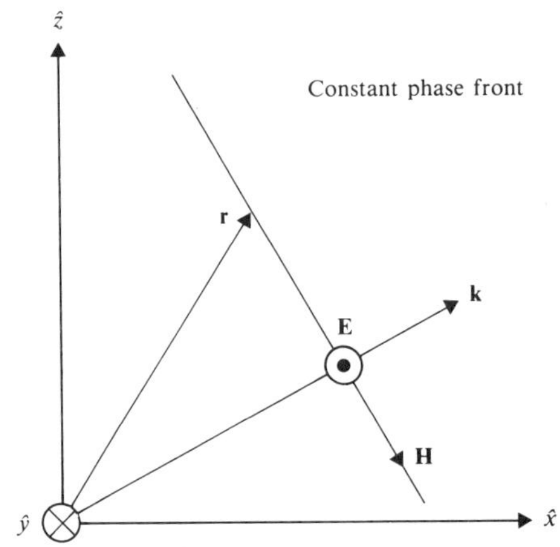

Figure 3-1 A uniform plane wave propagating through an isotropic medium in the direction, $\hat{k}$.

where, in the final expression of Eq. 3-13, we have changed the meaning of μ_1 to be *relative* permeability (i.e., normalized to free space permeability μ_0) and η_0 is the intrinsic impedance of free space (Section 1-5, Equation 1–52). **H** is perpendicular to both **E** and **k**, since $\mathbf{H}\cdot\mathbf{E}=0$ and $\mathbf{H}\cdot\mathbf{k}=0$. Moreover,

$$\mathbf{E}\times\mathbf{H}^* = \frac{(\hat{x}k_x+\hat{z}k_z)\,|E_0|^2}{k_0\omega_0\mu_1},$$

which is the direction of **k**.

3-2-1 Perpendicular Polarization

For an incident plane wave impinging on a plane boundary at an oblique angle, the normal to the plane boundary and the incidence direction form a plane called the *incidence plane.* Let **E** be perpendicular to the incidence plane and choose a coordinate system such that the boundary is $z=0$; we write (Fig. 3-2):

$$\begin{aligned}\mathbf{E}^i &= \hat{y}E_0\exp(-jxk_x-jzk_z),\\ \mathbf{H}^i &= (-\hat{x}k_z+\hat{z}k_x)\left(\frac{E_0}{k_0\omega_0\mu_1}\right)\exp(-jxk_x-jzk_z).\end{aligned} \tag{3-14}$$

A reflected wave is given by

$$\begin{aligned}\mathbf{E}^r &= \hat{y}R_{\text{per}}\,E_0\exp(-jk_{rx}x+jk_{rz}z)\\ \mathbf{H}^r &= (\hat{x}k_{rz}+\hat{z}k_{rx})\frac{R_{\text{per}}E_0}{k_0\omega_0\mu_1}\exp(-jk_{rx}x+jk_{rz}z)\end{aligned} \tag{3-15}$$

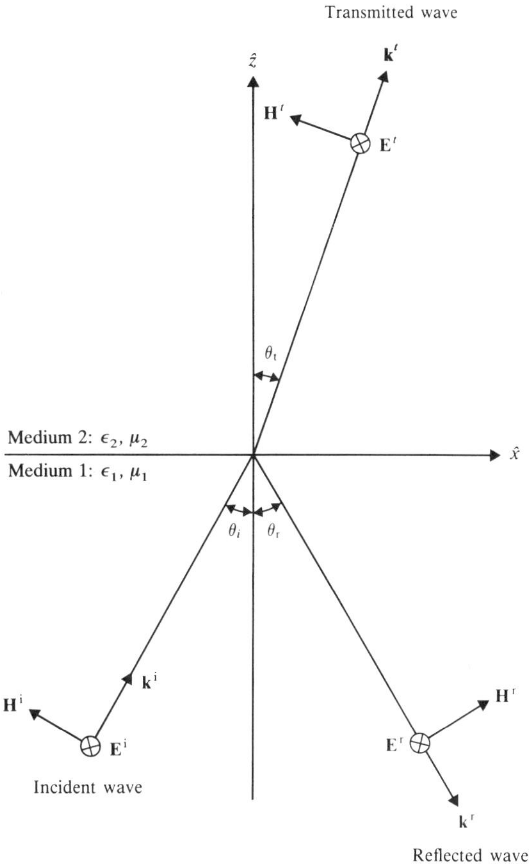

Figure 3-2 Transmission and reflection of a perpendicular polarized plane wave incident on a plane boundary between two media.

where R_{per} is the reflection coefficient for the perpendicularly polarized wave and

$$\mathbf{k} = \hat{x}k_{rx} - \hat{z}k_{rz} \tag{3-16}$$

its wave vector. (It propagates in the $+\hat{x}, -\hat{z}$–direction; see Fig. 3-2).

For the transmitted wave, with coefficient T_{per},

$$\begin{aligned}
\mathbf{E}^t &= \hat{y}T_{\mathrm{per}}E_0 \exp(-jkt_x x - jkt_z z), \\
\mathbf{H}^t &= (-\hat{x}k_{tz} + \hat{z}k_{tx})\left(\frac{T_{\mathrm{per}}E_0}{k_0\omega_0\mu_2}\right)\exp(-jk_{tx}x - jk_{tz}z),
\end{aligned} \tag{3-17}$$

where μ_2 is the relative permeability of medium 2. Assume that neither medium 1 nor medium 2 is a perfect conductor. The boundary conditions for **E** and **H** require that the tangential electric and magnetic field components be continuous at $z = 0$. Therefore,

$$\exp(-jk_x x) + R_{\text{per}} \exp(-jk_{rx} x) = T_{\text{per}} \exp(-jk_{tx} x)$$

$$\frac{-k_z}{k_0\omega_0\mu_1} \exp(-jk_x x) + \frac{k_{rz} R_{\text{per}}}{k_0\omega_0\mu_1} \exp(-jk_{rx} x) = -\frac{k_{tz} T_{\text{per}}}{k_0\omega_0\mu_2} \exp(-jk_{tx} x) \quad (3\text{-}18)$$

for *all* x in medium 1. So

$$k_x = k_{rx} = k_{tx}; \quad (3\text{-}19)$$

that is, the tangential components of $\mathbf{k}$, $\mathbf{k}_r$, and $\mathbf{k}_t$ are equal. This condition is called the *phase-matching* condition. The magnitudes of the three wave vectors can be obtained by substituting $\mathbf{E}^i$, $\mathbf{E}^r$, and $\mathbf{E}^t$ into the wave equations that hold in each medium:

$$\begin{aligned} (\nabla^2 + k_0^2\mu_1\epsilon_1)(\mathbf{E}^i, \mathbf{E}^r) &= 0, \\ (\nabla^2 + k_0^2\mu_2\epsilon_2)\mathbf{E}^t &= 0, \end{aligned} \quad (3\text{-}20)$$

where we note again that ϵ_i, μ_i, $i = 1, 2$, are *relative* constitutive parameters. The result of the substitutions is

$$\begin{aligned} k_x^2 + k_z^2 &= k_0^2\epsilon_1\mu_1 = k_1^2, \\ k_{rx}^2 + k_{rz}^2 &= k_0^2\epsilon_1\mu_1 = k_1^2, \\ k_{tx}^2 + k_{tz}^2 &= k_0^2\epsilon_2\mu_2 = k_2^2. \end{aligned} \quad (3\text{-}21)$$

But $k_x = k_{rx} = k_{tx}$; so $k_z = k_{rz}$, but the z-components of $\mathbf{k}_r$ and $\mathbf{k}_i$ point in opposite directions. Now, using the phase-matching condition (Eq. 3-19) with Eq. 3-18, we get

$$\begin{aligned} 1 + R_{\text{per}} &= T_{\text{per}} \\ 1 - R_{\text{per}} &= \frac{\mu_1 k_{tz} T_{\text{per}}}{\mu_2 k_z} \end{aligned} \quad (3\text{-}22)$$

and from these,

$$\begin{aligned} R_{\text{per}} &= \frac{\mu_2 k_z - \mu_1 k_{tz}}{\mu_2 k_z + \mu_1 k_{tz}} \\ T_{\text{per}} &= \frac{2\mu_2 k_z}{\mu_2 k_z + \mu_1 k_{tz}} \end{aligned} \quad (3\text{-}23)$$

which are the *Fresnel* reflection and transmission coefficients, respectively.

If we use angles θ of incidence, θ_r of reflection, and θ_t of transmission, we have

$$\begin{aligned} k_x &= k_1 \sin\theta, \\ k_{rx} &= k_1 \sin\theta_r, \\ k_{tx} &= k_2 \sin\theta_t. \end{aligned} \quad (3\text{-}24)$$

Using Eq. 3-24 in the phase-matching condition (Eq. 3-18) gives

$$k_1 \sin\theta = k_1 \sin\theta_r = k_2 \sin\theta_t \tag{3-25}$$

of which the first equality implies $\theta = \theta_r$; that is, angle of incidence = angle of reflection. The latter equality then produces *Snell's law*:

$$N_1 \sin\theta = N_2 \sin\theta_t, \tag{3-26}$$

where we have defined refractive indices

$$N_1 = \sqrt{\epsilon_1\mu_1},\ N_2 = \sqrt{\epsilon_2\mu_2} \tag{3-27}$$

for media 1 and 2, respectively. (We remind the reader again that ϵ_i, μ_i, $i = 1, 2$, are *relative* constitutive parameters.)

The perpendicular polarization reflection coefficient can be given explicitly in terms of the incidence angle θ and the constitutive parameters of the two media by using Eqs. 3-26, 3-24, and 3-21 in Eq. 3-23:

$$R_{\text{per}} = \frac{N_1\mu_2\cos\theta - \mu_1\sqrt{N_2^2 - (N_1\sin\theta)^2}}{N_1\mu_2\cos\theta + \mu_1\sqrt{N_2^2 - (N_1\sin\theta)^2}} \tag{3-28}$$

If Snell's law is rewritten,

$$\frac{N_1}{N_2}\sin\theta = \sin\theta_t \tag{3-29}$$

we see that if $N_1/N_2 > 1$ (with N_1 and N_2 real), then θ must be smaller than some θ_c such that

$$\theta < \arcsin\left(\frac{N_2}{N_1}\right) = \theta_c;$$

otherwise, θ_t is complex. In this circumstance, $k_{tz}^2 = k_2^2 - k_x^2 < 0$ which implies that $k_{tz} = \pm j\alpha$, α real. We choose the minus sign so that

$$\mathbf{E}^t = \hat{y}T_{\text{per}}E_0\exp(-jk_{tz}z - jk_{tx}x) = \hat{y}T_{\text{per}}E_0\exp(-\alpha z - jk_{tx}x) \tag{3-30}$$

which decays exponentially in the $+z$-direction. Since the magnitude decays so rapidly away from the interface, the wave is called a *surface wave* which propagates *along* the boundary. The refractive indices N_1 and N_2 may themselves be complex in which case k_{tz} will also be a complex number $\gamma \pm j\alpha$. The quantity α will then affect wave amplitude decay, as above, while γ will be a propagation constant in the z-direction.

3-2-2 Parallel Polarization

We can write the incident, reflected, and transmitted fields for this case from the perpendicular polarization results by replacing **H** by **E**, **E** by $-$**H**,

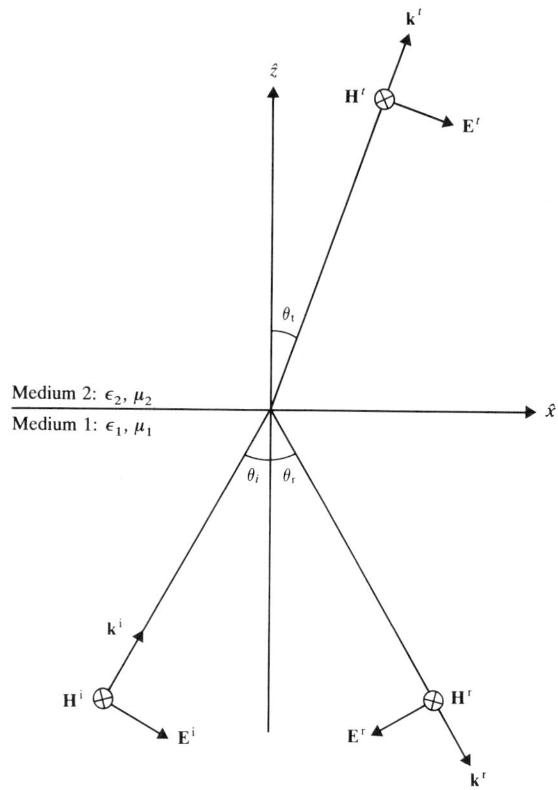

Figure 3-3 Transmission and reflection of a parallel polarized plane wave incident on a plane boundary between two media.

μ by ϵ, and ϵ by μ. In summary we have (Fig. 3-3):

$$\begin{aligned} \mathbf{H}^i &= \hat{y} H_0 \exp(-jk_x x - jk_z z); \\ \mathbf{E}^i &= (\hat{x}k_z - \hat{z}k_x)\left(\frac{H_0\eta_0}{k_0\epsilon_1}\right)\exp(-jk_x x - jk_z z); \end{aligned} \tag{3-31}$$

$$\begin{aligned} \mathbf{H}^r &= \hat{y} R_{\text{par}} H_0 \exp(-jk_{rx} x + jk_{rz} z); \\ \mathbf{E}^r &= (-\hat{x}k_{rz} - \hat{z}k_{rx})\left(\frac{R_{\text{par}} H_0\eta_0}{k_0\epsilon_1}\right)\exp(-jk_{rx} x + jk_{rz} z); \\ \mathbf{H}^t &= \hat{y} T_{\text{par}} H_0 \exp(-jk_{tx} x - jk_{tz} z); \\ \mathbf{E}^t &= (\hat{x}k_{tz} - \hat{z}k_{tx})\left(\frac{T_{\text{par}} H_0\eta_0}{k_0\epsilon_2}\right)\exp(-jk_{tx} x - jk_{tz} z). \end{aligned} \tag{3-32}$$

The reflection and transmission coefficients are

$$R_{\text{par}} = \frac{\epsilon_2 k_z - \epsilon_1 k_{tz}}{\epsilon_2 k_z + \epsilon_1 k_{tz}}; \; T_{\text{par}} = \frac{2\epsilon_2 k_z}{\epsilon_2 k_z + \epsilon_1 k_{tz}}. \tag{3-33}$$

3-2-3 Impedances

The perpendicular and parallel reflection coefficients can be manipulated and expressed in a variety of forms. One of these, which is very useful in later chapters (see, especially, Chapter 16), is the impedance concept, which can be conveniently introduced in connection with Eq. 3-23 and 3-33 above.

Consider the perpendicular case first. Rewrite Eq. 3-23 as

$$R_{\text{per}} = \frac{\mu_2 N_1 \cos\theta_1 - \mu_1 N_2 \cos\theta_2}{\mu_2 N_1 \cos\theta_1 + \mu_1 N_2 \cos\theta_2}. \tag{3-34}$$

If we introduce the ratios

$$\eta_i = \sqrt{\frac{\mu_i}{\epsilon_i}} \tag{3-35}$$

where i refers to a particular layer, then we can write Eq 3-34 in the form:

$$R_{\text{per}} = \frac{\eta_2 \cos\theta_1 - \eta_1 \cos\theta_2}{\eta_2 \cos\theta_1 + \eta_1 \cos\theta_2} \tag{3-36}$$

or, defining the perpendicular polarization impedances for the ith layer

$$z_{\text{per},i} = \frac{\eta_i}{\cos\theta_i}, \tag{3-37}$$

as

$$R_{\text{per}} = \frac{z_{\text{per},2} - z_{\text{per},1}}{z_{\text{per},2} + z_{\text{per},1}}. \tag{3-38}$$

For the parallel polarization case, we rewrite Eq. 3-33 as

$$R_{\text{par}} = \frac{\epsilon_2 N_1 \cos\theta_1 - \epsilon_1 N_2 \cos\theta_2}{\epsilon_2 N_1 \cos\theta_1 + \epsilon_1 N_2 \cos\theta_2}. \tag{3-39}$$

The parallel polarization impedances we define as

$$z_{\text{par},i} = \eta_i \cos\theta_i, \tag{3-40}$$

whence

$$R_{\text{par}} = \frac{\eta_1 \cos\theta_1 - \eta_2 \cos\theta_2}{\eta_1 \cos\theta_1 + \eta_2 \cos\theta_2} = \frac{z_{\text{par},1} - z_{\text{par},2}}{z_{\text{par},1} + z_{\text{par},2}}. \tag{3-41}$$

3-2-4 Special Case: $\theta = 0$

If $\theta = 0 = \theta_t$, then we find from Eqs. 3-23 and 3-33 that the perpendicular and parallel polarization reflection coefficients become, respectively,

$$R_{\text{per}} = \frac{\sqrt{\epsilon_1\mu_2} - \sqrt{\mu_1\epsilon_2}}{\sqrt{\epsilon_1\mu_2} + \sqrt{\mu_1\epsilon_2}} \tag{3-42}$$

$$R_{\text{par}} = \frac{\sqrt{\epsilon_2\mu_1} - \sqrt{\epsilon_1\mu_2}}{\sqrt{\epsilon_2\mu_1} + \sqrt{\epsilon_1\mu_2}} \tag{3-43}$$

It appears from Eqs. 3-42 and 3-43 that, at $\theta=0$, $R_{\text{per}}=R_{\text{par}}$, or that the two reflected fields would be 180° out of phase. The apparent discrepancy lies in the relation between incident and reflected fields, made clear by Figs. 3-2 and 3-3. In Fig. 3-2 the two are equal, in magnitude and direction, for $\theta=0$. But in Fig. 3-3 the two are equal in magnitude and 180° out of phase. The sign of R_{par} rectifies this seeming difficulty.

3-2-5 Brewster Angle

Let $\mu_1=\mu_2$. Consider $R_{\text{per}}=0$. Then $k_z=k_{tz}$ or

$$\omega\sqrt{\mu_1\epsilon_1}\cos\theta=\omega\sqrt{\mu_1\epsilon_2}\cos\theta_t.$$

By phase matching we also have

$$\omega\sqrt{\mu_1\epsilon_1}\sin\theta=\omega\sqrt{\mu_1\epsilon_2}\sin\theta_t.$$

To satisfy these two equations simultaneously we must have

$$\theta=\theta_t \qquad \text{and} \qquad \epsilon_2=\epsilon_1;$$

that is, the two media must be the same medium. But $R_{\text{par}}=0$ implies

$$\omega\sqrt{\mu_1\epsilon_2}\cos\theta_b=\omega\sqrt{\mu_1\epsilon_1}\cos\theta_t$$

and from phase matching

$$\omega\sqrt{\mu_1\epsilon_1}\sin\theta_b=\omega\sqrt{\mu_1\epsilon_2}\sin\theta_t.$$

These equations are satisfied simultaneously if

$$\theta_b+\theta_t=\frac{\pi}{2} \qquad \text{and} \qquad \theta_b=\arctan\sqrt{\frac{\epsilon_2}{\epsilon_1}}.$$

The angle θ_b is called the *Brewster angle.*

If an arbitrarily polarized wave is incident at the Brewster angle on a plane boundary between two media ($\mu_1=\mu_2$), the reflected wave comprises only a perpendicularly polarized component because the parallel polarized component is totally transmitted. So the Brewster angle may also be called the polarization angle.

4

GREEN'S FUNCTIONS

E can be obtained by operating on the current **J** with a dyad $\mathfrak{G}(\mathbf{r}, \mathbf{r}')$ called a Green's function. Set

$$\mathbf{E}(\mathbf{r}) = \int_\tau d\tau(\mathbf{r}')\mathfrak{G}(\mathbf{r}, \mathbf{r}') \cdot \mathbf{J}(\mathbf{r}') \tag{4-1}$$

where $\mathbf{r}$ is an observation point, and $\mathbf{r}'$ a source, or integration, point, and $d\tau$ a differential increment of the domain τ over which the argument $\mathbf{r}'$ of $\mathbf{J}(\mathbf{r}')$ ranges. Substituting **E** into the wave equation gives

$$\boldsymbol{\nabla} \times \boldsymbol{\nabla} \times \mathfrak{G}(\mathbf{r}, \mathbf{r}') - k^2\mathfrak{G}(\mathbf{r}, \mathbf{r}') = \mathfrak{I}\delta(\mathbf{r} - \mathbf{r}'), \tag{4-2}$$

since the delta function $\delta(\mathbf{r} - \mathbf{r}')$ has the property that

$$\int_\tau d\tau(\mathbf{r}')\mathbf{J}(\mathbf{r}')\,\delta(\mathbf{r} - \mathbf{r}') = \mathbf{J}(\mathbf{r}).$$

$\mathfrak{G}$ can be determined from a scalar Green's function $g(\mathbf{r}, \mathbf{r}')$, defined by

$$\mathfrak{G}(\mathbf{r}, \mathbf{r}') = (\mathfrak{I} + k^{-2}\boldsymbol{\nabla\nabla})g(\mathbf{r}, \mathbf{r}'), \tag{4-3}$$

where $g(\mathbf{r}, \mathbf{r}')$ satisfies the inhomogeneous Helmholtz equation

$$(\nabla^2 + k^2)g(\mathbf{r}, \mathbf{r}') = -\delta(\mathbf{r} - \mathbf{r}'). \tag{4-4a}$$

The manipulations necessary to verify Eq. 4-3 can be understood most easily by using matrix representations for the operators $\boldsymbol{\nabla}$ and $\boldsymbol{\nabla}\times$. Apply $\boldsymbol{\nabla} \times \boldsymbol{\nabla}\times - k^2$ to the expression for $\mathfrak{G}$ given by Eq. 4-3:

$$\begin{aligned}(\boldsymbol{\nabla} \times \boldsymbol{\nabla} \times - k^2)\mathfrak{G} &= \mathfrak{I}\nabla^2 g + k^{-2}\nabla^2\boldsymbol{\nabla\nabla} g + k^2\mathfrak{I}g \\ &\quad + \boldsymbol{\nabla\nabla} g - \boldsymbol{\nabla\nabla} g - k^{-2}\boldsymbol{\nabla\nabla\nabla\nabla} g \\ &= \mathfrak{I}(\nabla^2 + k^2)g + k^{-2}(\mathfrak{I}\nabla^2 - \boldsymbol{\nabla\nabla})\boldsymbol{\nabla\nabla} g.\end{aligned}$$

If we choose g so that Eq. 4-4a is satisfied, then for $\mathfrak{G}$ to be a solution of Eq. 4-2 we must show that the operator

$$(\mathfrak{I}\nabla^2 - \nabla\nabla)\nabla\nabla = -(\nabla\times\nabla\times)\nabla\nabla \tag{4-4b}$$

is *identically* zero.

First, we recall from Section 1 that $\nabla\times$ can be represented by a tensor of rank 2, that is, the operator (which operates on vectors, of course)

$$\nabla\times = \begin{pmatrix} 0 & -\partial/\partial z & \partial/\partial y \\ \partial/\partial z & 0 & -\partial/\partial x \\ -\partial/\partial y & \partial/\partial x & 0 \end{pmatrix}. \tag{4-4c}$$

So $\nabla\times\nabla\times$ can be represented as a simple matrix product.

Second, the expression $\nabla\nabla$ (or **grad grad**) can also be written as a matrix operator (Section 1):

$$\nabla\nabla = \begin{pmatrix} \partial/\partial x \\ \partial/\partial y \\ \partial/\partial z \end{pmatrix}(\partial/\partial x \ \partial/\partial y \ \partial/\partial z) = \begin{pmatrix} \partial^2/\partial x^2 & \partial^2/\partial x\partial y & \partial^2/\partial x\partial z \\ \partial^2/\partial y\partial x & \partial^2/\partial y^2 & \partial^2/\partial y\partial z \\ \partial^2/\partial z\partial x & \partial^2/\partial z\partial y & \partial^2/\partial z^2 \end{pmatrix}. \tag{4-4d}$$

Now we need only make the final assumption that any function of g, to which operators $\nabla\times\nabla\times$ and $\nabla\nabla$ are applied, has derivatives that are well behaved enough to permit exchanges in order of application of all derivatives up to order four. In such cases we may combine Eqs. 4-4c and 4-4d to show

$$\nabla\times\nabla\times\nabla\nabla = \mathfrak{O} \tag{4-4e}$$

where $\mathfrak{O}$ here stands for the 3×3 zero matrix operator.

It is useful to observe further some additional operations required to derive **H** from Eq. 4-1 above for **E**. We start with

$$\text{F: } \mathbf{H} = \left(\frac{j}{\omega\mu}\right)\nabla\times\mathbf{E},$$

and investigate $\nabla\times(\mathfrak{G}\cdot\mathbf{J})$ where $\nabla\times$ operates on the $\mathbf{r}$ argument of $\mathfrak{G}(\mathbf{r},\mathbf{r}')$:

$$\nabla\times(\mathfrak{I}g + k^{-2}\nabla\nabla g) = \nabla\times\mathfrak{I}g$$

since $\nabla\times\nabla\nabla = 0$. Then

$$\nabla\times(\mathfrak{G}\cdot\mathbf{J}) = (\nabla\times\mathfrak{I}g)\cdot\mathbf{J} = \nabla g\times\mathbf{J}.$$

This form is similar to the one we encounter later (Chapter 8) in our development of the physical optics approximation for the scattered field.

4-1 ASIDE ON THE δ-FUNCTION AND THE FOURIER TRANSFORM

It is useful to insert here some brief remarks on the δ-function and the Fourier transform. The interested reader may wish to consult more extensive treatments of these subjects given by Lighthill (1959), Carslaw (1930), Morse & Feshbach (1953), and Tai (1971).

4-1-1 The δ-Function

A δ-function $\delta(x)$ is a "generalized function" with the property that

$$\int_{-\infty}^{\infty} \delta(x)F(x)\,dx = F(0),$$

where $F(x)$ is any suitably continuous function of x. Continued parts integration can be used to show that

$$\int_{-\infty}^{\infty} \delta^{(n)}(x)f(x)\,dx = (-1)^n F^{(n)}(0),$$

where now F must have n derivatives.

The δ-function properties suggest certain physical idealizations. For example, the δ-function itself can be used to denote the distribution of charge along the x-axis called a point charge. And the δ-function derivative $\delta'(x)$ then corresponds to a dipole of unit electric moment, since

$$\int_{-\infty}^{\infty} x\delta'(x)\,dx = -1.$$

The notion of a generalized function $f(x)$ can be put on a firm basis by defining it as the equivalent of a sequence of analytic functions $f_n(x)$ such that

$$\lim_{n\to\infty} \int_{-\infty}^{\infty} f_n(x)F(x)\,dx = \int_{-\infty}^{\infty} f(x)F(x)\,dx$$

exists. For example, the sequence $\{\sqrt{n/\pi}\exp(-nx^2)\}$ defines a $\delta(x)$ such that

$$\int_{-\infty}^{\infty} \delta(x)F(x)\,dx = F(0).$$

We note that the sign function $\operatorname{sgn}(x) = 1,\ x > 0,\ \ = -1,\ x < 0,$ yields $(d/dx)\operatorname{sgn}(x) = 2\delta(x)$.

In three dimensions, we have, in three coordinate systems

$$\delta(\mathbf{r}) = \delta(x)\,\delta(y)\,\delta(z) \text{ (in rectangular coordinates);} \tag{4-5a}$$

$$= \frac{1}{2\pi\rho}\,\delta(\rho)\,\delta(z) \text{ (in cylindrical coordinates);} \tag{4-5b}$$

$$= \frac{1}{4\pi r^2}\,\delta(r) \text{ (in spherical coordinates)†;} \tag{4-5c}$$

since

$$\int \delta(\rho)\,d\rho = 1 = \int \delta(x)\,\delta(y)\,dx\,dy = \int \delta(x)\,\delta(y)\,(2\pi\rho)\,d\rho;$$

and

$$\int \delta(r)\,dr = 1 = \int \delta(x)\,\delta(y)\,\delta(z)\,dx\,dy\,dz = \int \delta(x)\,\delta(y)\,\delta(z)\,4\pi r^2\,dr.$$

4-1-2 The Fourier Transform

The Fourier transform $h(y)$ of $f(x)$ is defined as (using the standard engineering $j = \sqrt{-1}$)

$$h(y) = FT\{f(x)\} = \int_{-\infty}^{\infty} f(x)e^{-2\pi jxy}\,dx. \tag{4-6a}$$

Some authors use a variable change to bring out the factor 2π as a divisor of the integral. An inversion theorem may be based on this result and stated as follows:

$$\text{If } h(y) = FT\{f(x)\}, \text{ then } f(y) = FT\{h(-x)\}$$

$$= \int_{-\infty}^{\infty} h(-x)e^{-2\pi jxy}\,dx = \int_{-\infty}^{\infty} h(x)e^{2\pi jxy}\,dx. \tag{4-6b}$$

The generalized functions and, in particular, δ-functions can be combined with the Fourier transform theory. For example, we have

$$FT\{\delta(x-c)\} = e^{-2\pi jcy};$$

$$FT\{e^{2\pi jcx}\} = \delta(y-c).$$

And finally, a connection of the δ-function and Fourier transform concepts can be made with the Green's function concept. It can be shown (Tai, 1971) that a three-dimensional Fourier integral transform can be

† In Eq. 4-4a above some authors incorporate a factor of 4π with the δ-function on the right-hand side, obtaining a free space Green's function $(1/R)\exp(-jkR)$; others choose to keep the right-hand side as in Eq. 4-4a and use $(1/4\pi R)\exp(-jkR)$ as the free space Green's function.

applied to Eq. 4-4a to obtain the form of the free space Green's function,

$$g(\mathbf{r}, \mathbf{r}') = \frac{\exp(-jk \parallel \mathbf{r} - \mathbf{r}' \parallel)}{4\pi \parallel \mathbf{r} - \mathbf{r}' \parallel}. \tag{4-7}$$

This expression in turn can be substituted into Eq. 4-1 to give a representation for the electric field:

$$\mathbf{E}(\mathbf{r}) = -j\omega\mu(\mathfrak{I} + k^{-2}\nabla\nabla) \cdot \int d\tau(\mathbf{r}') \frac{\exp(-jk \parallel\mathbf{r} - \mathbf{r}'\parallel)}{4\pi \parallel\mathbf{r} - \mathbf{r}'\parallel} \mathbf{J}(\mathbf{r}'), \tag{4-8}$$

where the dyadic has been removed from the integral since it operates on $\mathbf{r}$ (not $\mathbf{r}'$).

The Green's function idea is treated from a somewhat simpler viewpoint in Section 9-2.

4-2 FAR APPROXIMATION

Let $r \gg r'$. Then $\|\mathbf{r} - \mathbf{r}'\| \approx r - \mathbf{r}' \cdot \hat{r}$, where $\mathbf{r} = r\hat{r}$. If $kr \gg 1$, we can define a radiation vector

$$\mathbf{A} = \int d\tau(\mathbf{r}') \frac{\exp(-jk \parallel\mathbf{r} - \mathbf{r}'\parallel)}{4\pi \parallel\mathbf{r} - \mathbf{r}'\parallel} \mathbf{J}(\mathbf{r}') \approx \frac{\exp(-jkr)}{4\pi r} \int d\tau(\mathbf{r}')\mathbf{J}(\mathbf{r}') \exp(j\mathbf{k} \cdot \mathbf{r}') \tag{4-9}$$

where we notice that $-\mathbf{r}' \cdot \hat{r}$ is neglected in the denominator but *not* in the exponent. The integral Eq. 4-9 has the form of a Fourier transform (see remarks below). Since the $\mathbf{k}$-vector is in the direction $\mathbf{r}$, we can replace ∇ by $-jk$ and write

$$\mathbf{E}(\mathbf{r}) \approx \left(-\frac{j}{\omega\epsilon}\right)(k^2\mathfrak{I} - \mathbf{kk}) \cdot \mathbf{A} = -j\omega\mu(\hat{\theta}A_\theta + \hat{\phi}A_\phi) \tag{4-10}$$

$$\mathbf{H}(\mathbf{r}) \approx -jk(\hat{\phi}A_\theta - \hat{\theta}A_\phi). \tag{4-11}$$

$\mathbf{A}$ then is the vector to find, given the source distribution $\mathbf{J}(\mathbf{r}')$. (Refer to Chapter 16 for a somewhat different viewpoint on this subject.)

4-3 HERTZIAN DIPOLE

Consider an infinitesimal dipole antenna composed of a current element with harmonic time dependence $\exp(j\omega t)$. Suppose it is situated at the origin and pointed in the $+\hat{z}$-direction.

This dipole can be viewed as being composed of two opposite charges $\pm q$ separated by distance s such that as $q \to \infty$, $s \to 0$, the product $p = qs$ remains finite. The dipole oscillates with frequency ω, of course.

The current is $I = j\omega q$. So the moment is $Is = j\omega p$. Such a dipole is called a Hertzian dipole:

$$\mathbf{J} = \hat{z} Is\delta(r'). \tag{4-12}$$

Substitute Eq. 4-12 into the **A** expression:

$$\mathbf{A} = \frac{Is \exp(-jkr)}{4\pi r}\hat{z}.$$

So

$$\mathbf{E}(\mathbf{r}) = -j\omega\mu(\Im + k^{-2}\nabla\nabla) \cdot \hat{z}\frac{Is \exp(-jkr)}{4\pi r} \tag{4-13}$$

where we have used

$$\frac{\partial}{\partial z}\frac{\exp(-jkr)}{r} = -\left(jk + \frac{1}{r}\right)\frac{\exp(-jkr)}{r}\cos\theta.$$

The electric field, we see, has *no* circumferential component. The magnetic field

$$\mathbf{H} = \hat{\phi}\frac{jkIs \exp(-jkr)}{4\pi r}\sin\theta = \hat{\phi}\frac{1}{\eta}E_\theta, \qquad \eta = \sqrt{\frac{\mu}{\epsilon}}, \tag{4-14}$$

provided, for $kr \gg 1$, only terms of order $1/r$ are important which implies

$$\mathbf{E} \approx \hat{\theta}\frac{j\omega\mu Is \exp(-jkr)}{4\pi r}\sin\theta. \tag{4-15}$$

So **H** has *only* a circumferential component. Therefore **E** is normal to **H**.

If the dipole is enclosed in a sphere of radius much larger than λ, then **E** is tangent to the sphere. The radiation pattern **E** consists of two circles on each side of the dipole axis in a plane containing the dipole axis. Since **E** is independent of ϕ, the three-dimensional pattern is obtained by rotating the circles about the dipole axis.

The radiation pattern is a plot of

$$\langle \mathbf{S} \rangle = \frac{1}{2}Re(\mathbf{E} \times \mathbf{H}^*) = \hat{r}\frac{\eta}{2}\left(\frac{kIs}{4\pi r}\right)^2 \sin^2\theta. \tag{4-16}$$

In the static limit, that is $kr \ll 1$, we keep only terms of order $1/r^3$:

$$\mathbf{E} \approx \left(\frac{p}{4\pi\epsilon r^3}\right)(\hat{r}\, 2\cos\theta + \hat{\theta}\sin\theta) \tag{4-17}$$

which is the solution for a static electric dipole.

4-4 DUALITY

A cursory examination of Maxwell's symmetric set of equations reveals the validity of a transformation called duality whose application leaves

Maxwell's equations unchanged. The transformation is defined in terms of replacement: $A \rightarrow B$, read A is replaced by B (or A is mapped onto B). The complete duality transformation encompasses the following set of replacements:

$$\mathbf{E} \rightarrow \mathbf{H} \qquad \mathbf{H} \rightarrow -\mathbf{E} \qquad \epsilon \rightarrow \mu$$

$$\mathbf{D} \rightarrow \mathbf{B} \qquad \mathbf{B} \rightarrow -\mathbf{D} \qquad \mu \rightarrow \epsilon$$

$$\mathbf{J} \rightarrow \mathbf{M} \qquad \mathbf{M} \rightarrow -\mathbf{J}$$

$$\rho \rightarrow m \qquad m \rightarrow -\rho$$

Carrying this idea one step further by letting electric conductors $\rightarrow$ magnetic conductors, and sources $\rightarrow$ duals, we can obtain *Babinet's principle*:

Consider a plane screen of electrical conductors. In the absence of the screen, let electromagnetic fields be **E**, **H**. Denote electric sources as *Is*, magnetic sources as *Ks*. In the presence of the screen, let electromagnetic fields be denoted $\mathbf{E}_a$, $\mathbf{H}_a$. So boundary conditions are

$$\hat{n} \times \mathbf{E}_a = 0 \text{ on } S_m,$$

$$\hat{n} \times \mathbf{H}_a = \hat{n} \times \mathbf{H} \text{ on } S_a,$$

where S_m denotes conductor area, S_a aperture area.

Consider a second screen of magnetic conductors. The two screens are *complementary* if, by superposition, the two cover the whole plane with *no* overlay. Letting $\mathbf{E}_m$, $\mathbf{H}_m$ be fields in the presence of the magnetically conducting screen, the boundary conditions are

$$\hat{n} \times \mathbf{E}_m = \hat{n} \times \mathbf{E} \text{ over } S_m,$$

$$\hat{n} \times \mathbf{H}_m = 0 \text{ over } S_a.$$

These two screens are complementary. After superposition,

$$\hat{n} \times (\mathbf{E}_a + \mathbf{E}_m) = \hat{n} \times \mathbf{E} \text{ on } S_m,$$

$$\hat{n} \times (\mathbf{H}_a + \mathbf{H}_m) = \hat{n} \times \mathbf{H} \text{ on } S_a;$$

and so the magnetic current sheets corresponding to the original tangential **E** are specified on S_m while the electric current sheets corresponding to the original tangential **H** are specified on S_a. By uniqueness of solutions (below)

$$\mathbf{E}_a + \mathbf{E}_m = \mathbf{E},$$

$$\mathbf{H}_a + \mathbf{H}_m = \mathbf{H},$$

which is Babinet's principle.

4-5 UNIQUENESS

Assume that a set of sources inside a volume V produces two different field solutions: $\mathbf{E}_1$, $\mathbf{H}_1$ and $\mathbf{E}_2$, $\mathbf{H}_2$. Their differences $\delta\mathbf{E}$, $\delta\mathbf{H}$ satisfy Maxwell's equations and so to prove uniqueness of solution we must show that they are zero.

From the source-free Maxwell's equations,

$$\nabla \times \delta \mathbf{H} = j\omega(\boldsymbol{\epsilon} \cdot \delta \mathbf{E} + \boldsymbol{\xi} \cdot \delta \mathbf{H})$$
$$-\nabla \times \delta \mathbf{E} = j\omega(\boldsymbol{\zeta} \cdot \delta \mathbf{E} + \boldsymbol{\mu} \cdot \delta \mathbf{H}),$$

it can be shown that

$$\begin{aligned}\nabla \cdot (\delta \mathbf{E} \times \delta \mathbf{H}^*) &= j\omega(\delta \mathbf{E} \cdot \boldsymbol{\epsilon}^* \cdot \delta \mathbf{E}^* + \delta \mathbf{E} \cdot \boldsymbol{\xi}^* \cdot \delta \mathbf{H}^* \\ &\quad - \delta \mathbf{H}^* \cdot \boldsymbol{\zeta} \cdot \delta \mathbf{E} - \delta \mathbf{H}^* \cdot \boldsymbol{\mu} \cdot \delta \mathbf{H}) \\ \nabla \cdot (\delta \mathbf{E}^* \times \delta \mathbf{H}) &= j\omega(\delta \mathbf{E}^* \cdot \boldsymbol{\epsilon} \cdot \delta \mathbf{E} + \delta \mathbf{E}^* \cdot \boldsymbol{\xi} \cdot \delta \mathbf{H} \\ &\quad - \delta \mathbf{H} \cdot \boldsymbol{\zeta}^* \cdot \delta \mathbf{E}^* - \delta \mathbf{H} \cdot \boldsymbol{\mu}^* \cdot \delta \mathbf{H}^*).\end{aligned}$$

Adding and integrating over V:

$$\oiint \mathbf{dS} \cdot (\delta \mathbf{E} \times \delta \mathbf{H}^* + \delta \mathbf{E}^* \times \delta \mathbf{H})$$
$$= \iiint_V dV[\delta \mathbf{E}^* \;\; \delta \mathbf{H}^*] \cdot \left[-j\omega\, \delta \mathfrak{C}_{\mathbf{E}\,\mathbf{H}}\right] \cdot \begin{pmatrix} \delta \mathbf{E} \\ \delta \mathbf{H} \end{pmatrix}$$

where

$$\delta \mathfrak{C}_{\mathbf{E}\,\mathbf{H}} = \begin{pmatrix} \boldsymbol{\epsilon} - \boldsymbol{\epsilon}^+ & \boldsymbol{\xi} - \boldsymbol{\xi}^+ \\ \boldsymbol{\zeta} - \boldsymbol{\zeta}^+ & \boldsymbol{\mu} - \boldsymbol{\mu}^+ \end{pmatrix}$$

and $\mathbf{dS} = \hat{n}\, dS$, $\hat{n}$ being the surface normal.

We let $-j\omega\boldsymbol{\delta}\mathfrak{C}_{\mathbf{E}\,\mathbf{H}}$ be negative (or positive) definite, which means the medium is passive (or active). If the medium is not passive (or active), it can be treated as the limiting case of a passive (or active medium). For example, for a passive isotropic medium $\epsilon - \epsilon^* = -2j\epsilon''$ and $\mu - \mu^* = -2j\mu''$. So the volume integral becomes $-2\epsilon''\, \|\delta \mathbf{E}\|^2 - 2\mu''\, \|\delta \mathbf{H}\|^2$ and is therefore negative definite. Therefore, if the left-hand side $= 0$, we then get $\delta \mathbf{E} = \delta \mathbf{H} = 0$.

But the left-hand side $= 0$ if (1) $\hat{n} \times \mathbf{E}$ is specified over the whole boundary S, or (2) $\hat{n} \times \mathbf{H}$ is specified over the whole boundary of S. So the solution of Maxwell's equations is unique for a volume V bounded by a closed surface S if either $\hat{n} \times \mathbf{E}$ or $\hat{n} \times \mathbf{H}$ (but not both) is specified over the entire surface S.

4-6 EQUIVALENT SOURCE

When we are interested in a limited region of space, we can replace all uninteresting regions outside of this limited region by equivalent sources; or place equivalent sources in uninteresting regions; or place equivalent current sheets on boundaries of regions of interest.

Equivalent sources are not unique and there are many different ways of constructing them. But we must take care that

- Boundary conditions are satisfied; and
- Original sources and fields in the region of interest are preserved.

When two different representations give the same solution in the region of interest (but not necessarily outside of that region), they are said to be *equivalent.*

We can view a small current loop as a magnetic dipole if we enclose both in a small volume inside of which we have no interest. Only when we get inside of the volume can we distinguish the two. (In the interior, magnetic fields of a loop and a magnetic dipole point in the opposite direction.)

We may use the following building blocks:

- Electric dipoles for electric current sources;
- Magnetic dipoles for magnetic current sources.

Electric current sheets $\mathbf{J}_s$ are produced by discontinuities in the tangential magnetic field across a boundary (see Chapter 3):

$$\mathbf{J}_s = \hat{n} \times \delta\mathbf{H} \text{ (right-hand).} \tag{4-18}$$

Magnetic current sheets $\mathbf{K}_s$ are produced by discontinuities in the tangential electric field across a boundary:

$$\mathbf{K}_s = -\hat{n} \times \delta\mathbf{E} \text{ (left-hand).} \tag{4-19}$$

Two Dual Situations

1. A surface electric current sheet with surface current $\mathbf{J}_s$ at $z = 0$ flowing in the $-\hat{x}$-direction generates plane waves in both $\pm\hat{z}$-directions:

$$E_x = \frac{1}{2}\eta J_s\, e^{-jkz}, \qquad H_y = \frac{1}{\eta} E_x, \qquad z > 0; \tag{4-20}$$

$$E_x = \frac{1}{2}\eta J_s\, e^{jkz}, \qquad H_y = -\frac{1}{\eta} E_x, \qquad z < 0. \tag{4-21}$$

   All boundary conditions are satisfied; viz. tangential electric fields are continuous at $z = 0$ and the discontinuity in tangential magnetic fields on both sides is equal to the strength of the current sheet.

2. A surface magnetic current sheet with surface current $\mathbf{K}_s$ at $z = 0$ flowing in the $-\hat{y}$-direction generates plane waves in both $\pm\hat{z}$-directions:

$$H_y = \frac{1}{2\eta} K_s\, e^{-jkz}, \qquad E_x = \eta H_y, \qquad z > 0; \tag{4-22}$$

$$H_y = \frac{1}{2\eta} K_s\, e^{jkz}, \qquad E_x = -\eta H_y, \qquad z < 0. \tag{4-23}$$

Note that by suitable choices of current sheets we can generate a plane wave with arbitrary direction such as $\mathbf{J}_s = \hat{x} J_s e^{-jk_y y}$ gives wave vectors $\mathbf{k} = \hat{y} k_y + \hat{z} k_z$ and $\mathbf{k} = \hat{y} k_y - \hat{z} k_z$. If $k_y > k_z$, the waves are evanescent in the $\hat{z}$-directions.

Examples Let's look at equivalent situations for a plane wave propagating in the $\hat{z}$-direction. Let

$$\mathbf{E} = \hat{x} E_x e^{-jkz}, \qquad \mathbf{H} = \hat{y} \frac{1}{\eta} E_0 e^{-jkz}$$

with interesting region $z > 0$. Equivalences are:

1. Combine an electric current sheet with $\mathbf{J}_s = +\hat{x} \frac{1}{\eta} E_0$ and a magnetic current sheet with $\mathbf{K}_s = \hat{y} E_0$. The same field is preserved for $z > 0$. For $z < 0$, there is no field.
2. Choose an electric current sheet with $\mathbf{J}_s = -\hat{x} \frac{2}{\eta} E_0$. The same field results for $z > 0$. For $z < 0$, we have a plane wave with $\mathbf{E} = \hat{x} E_0\, e^{jkz}$.
3. Choose a magnetic current sheet with $\mathbf{K}_s = -\hat{y} 2E_0$. The same field results for $z > 0$. For $z < 0$, we have a plane wave with $\mathbf{H} = \hat{y} \frac{1}{\eta} E_0\, e^{jkz}$.
4. Let $z < 0$ be a perfect conductor. Place in front of the conductor an electric current sheet with $\mathbf{J}_s = \hat{x} \frac{1}{\eta} E_0$ and a magnetic current sheet with $\mathbf{K}_s = -\hat{y} E_0$. The electric current sheet generates no field because an equal and opposite electric current sheet is induced on the surface of the conductor which cancels $\mathbf{J}_s$.
5. For the dual case, we introduce the concept of magnetic conductor on the surface of which tangential magnetic fields vanish; suppose this condition extends throughout the region $z < 0$. Place the same electric and magnetic current sheets as in example 4. Then the magnetic current generates no field; the electric current sheet generates the plane wave for $z > 0$.

We must distinguish between *impressed* and *induced* current sheets. When a layer of charge or current is impressed on a surface, induced surface charge and current sheets are generated at the surface so that boundary conditions are satisfied.

Consider a plane wave incident on a perfectly conducting half-space bounded by $z = 0$. Let the electric field be

$$\mathbf{E} = \hat{x}E_0 e^{-jkz}.$$

This field induces an electric current sheet $\mathbf{J}_s = \hat{x}2E_0/\eta$ on the conducting surface. The conductor can be replaced by the current sheet, which radiates into both $z > 0$ and $z < 0$ half-spaces. It generates a reflected field

$$\mathbf{E} = -\hat{x}E_0 e^{jkz}$$

so that at the boundary the total field is zero. The induced current sheet generates a plane wave

$$\mathbf{E} = -\hat{x}E_0 e^{-jkz}$$

in the region $z > 0$, which combines with the incident wave to produce a zero field inside the conductor.

Magnetic sources as a concept are useful, even though they do not exist. So we add an equivalent magnetic source to Faraday's law (and repeat Ampère's law for comparison):

$$\text{F: } -\nabla \times \mathbf{E} = j\omega\mathbf{B} + \mathbf{K};$$

$$\text{A: } \nabla \times \mathbf{H} = j\omega\mathbf{D} + \mathbf{J}.$$

So we have in addition to the boundary condition $\hat{n} \times \delta\mathbf{H} = \mathbf{J}_s$, a similar condition for tangential electric fields: $-\hat{n} \times \delta\mathbf{E} = \mathbf{K}_s$.

In addition, it is also necessary to include a magnetic charge term in Gauss's magnetic flux law (GB). These laws read:

$$\text{GB: } \nabla \cdot \mathbf{B} = m;$$

$$\text{GD: } \nabla \cdot \mathbf{D} = \rho.$$

The equivalence notion is useful in converting a problem into one where all uninteresting regions are replaced by equivalent sources. These sources are unknown until the problem is completely solved. In practice they are useful to obtain approximate solutions. If equivalent sources are given, we can readily construct solutions using Huygens's principle. (Baker and Copson, 1953).

4-7 HUYGENS'S PRINCIPLE

Huygens's principle expresses fields at an observation point in terms of fields at a boundary surface. Let S' be a surface enclosing a radiation source. Integrate the identity

$$\mathbf{P} \cdot \nabla \times \nabla \times \mathbf{Q} - \mathbf{Q} \cdot \nabla \times \nabla \times \mathbf{P} = \nabla \cdot (\mathbf{Q} \times \nabla \times \mathbf{P} - \mathbf{P} \times \nabla \times \mathbf{Q}) \quad (4\text{-}24)$$

over the volume V enclosed by S' and S_∞, the surface at infinity, letting

$\mathbf{P} = \mathbf{E}$ and $\mathbf{Q} = \mathfrak{G} \cdot \mathbf{a}$, $\mathbf{a}$ an arbitrary constant vector:

$$\iiint dV\{\mathbf{E}(\mathbf{r}) \cdot \nabla \times \nabla \times \mathfrak{G}(\mathbf{r}, \mathbf{r}') \cdot \mathbf{a} - [\nabla \times \nabla \times \mathbf{E}(\mathbf{r})] \cdot \mathfrak{G}(\mathbf{r}, \mathbf{r}') \cdot \mathbf{a}\}$$

$$= -\oint\!\!\oint dS\,\hat{s} \cdot \{[\nabla \times \mathbf{E}(\mathbf{r})] \times \mathfrak{G}(\mathbf{r}, \mathbf{r}') \cdot \mathbf{a} + \mathbf{E}(\mathbf{r}) \times \nabla \times \mathfrak{G}(\mathbf{r}, \mathbf{r}') \cdot \mathbf{a}\}, \quad (4\text{-}25)$$

where we have set $S = S' + S_\infty$. Equation (4-25) is the vector analog of Green's theorem. On its right-hand side the surface at infinity does not contribute; and on S', $\hat{s} = -\hat{n}$, where $\hat{n}$ is the outward normal to S'. On its left-hand side, we use the equation satisfied by $\mathfrak{G}(\mathbf{r}, \mathbf{r}')$; moreover, $\nabla \times \nabla \times \mathbf{E}(\mathbf{r}) = k^2\mathbf{E}(\mathbf{r})$, since $\mathbf{J}(\mathbf{r}) = 0$ in V. So

$$-j\omega\mu \iiint dV\, \delta(\mathbf{r} - \mathbf{r}')\mathbf{E}(\mathbf{r}) \cdot \mathbf{a} = \oint\!\!\oint dS\{[\hat{n} \times \nabla \times \mathbf{E}(\mathbf{r})] \cdot \mathfrak{G}(\mathbf{r}, \mathbf{r}') \cdot \mathbf{a}$$

$$+ [\hat{n} \times \mathbf{E}(\mathbf{r})] \cdot \nabla \times \mathfrak{G}(\mathbf{r}, \mathbf{r}') \cdot \mathbf{a}\}. \quad (4\text{-}26)$$

Cancel $\mathbf{a}$ from both sides, replace $\nabla \times \mathbf{E}$ by $-j\omega\mu\mathbf{H}$, and interchange primed and unprimed variables to get

$$-j\omega\mu\mathbf{E}(\mathbf{r}) = \iint_{S'} dS\{-j\omega\mu\mathfrak{G}(\mathbf{r}, \mathbf{r}') \cdot [\hat{n} \times \mathbf{H}(\mathbf{r}')]$$

$$+ \nabla \times \mathfrak{G}(\mathbf{r}, \mathbf{r}') \cdot [\hat{n} \times \mathbf{E}(\mathbf{r}')]\}, \quad (4\text{-}27)$$

where the triple products in the surface integral have been rearranged. By duality

$$\mathbf{H}(\mathbf{r}) = \iint_{S'} dS\{j\omega\epsilon\mathfrak{G}(\mathbf{r}, \mathbf{r}') \cdot [\hat{n} \times \mathbf{E}(\mathbf{r}')]$$

$$+ \nabla \times \mathfrak{G}(\mathbf{r}, \mathbf{r}') \cdot [\hat{n} \times \mathbf{H}(\mathbf{r}')]\}. \quad (4\text{-}28)$$

In these equations we observe that $\hat{n} \times \mathbf{H} = \mathbf{J}$ and $-\hat{n} \times \mathbf{E} = \mathbf{K}$ are identifiable as current sheets. These concepts are examined again, in greater detail, and with emphasis directed toward computational methods in Section 9-2-1.

Example. Diffraction of a Plane Wave by a Slit. Let the slit be infinite in the $\hat{x}$-direction and have width w in the $\hat{y}$-direction, with $k\omega > 1$.

Let the plane wave be polarized linearly in the $\hat{x}$-direction and normally incident on the slit from the $z < 0$ region. The wave is diffracted into the half-space $z > 0$.

The solution uses the equivalence principle:

Assume that the field at the slit equals the original field. And, as we did in equivalence example 2, we assume at the aperture an electric current sheet $\mathbf{J}_s = 2\hat{z} \times \mathbf{H}$ and no magnetic current sheet:

$$\mathbf{H}(\mathbf{r}') = \hat{y}\left(\frac{E_0}{\eta}\right) U\left(\frac{w}{2} - |y'|\right), \quad (4\text{-}29)$$

where $U\left(\frac{w}{2}-|y'|\right)$ is a unit step function, $=1$ for $|y'|\leq w/2$, $=0$ for $|y'|\geq w/2$. So

$$\mathbf{E}(\mathbf{r}) = -2\iint_{S'} dS j\omega\mu \mathfrak{G}(\mathbf{r},\mathbf{r}')\cdot\hat{z}\times\mathbf{H}(\mathbf{r}'). \tag{4-30}$$

Using the two-dimensional Green's function (the Hankel function) and noting that $\partial/\partial x=0$, since the statements above are independent of x, we get

$$\mathbf{E}(y,z)=\hat{x}\frac{\omega\mu}{2\eta}\int_{-\infty}^{\infty} dy' E_0 U\left(\frac{w}{2}-|y'|\right)H_0^{(1)}[k\sqrt{(y-y')^2+z^2}]. \tag{4-31}$$

In the far zone we use

$$H_0^{(1)}(k\rho)\approx\sqrt{\frac{2}{-jk\pi\rho}}\,e^{-jk\rho}.$$

Assume that observation points are far from the aperture but near to the $\hat{z}$-axis. Then $|z|\gg|y-y'|$. So

$$k\sqrt{(y-y')^2+z^2}\approx kz\left[1+\frac{1}{2}\left(\frac{y-y'}{z}\right)^2+\cdots\right]$$

and

$$\mathbf{E}(y,z)\approx\hat{x}\frac{k}{2}\sqrt{\frac{-2}{jk\pi z}}\,e^{-jkz}\int_{-\infty}^{\infty} dy' E_0 U\left(\frac{w}{2}-|y'|\right)\exp\left(-\frac{jk(y-y')^2}{2z}\right) \tag{4-32}$$

by using the first two terms of the expansion above and the first term in the denominator of the Hankel function approximation. This latter is called the *Fresnel approximation.* The last integral can be evaluated in terms of Fresnel integrals (Magnus and Oberhettinger, 1954, p. 96).

The maximum value of y' is $w/2$. So if we neglect $ky'^2/2z$, we get

$$k\sqrt{(y-y')^2+z^2}\approx kz+\left(\frac{ky^2}{2z}\right)-\left(\frac{kyy'}{z}\right)+\cdots$$

and hence

$$\mathbf{E}(y,z)\approx\hat{x}\frac{k}{2}\sqrt{\frac{-2}{jk\pi z}}\exp\left[-jkz-\frac{jky^2}{2z}\right]$$
$$\times\int_{-\infty}^{\infty} dy' E_0 U\left(\frac{w}{2}-|y'|\right)\exp\left[j\left(\frac{ky}{z}\right)y'\right]. \tag{4-33}$$

This result, called the *Fraunhofer* approximation, is the Fourier transform of the step function and therefore

$$\mathbf{E}(y, z) = \hat{x}\frac{1}{2}E_0 k\omega\sqrt{\frac{-2}{j\pi kz}}\frac{\sin(kyw/2z)}{kyw/2z}\exp\left[-jkz - \frac{jky^2}{2z}\right]. \tag{4-34}$$

The observed field is the Fourier transform of the aperture field. According to distance from the aperture, the diffracted field can be divided into near zone, Fresnel zone, and Fraunhofer zone. The separation between Fresnel and Fraunhofer zones may be taken as $z_F = 2w^2/\lambda$.

5

RADAR CROSS SECTION (RCS)

Perhaps the best viewpoint from which to introduce RCS is the practical one embodied by the radar range equation. This equation can be made to show (1) the effect of a radar transmitter, with gain G_t, on a source P_t of power; (2) the effect of a target, with RCS σ, on the transmitted power incident upon it; and finally, (3) the effect of a radar receiver, with gain G_r, on the power returned by a target: to produce a received power P_r. Suppose the target is at a distance R from the transmitter and at a distance r from the receiver. Then we have (Ridenour, 1947)

$$\underbrace{P_r}_{\substack{\text{power}\\ \text{received}}} = \underbrace{\underbrace{\frac{P_t G_t}{4\pi R^2}}_{\substack{\text{incident}\\ \text{power density}\\ \text{at the target}}} \cdot \overbrace{\frac{\sigma}{4\pi r^2}}^{\substack{\text{gain imposed}\\ \text{by the target}}}}_{\substack{\text{incident power}\\ \text{density at receiver}}} \cdot \underbrace{\frac{G_r \lambda^2}{4\pi}}_{\substack{\text{effective}\\ \text{receiving}\\ \text{cross-section}}} \tag{5-1}$$

Note that σ depends not only on the frequency of the transmitter but also on the target aspect with respect to both transmitter and receiver. The latter need not be colocated, in which case σ is called bistatic RCS. The angle β, subtended at the target, by which transmitter and receiver are separated is called the bistatic angle. Figure 5-1 is a cartoon depicting the factors of Eq.

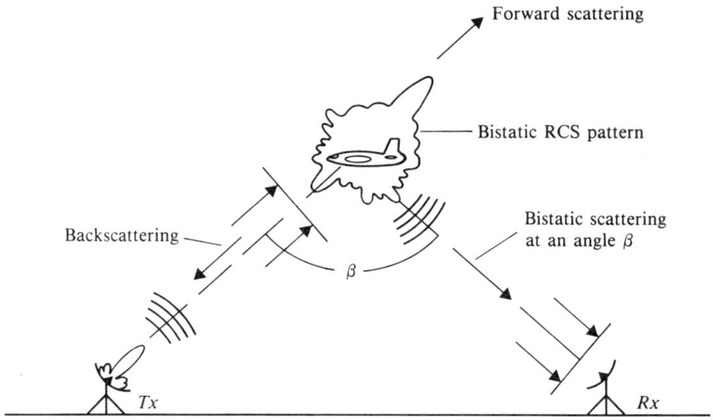

Figure 5-1 Cartoon of a bistatic RCS pattern for a target illuminated by a plane wave.

5-1. RCS σ depends as well on the polarizations of both transmitter and receiver because although RCS is defined in terms of a target gain, it can also be interpreted as a function of the electromagnetic fields that we established in preceding chapters.

Suppose we have a *plane wave* specified by $\mathbf{E} = \hat{x}Ee^{-jkz}$, $\mathbf{H} = \hat{y}\dfrac{E}{\eta}e^{-jkz}$ incident on a target. (Our use of a plane wave implies the assumption that the transmitter is far (in wavelengths) from the target.) Then the Poynting vector is

$$\mathbf{S} = \mathbf{E} \times \mathbf{H}^* = \frac{E^2}{\eta}\hat{z}; \; \|\mathbf{S}\| = \frac{1}{\eta}|E|^2. \tag{5-2}$$

The fields scattered by the target can also be assumed plane if the target is sufficiently far from the receiver. Denote these field and Poynting vectors with the superscript *s*. A standard definition of RCS σ is (Kerr, 1951, p. 33): σ is the area intercepting that amount of power which, when scattered isotropically, produces an echo equal to that observed at the target. Then $\mathbf{S}^s = (\sigma/4\pi r^2)\ \mathbf{S}$, where we have assumed the monostatic condition of coincident transmitter and receiver locations at distance r from the target. The isotropic condition produces the solid angle factor of 4π. Thus,

$$\sigma = 4\pi \lim_{r\to\infty} r^2 \|\mathbf{S}^s\|/\|\mathbf{S}\|$$

$$= 4\pi \lim_{r\to\infty} r^2\|\mathbf{E}^s\|^2/\|\mathbf{E}\|^2 = 4\pi \lim_{r\to\infty} r^2 \|\mathbf{H}^s\|^2/\|\mathbf{H}\|^2. \tag{5-3}$$

The limit operation is the result of our requirement that the target be far from the receiver.

Example: Backscattering RCS. Consider the dielectric sphere example discussed above.

$$\sigma(\theta) = 4\pi r^2 \langle \mathbf{S}^s \rangle \cdot \hat{r} / (\tfrac{1}{2} E_0^2 / \eta)$$

$$= 4\pi \left(\frac{\epsilon - \epsilon_0}{\epsilon + 2\epsilon_0} \right)^2 k^4 a^6 \sin^2 \theta.$$

5-1 OTHER CROSS SECTIONS

There are at least three other types of cross section defined by various authors. They are absorption cross section; extinction cross section; and total, or scattering, cross section. (For the first two, see Morse and Feshbach, 1953; for the latter, see Shen and Kong, 1983.)

5-1-1 Absorption Cross Section

Let the index of refraction of the scattering body be complex, say $n - jq$, so that the phase of the scattered wave has the form $-jk(n - jq)z$. Then if $q > 0$, energy will be attenuated (absorbed) as it passes through the body. An absorption cross section can then be defined as the power absorbed by the body per unit incident intensity, that is, take the difference between the amount of power radiated, or scattered, by the body with no absorption and subtract from it the amount of power radiated with absorption.

5-1-2 Extinction Cross Section

The amount of energy lost by the plane wave by virtue of scattering *and* by absorption is then called extinction cross section.

5-1-3 Total Cross Section

Finally, then, we have total cross section, which we define as the ratio of the *total* scattered power to the incident power density, that is, we are obliged to integrate the scattered Poynting vector over all space to find the total scattered power. Shen and Kong (1983) give an interesting example for scattering of X-band energy by a raindrop. Letting the diameter of the drop be 3 mm and taking a representative X-band wavelength to be 3 cm, = 10 diameters, we have a Rayleigh scattering situation to which we may apply previous results. Taking ϵ for water to be $61\epsilon_0$, we find (after Shen and Kong) $\sigma_{\text{total}} = 0.006(\pi d^2)$. For backscattering the result is $\sigma = 0.035[\pi(d/2)^2]$. That is to say, the backscattering RCS of the raindrop is 0.035 as large as its geometric cross section.

5-2 TWO- AND THREE-DIMENSIONAL RCS

In three dimensions the far-scattered field, for plane wave incidence, behaves like

$$\mathbf{E}^s \sim \frac{\exp(-jkr)}{kr} S\hat{e}, \qquad r \to \infty, \tag{5-4}$$

where $S = S(\theta, \phi)$ is the scattering coefficient of the target. So RCS may be defined as

$$\sigma(\theta, \phi) = \lim_{r \to \infty} 4\pi r^2 \, \|\mathbf{E}^s\|^2 / \|\mathbf{E}^i\|^2 = (4\pi/k^2) \, |S(\theta, \phi)|^2. \tag{5-5}$$

In two dimensions, that is, for infinitely long cylindrical bodies with cylindrical axis parallel to the z-axis, the z-component of the far-scattered field is proportional to (using a cylindrical coordinate system)

$$P(\phi) \sqrt{\frac{2}{\pi \rho k}} \exp\left(-jk\rho + \frac{j\pi}{4} \right), \ \rho \to \infty. \tag{5-6}$$

So RCS (actually, RCS per unit length which is called radar cross-width, or RCW, by some authors) may be defined as

$$\sigma(\phi) = \lim 2\pi\rho \, |V^s|^2 / |V^i|^2 = (4/k) \, |P(\phi)|^2, \tag{5-7}$$

where

$$\begin{aligned} V^{s,i} &= E_z \qquad \text{if } \mathbf{E} \text{ is parallel to the } z\text{-axis;} \\ &= H_z \qquad \text{if } \mathbf{H} \text{ is parallel to the } z\text{-axis.} \end{aligned}$$

The results here are applicable for finite cylinders of sufficient length L (in wavelengths; usually at least 10). If RCW and RCS are normalized to λ and λ^2, respectively, then a conversion factor of $2(L/\lambda)^2$ may be applied to the first to obtain the second, that is,

$$\frac{\sigma(\theta, \phi)}{\lambda^2} = 2\left(\frac{L}{\lambda}\right)^2 \left(\frac{\sigma(\phi)}{\lambda}\right). \tag{5-8}$$

Example 1. For incidence perpendicular to a wedge edge the surface current can be assumed to be the same for a long length of wedge as it is in the two-dimensional case (i.e., an infinite wedge). If we apply the physical optics method, we get (with z-axis along the wedge edge)

$$\exp(-jk\sqrt{r^2 + z^2}) \, dz \approx \sqrt{\lambda r} \exp\left(-jkr + \frac{j\pi}{4} \right). \tag{5-9}$$

But for the 3-dimensional case the integral is

$$\exp(-jk\sqrt{r^2 + z^2}) \, dz \approx L \exp(-jkr). \tag{5-10}$$

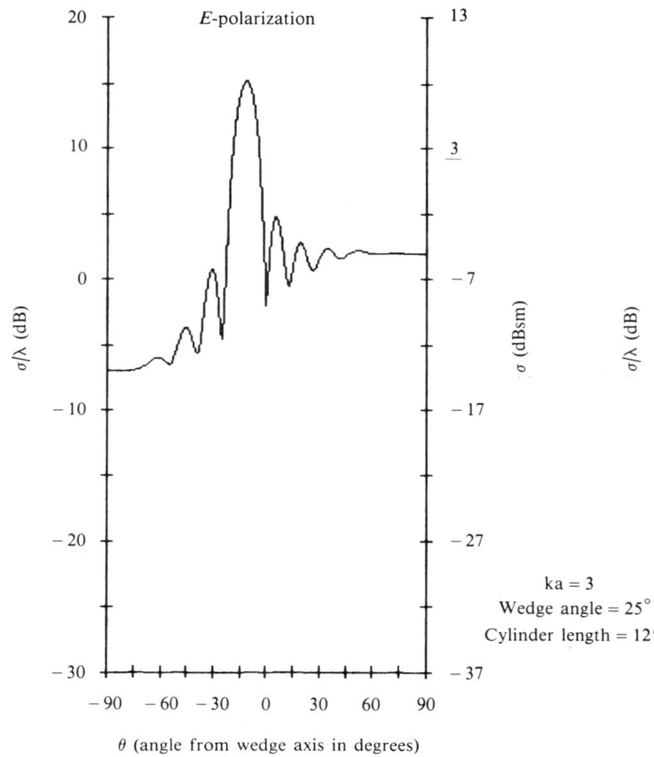

Figure 5-2 Backscattering RCS of a wedge cylinder in a plane normal to the cylinder axis: E-polarization.

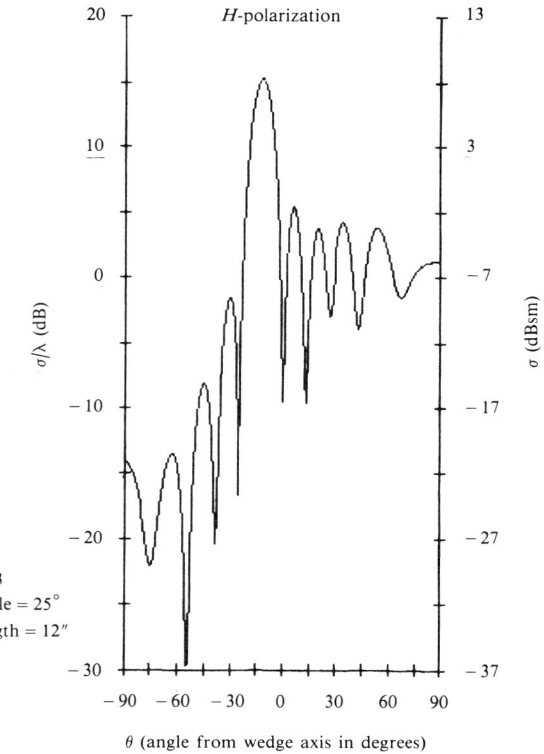

Figure 5-3 Backscattering RCS of a wedge cylinder in a plane normal to the cylinder axis: H-polarization.

Thus, the three-dimensional fields can be gotten from the two-dimensional fields by multiplying by $L \exp(j\pi/4)/\sqrt{\lambda r}$.

Example 2. Consider a wedge with a total interior angle of 25° to which is smoothly fitted a circular cylinder whose radius a is such that $ka = 3$ (which then amounts to an automatic choice of k or λ). In Figs. 5-2 and 5-3 we show RCW, normalized to λ (left ordinate scale), plotted against incidence/observation angle, beginning with an edge-on view. RCS, in dBsm, is also indicated for a finite length (12 in.) cylinder (right ordinate scale).

E-polarization means that $\mathbf{E}^i$ is parallel to the wedge edge; H-polarization means that $\mathbf{H}^i$ is parallel to the wedge edge.

Notice the pattern near the edge for H-polarization (Fig. 5-3). This effect is an important type of scattering called traveling wave return; it is treated in detail in Section 11-7.

Liepa et al. (1974) formulated a computer code, called RAMD, for arbitrary two-dimensional cylindrical bodies, based on solutions of integral equations. Its results are shown (Liepa *et al.*, 1974) to be in excellent agreement with data measured for an 18 in. cylinder at 7.5 GHz by the Air Force Avionics Laboratory. The code has been modified by the writer and H.W. Klimach for PC use and the modified version has been applied to obtain Figs 5-2 and 5-3. Additional discussion of scattering by two-dimensional bodies may be found in Sections 9-6, 11-2, and 16-6.

PART II

SCATTERING AND DIFFRACTION OF ELECTROMAGNETIC WAVES

The main thrust of the five chapters of Part II is the development of practical methods of RCS analysis, including, in particular, asymptotic methods (Chapters 7 and 8) and numerical methods and approximations (Chapter 9). These methods and the results obtained from them must be evaluated; the most universally accepted technique of evaluation is that of measurement of both full-scale and fractional scale models on a static radar range. Chapter 10 deals with types of radar ranges and accepted standards of measurement.

A most useful and beautiful way of approaching these ideas is to examine the problem of scattering of a plane wave by a half-plane. To sift out more easily the salient points we idealize this problem by assuming the half-plane to be a perfectly absorbing one; in this way we can ignore the reflected field and concentrate entirely on the incident and diffracted fields.

6

DIFFRACTION OF A PLANE WAVE BY A PERFECTLY ABSORBING HALF-PLANE†

To introduce the concept of scattering and diffraction by a plane-wave-illuminated object, we appeal to a fictitious situation for the reason that we can keep the mathematical work simple enough to avoid obscuring the principle to be developed.

We assume (using time dependence $e^{j\omega t}$) that we have a plane wave

$$\mathbf{E}^i = \hat{z}\exp(jkr\cos(\theta - \alpha_0)) \tag{6-1}$$

incident from the direction α_0 with the $+x$-axis on a perfectly absorbing (black) half-plane $y = 0$, $x > 0$ (see Fig. 6-1).

We shall obtain the field diffracted into the half-space $y < 0$ by using the surface current density idea. So we may deal with a pair E_x, E_z, or with a pair H_x, H_z, since tangential components are the quantitities that must be suitably prescribed on $y = 0$. In fact, in this fictitious problem, we shall postulate boundary values of tangential fields rather than face actual boundary conditions.

We could assume that we are dealing with:

1. A surface magnetic current density **K** that produces a discontinuity in tangential **E**; or
2. A surface electric current density **J** that produces a discontinuity in tangential **H**; or
3. A superposition of $p\mathbf{J}$ and $(1-p)\mathbf{K}$, with p a constant combination factor.

† The approach and results given here are adapted from treatments by Clemmow (1966) and Born and Wolf (1959). See also Booker and Clemmow (1950).

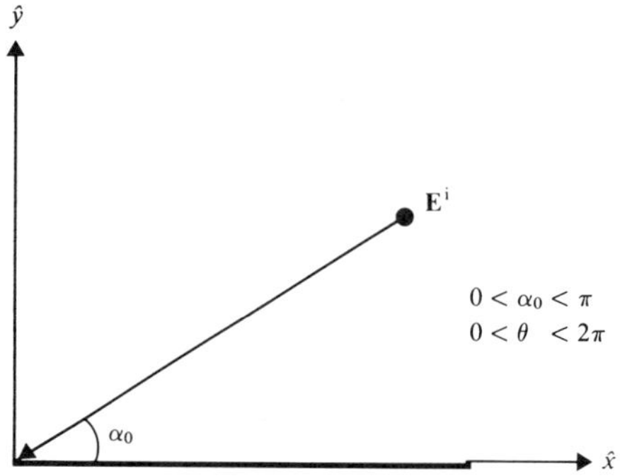

Figure 6-1 A plane wave incident on a half-plane at an angle α_0.

Each type of current density in $y = 0$ casts an image in the half-space $y > 0$ of the field it projects into $y < 0$. But the images in $y > 0$ are of opposite sign and so cancel. The choice of $p = 1/2$ therefore gives a unique surface current density on $y = 0$ but eliminates the need to worry about a reflected field in $y > 0$. We can put our attention *fully* on the incident and diffracted fields.

So we use both kinds of surface current density ideas (i.e., alternative 3 above; refer also to Section 4-6, example 1) and take the fields generated by them to be of the form of a plane wave spectrum representation:

$$E_z = \int_C P(\cos\alpha)\exp(-jkr\cos(\theta+\alpha))\,d\alpha \tag{6-2}$$

$$H_x = Y_0 \int_C \sin\alpha\, P(\cos\alpha)\exp(-jkr\cos(\theta+\alpha))\,d\alpha, \tag{6-3}$$

where $Y_0 = \sqrt{\epsilon_0/\mu_0}$ is the admittance of free space and the path C is described in Fig. 6-2b.

The boundary values we impose at $y = 0$ are

$$E_z = \begin{cases} 0, & x > 0, \\ \exp(jkx\cos\alpha_0), & x < 0; \end{cases} \tag{6-4}$$

and

$$H_x = \begin{cases} 0, & x > 0, \\ -Y_0\sin\alpha_0\exp(jkx\cos\alpha_0), & x < 0. \end{cases} \tag{6-5}$$

Remember then that our answer is 1/2 the sum of these two results. Equation 6-4 implies that

$$\int_{-\infty}^{\infty} \frac{P(\lambda)\exp(-jkx\lambda)\,d\lambda}{\sqrt{1-\lambda^2}} = \begin{cases} 0, & x > 0, \\ \exp(jkx\lambda_0), & x < 0. \end{cases} \tag{6-6}$$

This latter is a Fourier transform (FT) (see Chapter 4), from which we obtain

$$\frac{P(\lambda)}{\sqrt{1-\lambda^2}} = \frac{1}{2\pi j}\frac{1}{\lambda+\lambda_0} \tag{6-7}$$

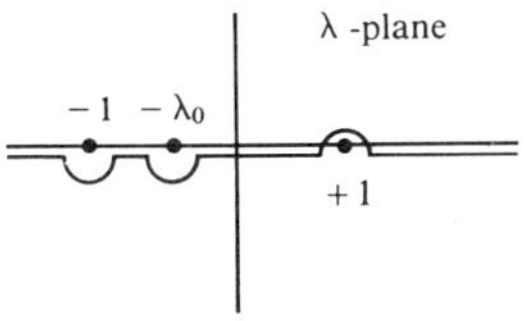

Figure 6-2a Integration path in the λ-plane. Reprinted with permission from Clemmow (1966), Copyright 1966, Pergamon Books Ltd.

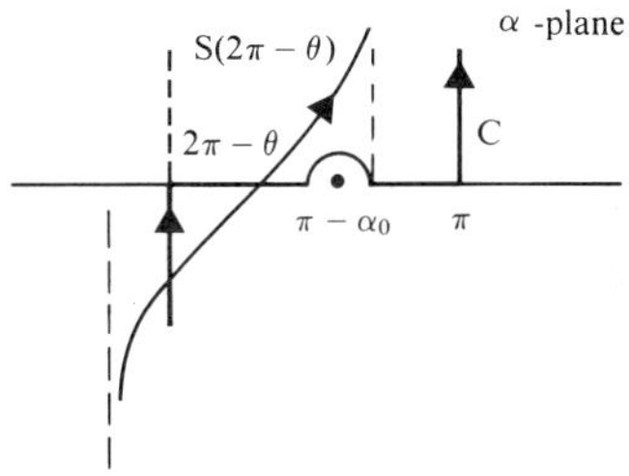

Figure 6-2b Integration path C and deformed path S in the α-plane.

The integration path is shown in Figs. 6-2a and 6-2b for both α and λ. Thus, from Eq. 6-7,

$$P(\cos\alpha) = \frac{1}{2\pi j}\frac{\sin\alpha}{\cos\alpha + \cos\alpha_0}. \tag{6-8}$$

and, from Eq. 6-5,

$$P(\cos\alpha) = \frac{1}{2\pi j}\frac{\sin\alpha_0}{\cos\alpha + \cos\alpha_0} \tag{6-9}$$

Therefore, using our combination rule, the field in $y<0$ is given by Eq. 6-2 with

$$P(\cos\alpha) = \frac{1}{4\pi j}\frac{\sin\alpha + \sin\alpha_0}{\cos\alpha + \cos\alpha_0} \tag{6-10}$$

that is, by

$$E_z = \frac{1}{4\pi j}\int_C \tan\frac{1}{2}(\alpha + \alpha_0)\exp(-jkr\cos(\theta + \alpha))\,d\alpha \tag{6-11}$$

Now we must evaluate Eq. 6-11. To do so, we assume $kr \gg 1$.

6-1 SOME DETAILS ABOUT THE PATH *C*

Any two-dimensional field can be regarded as a superposition of **E**-fields (i.e., only E_z, H_x, H_y, $J_z \neq 0$) of the form

$$\mathbf{E} = -\frac{1}{2}Z_0 J_z \hat{z}\exp(-jkr\cos(\theta \pm \alpha))$$

$$\mathbf{H} = \frac{1}{2}J_z(\pm\hat{x}\sin\alpha + \hat{y}\cos\alpha)\exp(-jkr\cos(\theta \pm \alpha)),$$

where the lower (upper sign) indicates an outgoing wave for $y<0$ ($y>0$)

from the plane $y = 0$, arising from an electric surface current density

$$J_z = J_0 \sin\alpha \exp(-jkx\cos\alpha)$$

located there. This notion was observed at the outset of this discussion on the perfectly absorbing half-plane. We can, in fact, use suitable imaginary α to include inhomogeneous plane waves. This process leads us to the concept of an *angular spectrum* of plane waves which we can write using FT theory. Thus,

$$\mathbf{E} = \hat{z} \int_C P(\cos\alpha) e^{-jkr\cos(\theta \pm \alpha)}\, d\alpha$$

$$\mathbf{H} = Y_0 \int_C (-\hat{x}\sin\alpha - \hat{y}\cos\alpha) P(\cos\alpha) e^{-jkr\cos(\theta\pm\alpha)}\, d\alpha$$

would arise from

$$J_z(x) = -2Y_0 \int_C \sin\alpha\, P(\cos\alpha) e^{-jkx\cos\alpha}\, d\alpha.$$

For these representations the integration path C must be chosen such that $\cos\alpha$ remains real (see Fig. 6-2b for path C). In somewhat more detail, we have

C: $\cos\alpha$ real covers the range $-\infty, \infty$ (for the FT):

(a) α: range $-i\beta$, $0<\beta<\infty$(negative imaginary axis) implies $1<\cosh\beta<\infty$.

(b) α: range $0\le\alpha\le\pi$ implies $1>\cos\alpha>-1$.

(c) α: $\pi + i\beta$, $0<\beta<\infty$ implies $-\infty<\cosh\beta<-1$.

Equation 6-11 has a saddle point at $\alpha = 2\pi - \theta$ and a pole at $\alpha = \pi - \alpha_0$. These two coalesce at $\theta = \pi + \alpha_0$. (Recall: $\pi<\theta<2\pi$.) So if we evaluate Eq. 6-11 by the saddle point method by deforming C into $S(2\pi-\theta)$, the steepest descent path, we must worry whether we cross over the pole or not. This occurrence takes place in accordance with whether θ is less than or greater than $\pi + \alpha_0$. Physically this dichotomy indicates whether we are in the illuminated region or the shadowed region. If the pole is included in the region bounded by C and S, then its residue, which is the geometrical optics incident field, must be added to the diffracted field.

On the account of the pole inclusion in the region bounded by C and S, we can write our field component as a sum consisting of an incident part (geometrical optics) E_z^g and a diffracted part E_z^d:

$$E_z = E_z^g + E_z^d, \tag{6-12}$$

with

$$E_z^g = \begin{cases} \exp(jkr\cos(\theta - \alpha_0)), & \pi<\theta<\pi+\alpha_0, \\ 0, & \pi+\alpha_0<\theta<2\pi; \end{cases} \tag{6-13}$$

and

$$E_z^d = \frac{1}{4\pi j} \int_{S(2\pi-\theta)} \tan \frac{1}{2} (\alpha + \alpha_0) e^{-jkr\cos(\theta+\alpha)} \, d\alpha. \tag{6-14}$$

Note that the integration path has been deformed (see Section 6-1) from C to the steepest descent path $S(2\pi - \theta)$ on which $P(\cos\theta)$ is real and decreases as we move away from $2\pi - \theta$. (It increases in the orthogonal direction.)

Then the evaluation of Eq. 6-14 becomes

$$E_z^d = \mp \frac{\exp(j\pi/4)}{\sqrt{\pi}} \sin \frac{1}{2} (\theta - \alpha_0) \, e^{-jkr} F\left[\pm \sqrt{2kr} \cos \frac{1}{2} (\theta - \alpha_0) \right], \tag{6-15}$$

with upper (lower) sign denoting $\theta \lessgtr \pi + \alpha_0$. $F[\cdot]$ is a Fresnel integral (Magnus and Oberhettinger, 1954, p. 96).

If we use the asymptotic expansion for the Fresnel integral, for which we must have

$$|\sqrt{2kr} \cos \tfrac{1}{2} (\theta - \alpha_0)| \gg 1, \tag{6-16}$$

we obtain

$$E_z^d \approx - \frac{\exp(-j\pi/4)}{2\sqrt{2\pi}} \tan \frac{1}{2} (\theta - \alpha_0) \frac{\exp(-jkr)}{\sqrt{kr}}. \tag{6-17}$$

This result is obtainable by replacing the integrand factor

$$\tan \frac{1}{2} (\alpha + \alpha_0)$$

in Eq. 6-14 by its value at the saddle point $\alpha = 2\pi - \theta$. (This answer is essentially the one we could also get using the method of stationary phase.)

Equation 6-17 is called an "edge wave" or "fringe wave" approximation for the diffracted field. By Eq. 6-16 it is valid outside of some parabola

$$r \cos(\theta - \alpha_0) + r = \text{constant},$$

whose focus is at the diffracting edge with axis along the shadow line $\pi + \alpha_0$.

A physical interpretation for the saddle point is that for the diffracted field it selects from the plane wave spectrum those waves whose combination is cylindrical (Eq. 6-17). If we construct higher order solutions, then other wave types will be included.

Note that these cylindrical waves must change *continuously* from a plane wave as the shadow boundary is crossed. (See Fig. 6-3.)

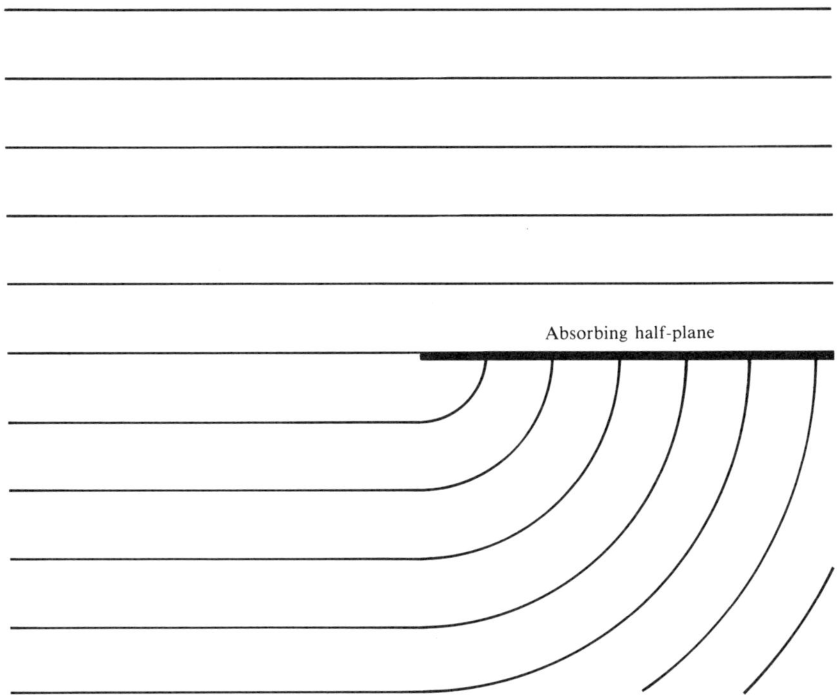

Figure 6-3 Phase contours of a normally incident H-polarized plane wave diffracted by a perfectly absorbing half-plane.

6-2 INTEGRATION BY STATIONARY PHASE AND STEEPEST DESCENTS METHODS

Given an integral of the form

$$I = \int_C P(\cos\alpha) \exp[-jkr\cos(\theta-\alpha)]\, d\alpha.$$

Stationary phase assumes that, if $kr \gg 1$, the phase of the waves will vary rapidly, canceling each other by destructive interference—except for those cases when $\theta = \alpha$. Here the phases vary slowly and so contribute most to the integral. An approximation is

$$I \approx P(\cos\theta) \int_C e^{-jkr\cos(\theta-\alpha)}\, d\alpha$$

$$\approx \sqrt{2\pi}\, e^{j\pi/4}\, P(\cos\theta) \frac{\exp(-jkr)}{\sqrt{kr}}$$

For steepest descents, the integration path C is deformed into one on which

$$jkr[1 - \cos(\theta - \alpha)]$$

is real. That is, setting $\alpha = \alpha_r + j\alpha_i$,

$$\begin{aligned} -jkr\cos(\alpha - \theta) = -jkr\cos(\alpha_r - \theta)\cosh\alpha_i \\ -kr\sin(\alpha_r - \theta)\sinh\alpha_i. \end{aligned}$$

So, for the last term to be real, the stated condition must be satisfied. That is, $\cos(\alpha_r - \theta)\cosh\alpha_i$ *must* $= 1$.

If $\alpha = \alpha_r + j\alpha_i$, then $S(2\pi - \theta)$ is given by

$$\cos(\alpha_r - \theta)\cosh\alpha_i = 1$$

and has asymptotes

$$\alpha_r = \frac{5\pi}{2} - \theta$$

and

$$\alpha_r = \frac{3\pi}{2} - \theta \qquad (\pi < \theta < 2\pi).$$

In carrying out the above manipulations, we note that

$$\cos(\theta + \alpha_r + j\alpha_i) = \cos(\theta + \alpha_r)\cosh\alpha_i - j\sin(\theta + \alpha_r)\sinh\alpha_i,$$

which implies that

$$2\pi - \theta < \alpha_r < 3\pi - \theta,\ \alpha_i > 0,$$

$$\pi - \theta < \alpha_r < 2\pi - \theta,\ \alpha_i < 0.$$

From these statements the α_r and α_i regions can be shown to contain a path on which the real part of

$$-jkr\cos(\theta - \alpha)$$

is stationary, without being a maximum or minimum, at some point. From this point a steepest descent path can be chosen so that

$$\text{Re}[-jkr\cos(\theta - \alpha)]$$

decreases most rapidly. So, for $kr \gg 1$, only the part of this path in the vicinity of the saddle point contributes significantly to the integral I.

7

ASYMPTOTIC APPROXIMATION METHODS

If we turn briefly to a rigorous scattering solution† for a sphere, we get an indication of the analytical difficulties that arise in scattering solutions for even the simplest bodies. We can avoid some of these analytical difficulties by turning to asymptotic methods involving wavelengths λ that are either large or small in respect to some major linear dimension d of the scattering body. The first, $d \ll \lambda$, is Rayleigh scattering, (long-wavelength case), after Lord Rayleigh's early work with this type of analysis. The second, $\lambda \ll d$, is optics scattering (short-wavelength case), so called for the resemblance of its rules of reflection, refraction, and diffraction to the analogous rules for optics. This topic will extend into Chapter 13 in pursuit of special extensions of simple methods of optics to describe adequately the phenomena of diffraction.

These asymptotic methods include assumptions about the wavelength λ that are in addition to the assumption already made regarding the observer/illuminator distance r from the scatterer, namely, that $r \gg \lambda$. This assumption continues in force unless *explicitly* lifted for brief discussions.

7-1 RAYLEIGH THEORY

The electric and magnetic fields $\mathbf{E}^s$ and $\mathbf{H}^s$ scattered by a perfect conductor illuminated by the incident fields $\mathbf{E}^i$ and $\mathbf{H}^i$ (with suppressed time

† See Born and Wolf (1959, p. 630); or Bowman et al. (1969, p. 353).

dependence $e^{j\omega t}$ and Gaussian units) satisfy Maxwell's equations:

$$\begin{aligned} &(\text{F}): \quad \nabla \times \mathbf{E}^s = jk\mathbf{H}^s; \qquad &&(\text{A}): \quad \nabla \times \mathbf{H}^s = jk\mathbf{E}^s; \\ &(\text{GB}): \quad \nabla \cdot \mathbf{E}^s = 0; \quad \text{and} \qquad &&(\text{GD}): \quad \nabla \cdot \mathbf{H}^s = 0. \end{aligned} \tag{7-1}$$

If S is a smooth perfect conductor, then the incident and scattered fields are related by boundary conditions at the surface of S:

$$\begin{aligned} &(\text{a}) \quad \hat{n} \times \mathbf{E}^s \,|_S = -\hat{n} \times \mathbf{E}^i \,|_S; \\ &(\text{b}) \quad \hat{n} \cdot \mathbf{H}^s \,|_S = -\hat{n} \cdot \mathbf{H}^i \,|_S. \end{aligned} \tag{7-2}$$

In addition, the scattered electric field must satisfy a radiation condition at infinity [Kleinman, 1965]:

$$\lim_{r\to\infty} \mathbf{r} \times (\nabla \times \mathbf{E}^s) - jkr\mathbf{E}^s = 0. \tag{7-3}$$

The scattered fields, $\mathbf{E}^s$, $\mathbf{H}^s$, exterior to the smooth perfectly conducting surface S illuminated by incident fields, $\mathbf{E}^i$, $\mathbf{H}^i$, can be represented as *convergent series* (with time dependence $e^{j\omega t}$ as remarked previously):

$$\mathbf{E}^s = \sum_{m=0}^{\infty} (-jk)^m \mathbf{E}^s_m, \tag{7-4}$$

$$\mathbf{H}^s = \sum_{m=0}^{\infty} (-jk)^m \mathbf{H}^s_m, \tag{7-5}$$

for suitably small wave number k (or large wavelength λ).

Terms of the series must be obtained iteratively from the Maxwell equations and boundary conditions given above. The *first* terms of the series require (comparing coefficients of like powers of k):

$$\begin{aligned} &(\text{F}): \quad \nabla \times \mathbf{E}^s_0 = 0; \qquad &&(\text{A}): \quad \nabla \times \mathbf{H}^s_0 = 0; \\ &(\text{GB}): \quad \nabla \cdot \mathbf{E}^s_0 = 0; \qquad &&(\text{GD}): \quad \nabla \cdot \mathbf{H}^s_0 = 0; \end{aligned} \tag{7-6}$$

$$(\text{a}) \quad \hat{n} \times \mathbf{E}^s_0 \,|_S = -\hat{n} \times \mathbf{E}^i_0 \,|_S; \qquad (\text{b}) \quad \hat{n} \cdot \mathbf{H}^s_0 \,|_S = \hat{n} \cdot \mathbf{H}^i_0 \,|_S; \tag{7-7}$$

$$(\text{a}) \quad \lim_{r\to\infty} r\mathbf{E}^s_0 = 0; \quad \text{and} \quad (\text{b}) \quad \lim_{r\to\infty} r\mathbf{H}^s_0 = 0. \tag{7-8}$$

These equations are satisfied if we take

$$\mathbf{E}^s_0 = \nabla \phi_0, \tag{7-9}$$

$$\mathbf{H}^s_0 = \nabla \psi_0, \tag{7-10}$$

with

$$\begin{aligned} &(\text{a}) \quad \nabla^2 \phi_0 = 0, \\ &(\text{b}) \quad \hat{n} \times \nabla \phi_0 \,|_S = -\hat{n} \times \mathbf{E}^i_0 \,|_S, \\ &(\text{c}) \quad \lim_{r\to\infty} r\left(\frac{\partial \phi_0}{\partial r}\right) = 0; \end{aligned} \tag{7-11}$$

and with

$$
\begin{aligned}
&\text{(a)} \quad \nabla^2 \psi_0 = 0, \\
&\text{(b)} \quad \left.\frac{\partial \psi_0}{\partial n}\right|_S = -\hat{n} \cdot \mathbf{H}_0^i \,|_S, \\
&\text{(c)} \quad \lim_{r \to \infty} r\left(\frac{\partial \psi_0}{\partial r}\right) = 0.
\end{aligned} \tag{7-12}
$$

The first of these boundary value problems is referred to as a *Dirichlet* problem, the second as a *Neumann* problem, since they were treated by these two mathematicians more than 100 years ago.

The solutions are of the following form:

$$\phi_0 = \frac{-\mathbf{p} \cdot \hat{r}}{r^2}, \tag{7-13}$$

$$\psi_0 = \frac{-\mathbf{m} \cdot \hat{r}}{r^2}, \tag{7-14}$$

where $\mathbf{p}$ and $\mathbf{m}$ are electric and magnetic dipole moments, respectively, of dipoles situated at the origin. The fields due to the dipole moments $\mathbf{p}$ and $\mathbf{m}$ may be written (Kleinman, 1965):

$$\mathbf{E} = \nabla \times \left(\nabla \times \frac{e^{-jkr}}{4\pi\epsilon r}\mathbf{p}\right) - jkZ\nabla \times \frac{e^{-jkr}}{4\pi r}\mathbf{m}; \tag{7-15}$$

and

$$\mathbf{H} = jkY\nabla \times \frac{e^{-jkr}}{4\pi\epsilon r}\mathbf{p} + \nabla \times \left(\nabla \times \frac{e^{-jkr}}{4\pi r}\mathbf{m}\right); \tag{7-16}$$

in Eqs. 7-15 and 7-16 Z is the impedance of the medium through which $\mathbf{E}$ and $\mathbf{H}$ propagate and $Y = 1/Z$ is the admittance.

Thus, expanding the dipole fields $\mathbf{E}$ and $\mathbf{H}$ and e^{-jkr} in powers of k, and neglecting terms which vanish faster than $1/r$ (to get a *far-field* expression), Kleinman has shown that

$$
\begin{aligned}
\mathbf{E}_0^s &= \frac{e^{-jkr}}{4\pi r} k^2\left[\frac{\hat{r}}{\epsilon} \times (\mathbf{p} \times \hat{r}) - Z\hat{r} \times \mathbf{m}\right] \\
&= \frac{e^{-jkr}}{4\pi r} k^2\left[\frac{1}{\epsilon}\mathbf{p} - (\hat{r} \cdot \mathbf{p})\hat{r} - Z\hat{r} \times \mathbf{m}\right],
\end{aligned} \tag{7-17}
$$

$$\mathbf{H}_0^s = \frac{e^{-jkr}}{4\pi r} k^2\left[\hat{r} \times (\mathbf{m} \times \hat{r}) + \frac{Y}{\epsilon}\hat{r} \times \mathbf{p}\right]. \tag{7-18}$$

To be more explicit about the nature of these scattered fields, suppose we take the incident electric and magnetic fields to be

$$\mathbf{E}^i = \hat{a} \exp(-j\mathbf{k} \cdot \mathbf{r}), \tag{7-19}$$

$$\mathbf{H}^i = Y\hat{b} \exp(-j\mathbf{k} \cdot \mathbf{r}), \tag{7-20}$$

where $\mathbf{k} = k\hat{k}$, $\hat{k}$ being incidence direction and k wave number, and $\hat{b} = \hat{k} \times \hat{a}$, $\hat{a} \cdot \hat{k} = 0$. In these circumstances, the electric dipole moment $\mathbf{p}$ and the magnetic dipole moment $\mathbf{m}$ can be written in terms of electric and magnetic polarizability tensors (of rank 2) $\mathfrak{P}$ and $\mathfrak{M}$, respectively, as follows (Keller et al., 1972):

$$\mathbf{p} = \epsilon \mathfrak{P} \cdot \hat{a}; \tag{7-21}$$

$$\mathbf{m} = -Y\mathfrak{M} \cdot \hat{b}. \tag{7-22}$$

Keller et al. show that the entries P_{ij} of $\mathfrak{P}$ and M_{ij} of $\mathfrak{M}$ can be expressed as either surface or volume integrals over exterior potential functions.

For axially symmetric (about, say, the z-axis) bodies, it can be shown that (Keller et al., 1972)

$$P_{ij} = M_{ij} = 0, \qquad i \neq j; \tag{7-23}$$

$$P_{22} = P_{11}, \qquad M_{22} = M_{11}; \quad \text{and} \tag{7-24}$$

$$M_{33} = \frac{1}{2} P_{11}. \tag{7-25}$$

Thus, there are three independent quantities remaining in the electric and magnetic polarizability tensors $\mathfrak{P}$ and $\mathfrak{M}$ and these satisfy the fundamental inequalities

$$M_{11}, M_{33}, P_{33} \geq V, \tag{7-26}$$

where V is the volume of the axially symmetric scattering body. Furthermore, because of Eq. 7-25,

$$P_{11} \geq 2V. \tag{7-27}$$

Kleinman and Senior (1972) deduce the further inequalities:

$$\begin{aligned} &2P_{11} + P_{33} \geq 9V; \\ &2M_{11} + M_{33} \geq \frac{9V}{2}; \\ &(P_{11} - V)(P_{11} + 2P_{33} - 3V) \geq 12V^2; \\ &(M_{11} - V)(M_{11} + 2M_{33} - 3V) \geq \frac{3V^2}{4}; \end{aligned} \tag{7-28}$$

all of which are optimum in the sense that *equality* holds for at least one body, the sphere. In this case:

$$\begin{aligned} &P_{33} = P_{11} = 3V, \quad \text{and} \\ &M_{33} = M_{11} = \frac{1}{2} P_{11} = \frac{3V}{2}; \end{aligned} \tag{7-29}$$

so for the sphere, the electric and magnetic polarizability tensors have only one independent entry and that entry depends on the sphere volume V. We may write out the electric (far) scattered field for the sphere as

$$\mathbf{E}^s = \frac{3k^2Ve^{-jkr}}{4\pi r}\left[\hat{r}\times(\hat{a}\times\hat{r}) + \frac{1}{2}(\hat{b}\times\hat{r})\right].$$

If we take $\hat{k} = -\hat{z}$ and consider the case of backscattering $\hat{r} = \hat{z}$, then

$$\mathbf{E}^s = \frac{3k^2Ve^{-jkr}}{4\pi r}\left[\hat{a} + \frac{1}{2}\hat{a}\right] = \frac{9k^2Ve^{-jkr}}{8\pi r}\hat{a}. \tag{7-30}$$

The Rayleigh backscattering RCS for the sphere can now be written immediately as

$$\sigma = 4\pi r^2 \,\|\mathbf{E}^s\|^2/\|\mathbf{E}^i\|^2$$

$$= \frac{81k^4V^4}{16\pi}; \tag{7-31}$$

$$\frac{\sigma}{\lambda^2} = \frac{9\pi^5}{4}\left(\frac{2a}{\lambda^6}\right).$$

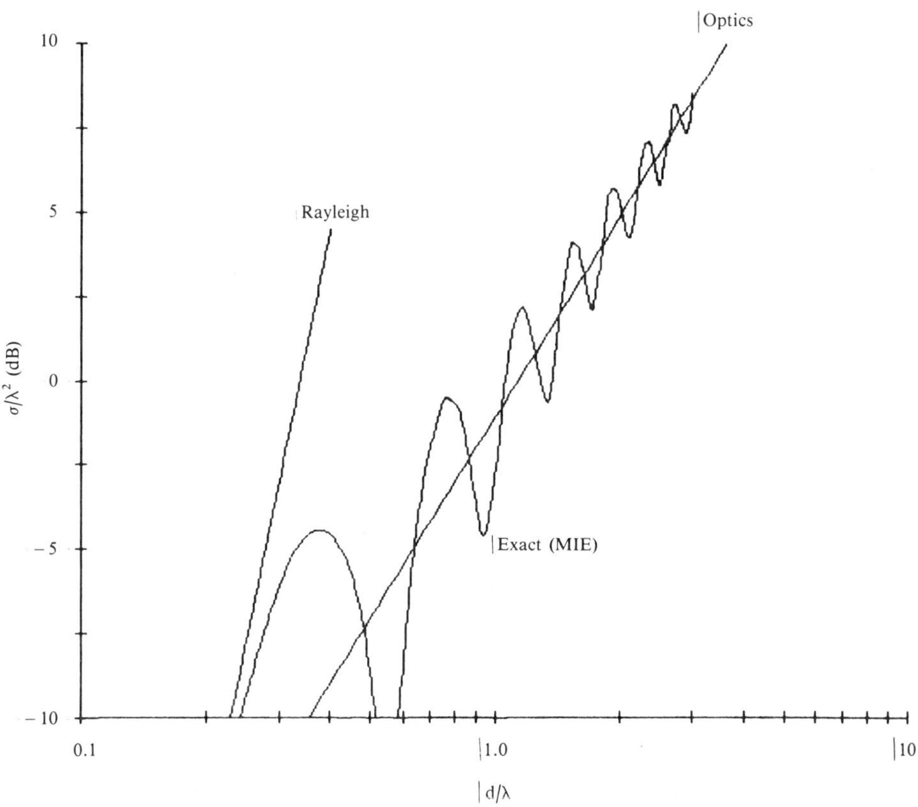

Figure 7-1a Sphere RCS normalized to λ^2 (exact, Rayleigh, and optics theory).

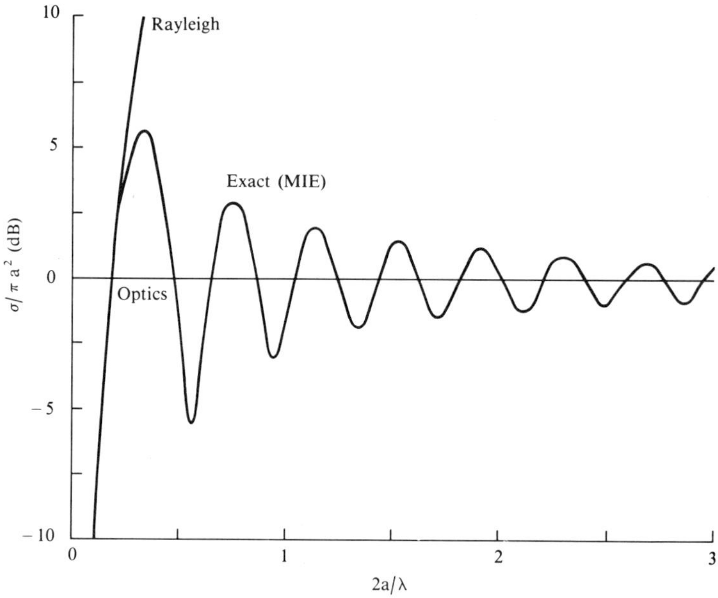

Figure 7-1b Sphere RCS normalized to geometrical optics RCS.

Equation 7-31 can be used to find a Rayleigh approximation to the exact Mie sphere RCS, as is shown in Figs. 7-1a and 7-1b, which show sphere RCS for two different normalizations.

7-2 GEOMETRICAL OPTICS

From Maxwell's equations, the electric field satisfies

$$\nabla^2\mathbf{E} + k^2\mathbf{E} = 0, \qquad \text{subject to } \nabla \cdot \mathbf{E} = 0. \tag{7-32}$$

For large ω, a solution (Luneberg–Kline) is of the form

$$\mathbf{E}(\mathbf{R}, \omega) = e^{-jk_0\psi(\mathbf{R})} \sum_{m=0}^{\infty} (j\omega)^{-m}\mathbf{E}_m(\mathbf{R}), \tag{7-33}$$

where $\mathbf{R}$ is a position vector, $k = \omega\sqrt{\epsilon\mu}$, and k_0 is a phase constant (or wave number) of free space (Kouyoumjian in Mittra, 1975).

Substitute the sum in the differential equation, getting

$$\|\nabla\psi\|^2 = n^2, \qquad n = \sqrt{\epsilon\mu} \qquad \text{(eikonal equation)} \tag{7-34}$$

together with

$$\left(\frac{\partial}{\partial s} + \frac{1}{2}\frac{\nabla^2\psi}{n}\right)\mathbf{E}_0 = 0 \quad \text{(transport equation, 1st order)}, \tag{7-35}$$

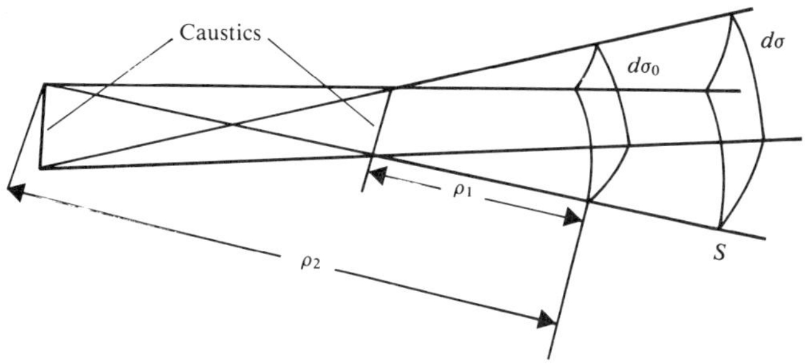

Figure 7-2 Wavefront with principal radii of curvature ρ_1 and ρ_2.

and

$$\hat{s} \cdot \mathbf{E}_0 = 0 \quad \text{(conditional equation, 1st order)}, \tag{7-36}$$

where $\hat{s} = \nabla\psi/n$ and s is distance along a ray path.

At high frequency (small wavelength),

$$\mathbf{E}(s) \approx \exp[-jk_0\psi(s)]\mathbf{E}_0(s). \tag{7-37}$$

Then integration of the transport equation yields

$$\mathbf{E}_0 \approx \mathbf{E}_0(0)\exp(-jk_0\psi(0))\sqrt{\rho_1\rho_2(\rho_1+s)^{-1}(\rho_2+s)^{-1}}\exp(-jks), \tag{7-38}$$

where ρ_1 and ρ_2 are principal radii of curvature of the wavefront at $s = 0$ (Fig. 7-2). The segments labeled "caustic" consist of the locus of points where either the expression $\rho_1 + s$ or the expression $\rho_2 + s$ can be zero; Eq. 7-38 for the scattered field is invalid because of this singular behavior and so must be treated in a special way (see Kouyoumjian in Mittra, 1975).

The magnetic field can be obtained from Maxwell's equations:

$$\mathbf{H}(s) \approx \frac{(\hat{s} \times \mathbf{E}(s))}{Z_c}, \tag{7-39}$$

where $Z_c = \sqrt{\mu/\epsilon}$ is the characteristic impedance of the medium.

7-2-1 Reflection

Set

$$\mathbf{E}^r(s) = \exp(-jk_0\psi^r(s))\mathbf{E}_0^r(s). \tag{7-40}$$

From the boundary condition

$$\hat{n} \times (\mathbf{E}^i + \mathbf{E}^r) = 0 \tag{7-41}$$

on a surface S (refer to Fig. 7-3 and compare with Fig. 3-3) at a point Q_R, it can be shown (Stratton, 1941) that

$$\mathbf{E}_0^r(Q_R) = \mathbf{E}_0^i(Q_R)\Re, \tag{7-42}$$

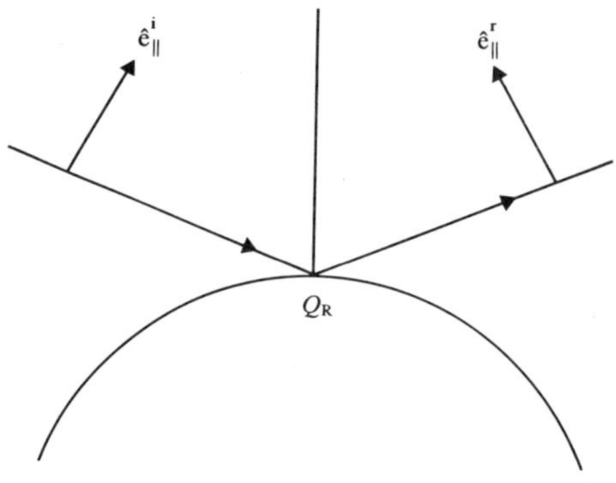

Figure 7-3 Incident and reflected fields at a surface S with principal radii of curvature a_1 and a_2.

where $\mathbf{E}_0(Q_R)$ has components along the unit vectors $\hat{e}_{\|}$ and $\hat{e}_{\perp}$, respectively, parallel and perpendicular to the incidence plane with $\hat{e}_{\|} = \hat{e}_{\perp} \times \hat{s}$.

$\mathfrak{R}$ is a dyadic, or matrix, reflection coefficient. In the case of reflection of a plane wave from a plane perfectly conducting surface,

$$\mathfrak{R} = \begin{pmatrix} R_{\text{per}} & 0 \\ 0 & R_{\text{par}} \end{pmatrix}$$

where R_{per} and R_{par} are perpendicular and parallel reflection coefficients, respectively (see Section 3-2).

7-2-2 Reflection Law

The phases of incident and reflected fields must be matched on S:

$$\psi^i(Q_R) = \psi^r(Q_R).$$

This fact, taken together with the previous observations regarding the total field at S, yields the *reflection law:*

$$\mathbf{E}^r(s) = \mathbf{E}^i(Q_R) \cdot \mathfrak{R}\sqrt{\rho_1^r \rho_2^r (\rho_1^r + s)^{-1} (\rho_2^r + s)^{-1}}\, e^{-jks}, \tag{7-43a}$$

where

$$\frac{1}{\rho_1^r} = \frac{1}{2}\left(\frac{1}{\rho_1^i} + \frac{1}{\rho_2^i}\right) + \frac{1}{f_1} \tag{7-43b}$$

and

$$\frac{1}{\rho_2^r} = \frac{1}{2}\left(\frac{1}{\rho_1^i} + \frac{1}{\rho_2^i}\right) + \frac{1}{f_2} \tag{7-43c}$$

(recall the lens and mirror formulas).

For an incident spherical wave ($\rho_1^i = \rho_2^i = \rho$)

$$\frac{1}{f_{1,2}} = \frac{1}{\cos\theta_i}\left(\frac{\sin^2\theta_2}{a_1} + \frac{\sin^2\theta_1}{a_2}\right) \pm \sqrt{\frac{1}{\cos^2\theta_i}\left(\frac{\sin^2\theta_2}{a_1} + \frac{\sin^2\theta_1}{a_2}\right)^2 - \frac{4}{a_1 a_2}} \tag{7-44}$$

where $\theta_{1,2}$ are, respectively, the angles between $\hat{s}$ and the principal directions associated with the principal radii $a_{1,2}$ of surface curvature.

For an incident plane wave ($\rho_{1,2}^i \rightarrow \infty$), the above formulas yield

$$\sqrt{\rho_1^r \rho_2^r} = \frac{1}{2}\sqrt{a_1 a_2}. \tag{7-45}$$

Thus, for $s \gg a_{1,2}$, using the reflected field above as the scattered field in the RCS formula $\sigma = \lim_{s\rightarrow\infty} 4\pi s^2 \, \|\mathbf{E}^s\| / \|\mathbf{E}^i\|$ yields the geometrical optics RCS,

$$\sigma = \pi a_1 a_2. \tag{7-46}$$

8

DIFFRACTION APPROXIMATION METHODS

From the last result of the previous chapter we see that the geometrical optics short-wavelength asymptotic scattering result leaves something to be desired. It is inapplicable to surfaces other than those having two finite radii of curvature. Phrased somewhat differently, the geometrical optics method treats only specular scattering from targets and fails to treat any diffraction effects (such as scattering by edges and points on the target).

So we must turn to other methods to account for such diffraction effects. From among several methods we choose two: physical optics (PO) and the method of equivalent currents (MEC). Our choice of these two methods is based on their tractability and the directness of their physical imagery.

We begin with PO which, for all of its limitations, might be called the workhorse of scattering theory.

8-1 PHYSICAL OPTICS APPROXIMATION FOR $\mathbf{H}^s$; THE BISTATIC THEOREM

Consider a plane wave incident on a target surface S that we take to be a closed perfect conductor. Assume that S is smooth, that is, its radii of curvature at all points are at least as large as a wavelength. In the following discussion we work with the magnetic field because its form is simpler than that of the electric field for perfectly conducting scatterers. (See Section 4-7, Eqs. 4-27, 4-28, and Section 9-2 as well.)

We assume that the tangential component of the total magnetic field $\mathbf{H}$ on S can be approximated by *twice* the tangential component of $\mathbf{H}^{inc}$ on the illuminated region of S (i.e., the tangential component of the incident field plus the tangential component of the reflected field on S), and by zero on its shadowed region, since there is no incident field or reflected field in the shadow. Notice then that the surface field [or current $\hat{n}' \times \mathbf{H}(\mathbf{r}')$] may be viewed as a geometrical optics field. And although we have placed a lower bound on the radii of curvature of S, we have placed no upper bound on them. Thus, we can use PO to find the approximate RCS of a flat plate, whereas we cannot so use GO. The resulting integral approximation for $\mathbf{H}^s$ is called the *physical optics approximation* for the scattered field:

$$\mathbf{H}^s = \frac{1}{4\pi}\int_S [\hat{n}' \times \mathbf{H}(\mathbf{r}')] \times \nabla g(R')\, dS$$
$$= \frac{1}{2\pi}\int_{S_{\text{illuminated}}} [\hat{n}' \times \mathbf{H}^{inc}(\mathbf{r}')] \times \nabla g(R')\, dS \qquad (8\text{-}1)$$

where, as before, $g(R')$ is a Green's function:

$$g = \frac{1}{R'} e^{-jkR'}.$$

(The factor $1/4\pi$ has been removed from the Green's function and placed outside the integral in Eq. 8-1 to emphasize the cancellation of the factor 2 when the surface field is replaced, that is, approximated, by twice the tangential component of the incident field at the surface.)

Let O be an origin of coordinates within or in the vicinity of S. Assuming that an observer Rx is far from O (Fig. 8-1), we make the following approximations:

$$\begin{aligned} R'^2 &= r'^2 + R^2 - 2\mathbf{r}' \cdot R\hat{r} \\ R' &\approx R - \mathbf{r}' \cdot \hat{r} \quad \text{for} \quad R \gg r'. \end{aligned} \qquad (8\text{-}2)$$

$$\mathbf{grad}\,\frac{e^{-jkR'}}{R'} \approx \frac{-jk}{R}\,\hat{r} e^{-jk(R - \mathbf{r}' \cdot \hat{r})}. \qquad (8\text{-}3)$$

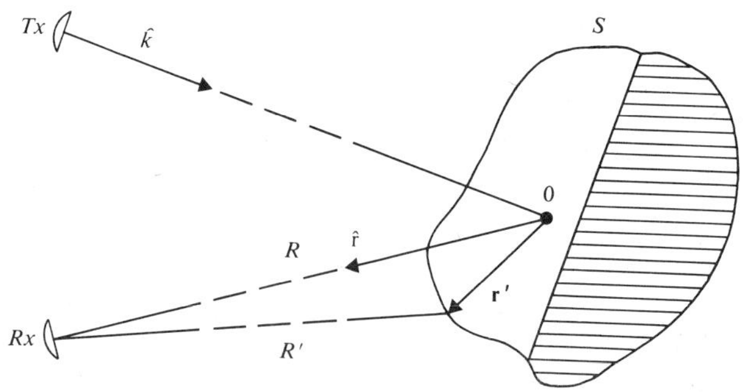

Figure 8-1 Definition of physical optics parameters.

Note that, in the gradient operation on the Green's function, we make a more careful approximation of the phase (in the exponential: $R' \approx R - \mathbf{r}' \cdot \hat{r}$) than we do of the amplitude (in the denominator: $R' \approx R$), since the former has a much greater effect on $\mathbf{H}^s$ than the latter.

As was indicated above in the introductory remarks, we take

$$\mathbf{H}(\mathbf{r}') = \begin{cases} 2\mathbf{H}_{\tan}^{inc}(\mathbf{r}') \text{ on } S_{\text{illuminated}} \text{ (set} = S_0), \\ 0, \quad \text{otherwise,} \end{cases} \tag{8-4}$$

whence

$$\hat{n}' \times 2\mathbf{H}_{\tan}^{inc}(\mathbf{r}') = \hat{n}' \times 2\mathbf{H}^{inc}(\mathbf{r}'). \tag{8-5}$$

Let

$$\mathbf{H}^{inc}(\mathbf{r}') = He^{-j\mathbf{k}\cdot\mathbf{r}'}\hat{h}.$$

Then

$$\begin{aligned} \mathbf{H}^s &= \frac{jk}{2\pi R} e^{-jkR} \int_{S_0} e^{jk\hat{r}\cdot\mathbf{r}'} \hat{r} \times (\hat{n}' \times H\hat{h}e^{-jk\hat{k}\cdot\mathbf{r}'})\, dS \\ &= \frac{jkH}{2\pi R} e^{-jkR} \int_{S_0} e^{jk(\hat{r}-\hat{k})\cdot\mathbf{r}'} [\hat{r} \times (\hat{n}' \times \hat{h})]\, dS. \end{aligned} \tag{8-6}$$

(Note: For backscattering, $\hat{r} = -\hat{k}$.) Therefore,

$$\mathbf{H}_{\text{PO}}^s = \frac{jkH}{2\pi R} e^{-jkR} \int_{S_0} e^{jk(\hat{r}-\hat{k})\cdot\mathbf{r}'} [(\hat{r} \cdot \hat{h})\hat{n}' - (\hat{r} \cdot \hat{n}')\hat{h}]\, dS. \tag{8-7}$$

Set

$$\mathbf{f} = \int_{S_0} \hat{n}' e^{jk(\hat{r}-\hat{k})\cdot\mathbf{r}'}\, dS = f_x\hat{x} + f_y\hat{y} + f_z\hat{z}. \tag{8-8}$$

Then

$$\mathbf{H}_{\text{PO}}^s = \frac{jkH}{2\pi R} e^{-jkR} [(\hat{r} \cdot \hat{h})\mathbf{f} - (\hat{r} \cdot \mathbf{f})\hat{h}], \tag{8-9}$$

and the RCS becomes

$$\sigma = \frac{4\pi}{\lambda^2} |(\hat{r} \cdot \hat{h})\mathbf{f} - (\hat{r} \cdot \mathbf{f})\hat{h}|^2. \tag{8-10}$$

If a receiver magnetic field direction is denoted by $\hat{d}$, then the effective RCS is

$$\sigma_e = \frac{4\pi}{\lambda^2} |(\hat{r} \cdot \hat{h})(\mathbf{f} \cdot \hat{d}) - (\hat{r} \cdot \mathbf{f})(\hat{h} \cdot \hat{d})|^2. \tag{8-11}$$

8-1-1 Special Cases, Bistatic

Let (with θ, or θ_0, denoting the usual polar angle measured from the z-axis)

$$\hat{k} = -\hat{x}\sin\theta_0 - \hat{z}\cos\theta_0 \qquad \text{(direction from transmitter to target);} \tag{8-12}$$

$$\hat{r} = \hat{x}\sin\theta + \hat{z}\cos\theta \qquad \text{(direction from target to receiver).} \tag{8-13}$$

Let $\hat{h} = \hat{y}$ denote parallel, or "vertical," polarization (since $\hat{e} = \hat{h} \times \hat{k}$ lies in the xz-plane); and let $\hat{h} = \hat{x}\cos\theta_0 - \hat{z}\sin\theta_0$ denote perpendicular, or "horizontal," polarization (since $\hat{e} = \hat{h} \times \hat{k}$ is perpendicular to the xz-plane). Then

$$\sigma_{VV} = \frac{4\pi}{\lambda^2}\{|f_x|^2\sin^2\theta + |f_z|^2\cos^2\theta + \mathrm{Re}(f_x f_z^*)\sin 2\theta\} \tag{8-14}$$

$$\sigma_{HH} = \frac{4\pi}{\lambda^2}\{|f_x|^2\sin^2\theta_0 + |f_z|^2\cos^2\theta_0 + \mathrm{Re}(f_x f_z^*)\sin 2\theta_0 + |f_y|^2\sin^2(\theta - \theta_0)\}, \tag{8-15}$$

where Re means "real part of" and $*$ means complex conjugate.

8-1-2 Special Cases, Monostatic

If $\theta_0 = \theta$ (which is the monostatic case, since $\hat{r} = -\hat{k}$), then $\sigma_{VV} = \sigma_{HH}$. Thus, physical optics gives *no polarization information in the monostatic case.* This shortcoming is, in fact, one of the drawbacks of the PO method.

If $\hat{k} = -\hat{z}$, and if the target is symmetric with respect to the xz-plane (i.e., the xz-plane mirrors each half of the target), then $f_y = 0$.

If $\hat{k} = -\hat{z}$, and if the target is a body of revolution about the z-axis, then $f_x = f_y = 0$, and

$$\sigma_{VV} = \sigma_{HH} = \frac{4\pi}{\lambda^2}|f_z|^2, \tag{8-16}$$

where

$$f_z = \int_{S_0} \hat{n}'_z e^{-2jk\hat{z}\cdot\mathbf{r}'}\,dS. \tag{8-17}$$

8-1-3 The Bistatic Theorem

In the short-wavelength limit, which is subsumed under our PO assumptions, it is possible to prove an interesting relationship between bistatic and monostatic RCS. The relationship is called the bistatic theorem. We state it only and refer the reader to Crispin and Maffett (1965) or to Kell (1965) for proofs.

Crispin and Maffett prove, using only PO current distribution assumptions, the theorem that bistatic RCS for a bistatic angle β is approximately equal to monostatic RCS at the bisector of β. From experimental data, one observes that β should not exceed 130° for the theorem to be valid.

Using evidence about surface current distribution over and above the PO assumptions, Kell proves the theorem that the bistatic RCS of a target for a given bistatic angle β and frequency f is equal to the monostatic RCS at the bisector of β at a frequency of $f/\cos(\beta/2)$. Once again, β is limited as suggested above. For moderate β-values, the two theorems are essentially alike.

8-1-4 Spurious Terms in Physical Optics

Write $\mathbf{H}^s = (\exp(-jkR')/kR)A\hat{h}^s$ and consider the expression A for several target bodies. We wish to illustrate the effect of the geometrical optics surface current assumption on the terms of the PO answer by showing that the artificial discontinuity in surface current at the shadow boundary produces spurious terms in the scattered field.

1. Consider a sphere, of radius a, with origin at sphere center: From the PO integral we find that

$$A = \underbrace{-\frac{ka}{2}\left(1 - \frac{-j}{2ka}\right)e^{2jka}}_{\substack{\text{front face reflection}\\ \text{(specular)}}} + \underbrace{\frac{j}{4}}_{\substack{\text{shadow boundary}\\ \text{(spurious)}}}.$$

   where the individual terms can be identified as arising from the designated locations. The shadow boundary does, as anticipated, produce a contribution to the answer, since the integration reflects the artificial jump there from geometrical optics in the illuminated region to zero in the shadow region.

2. Consider next a cone-sphere, of halfangle α and base radius a, with origin at the sphere center. Here we find from the PO integral that

$$A = \underbrace{-\frac{-j}{4}\tan^2\alpha\, e^{2jka\csc\alpha}}_{\text{tip reflection}} \underbrace{-\frac{j}{4}\sec^2\alpha\, e^{2jka\sin\alpha}}_{\text{join reflection}} + \underbrace{\frac{j}{4}}_{\substack{\text{shadow boundary}\\ \text{(spurious)}}} \quad \text{(nose-on)}.$$

   Again the PO integral produces a false term at the shadow boundary.

   Not only does the PO integral produce spurious terms, it also omits terms that should be included. For instance, in the above two examples, we find that there is missing from both a "creeping wave" term that is exponentially attenuated for large ka.

3. Next we examine the right circular cone, of halfangle α and vertical height h, with origin at tip. We find here that

$$A = \underbrace{\frac{j}{4}\tan^2\alpha}_{\text{tip reflection}} - \underbrace{\frac{j}{4}\tan^2(1+2jkh)e^{-2jkh}}_{\substack{\text{rim reflection} \\ \text{(both shadow boundary} \\ \textit{and}\text{ singularity)}}} \quad \text{(nose-on).}$$

Since the shadow boundary coincides with the rim which is a bonafide contributor, we must take further steps to identify the actual rim contribution. Here, we use a wedge model [refer to Section 3-2 below] for the cone base to obtain a valid expression for the rim reflection. It is:

$$A = \frac{j}{4}\tan^2\alpha - \frac{kh}{2n}\tan\alpha\csc\frac{2\pi}{n}e^{-2jkh},$$

$$\text{where} \quad n = \frac{3}{2} + \frac{\alpha}{\pi} \quad \text{(nose-on).}$$

From the above examples we see that we must always examine the PO answers carefully to be certain that the shadow boundary contribution does not introduce a significant error into the RCS.

8-2 METHOD OF EQUIVALENT CURRENTS (MEC)

For our discussion of edge diffraction we choose the MEC, which is an integrative technique dealing with radiation integrals. The diffracted field source is ascribed to fictitious equivalent currents, both electric and magnetic, flowing along the edge (cf. Chapter 4). Consequently, the MEC involves only line integrals. In this respect, it is a narrower theory than the Ufimtsev (1971) physical theory of diffraction (PTD) which uses surface integrals. On the other hand, it is a wider theory than the Keller (1962) geometrical theory of diffraction (GTD) since, as we shall see below, the MEC encompasses situations other than those described by the Keller cone of diffracted rays.

Equivalent edge currents have been discussed by Clemmow (1966), Millar (1956, 1957), Ryan and Peters (1970), Knott and Senior (1974), James (1976), and, most recently, by Michaeli (1984, 1986a, 1986b, 1987), by Knott (1985), and by Knott, Shaeffer, and Tuley, 1986, all of whose results we draw upon for the following discussion.

8-2-1 Edge-Diffracted Field

The diffracted field $\mathbf{E}^d$ from an edge discontinuity C, for the Fresnel and far zones, has the integral representation (James, 1976):

$$\mathbf{E}^d \approx jk \int_C [ZJ(\mathbf{r}')\hat{s} \times (\hat{s} \times \hat{t}) + K(\mathbf{r}')\hat{s} \times \hat{t}]g(\mathbf{r}, \mathbf{r}')\, dc \tag{8-18}$$

where we have employed the following notation:

Z = impedance of the medium,
$\mathbf{r}$ = position vector of the observation point,
$\mathbf{r}'$ = position vector of the integration point on C,
$dc = \|d\mathbf{r}'\|$: increment of arclength c along C,
$\hat{t} = d\mathbf{r}'/dc$: unit tangent vector,
$\hat{s} = \mathbf{s}/s = (\mathbf{r} - \mathbf{r}')/\|\mathbf{r} - \mathbf{r}'\|$,
$= \sin\beta \cos\phi \hat{x} + \sin\beta \sin\phi \hat{y} + \cos\beta\, \hat{z}$
(β and ϕ are polar and azimuth angles, respectively; see Fig. 8-2. The incidence direction is defined by the polar pair β, ϕ.)
$g(\mathbf{r}, \mathbf{r}') = \dfrac{\exp(-jks)}{4\pi s}$,
$\mathbf{J}(\mathbf{r}') = J(\mathbf{r}')\hat{t}$: electric equivalent current, and
$\mathbf{K}(\mathbf{r}') = K(\mathbf{r}')\hat{t}$: magnetic equivalent current;

$J(\mathbf{r}')$ and $K(\mathbf{r}')$ are assumed to depend linearly on the incident electric and magnetic field intensities E^i and H^i at $\mathbf{r}'$. As a result of this linear dependence, the phase factor $\exp[-jk\psi(\mathbf{r}')]$ describing spatial variation of E^i and H^i enters the integrand of Eq. 8-18 via J and K. Thus, the total phase variation of the integrand along C is contained in $\psi(\mathbf{r}') + |\mathbf{r} - \mathbf{r}'|$.

As $k \to \infty$, Eq. 8-18 has an asymptotic evaluation obtainable by the stationary phase method [assuming $\mathbf{r}$ not to lie on a caustic; the definition here is the same as it was for geometrical optics in Section 7-2 (see Fig. 7-2)], wherein the stationary phase points are obtained as solutions of

$$\hat{t} \cdot \nabla_{r'}[\psi(\mathbf{r}') + \|\mathbf{r} - \mathbf{r}'\|] = 0 \tag{8-19}$$

or

$$\beta = \beta', \tag{8-20}$$

where

$$\beta' = \cos^{-1}(\hat{t} \cdot \hat{s}'), \tag{8-21a}$$

$$\hat{s}' = \nabla_{r'}\psi, \tag{8-21b}$$

and

$$\beta = \cos^{-1}(\hat{t} \cdot \hat{s}). \tag{8-21c}$$

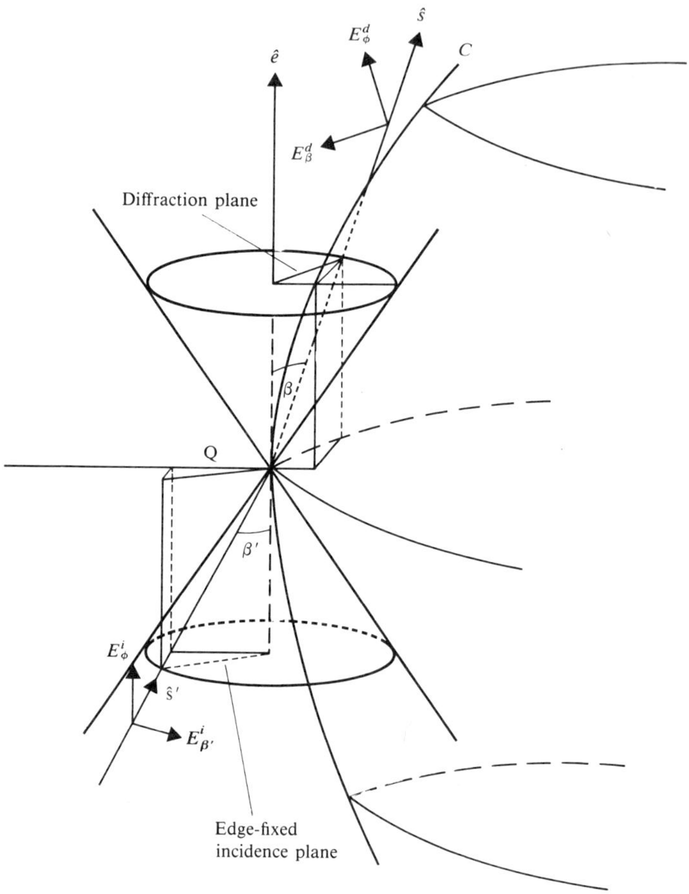

Figure 8-2 Diffraction by a curved edge.

8-2-2 Keller's Geometrical Theory of Diffraction (GTD)

Equation 8-20 is Keller's law of diffraction; it defines a forward cone of observation directions or Keller's cone of diffracted rays. So the MEC radiation integral reduces, by a stationary phase evaluation, to a sum of ray field contributions from (isolated) diffraction points on C satisfying Eq. 8-20.

These contributions are known from GTD and so J and K can be obtained therefrom:

$$J = -\frac{2(\mathbf{E}^i \cdot \hat{t})D_s}{jkZ \sin \beta'}; \qquad K = \frac{2Z(\mathbf{H}^i \cdot \hat{t})D_h}{jk \sin \beta'},$$

where $D_{s(h)}$ are the soft (hard) scalar diffraction coefficients from GTD. So

GTD is a corollary of the MEC (when the observation point is not near a caustic). But the MEC includes situations where the integral (Eq. 8-18) need not be evaluated by stationary phase.

Before looking at these situations, we turn aside briefly to look at the GTD.

8-2-3 Postulates for GTD

1. A diffracted field propagates along rays determined by a *generalized Fermat's principle* (which includes surface boundary points in a ray trajectory).
2. Diffraction, like reflection and transmission, is a *local* phenomenon at high frequencies.
3. A diffracted wave propagates along its ray in such a way that
   (a) power is conserved in a tube of rays, and
   (b) the phase delay along a path is equal to the product of medium wave number and distance in the medium.

Diffracted rays are initiated at boundary surface points where the incident geometrical optics field is discontinuous (i.e., at shadow or reflection boundary points).

Rays follow paths that make the optical distance between source and field points an extremum (usually a minimum). (On a surface, such a path is called a geodesic, e.g., the creeping wave on a sphere.)

For a wedge of exterior angle $\alpha = n\pi$, the scalar diffraction coefficients are, for the soft (s) and hard (h) cases:

$$D_{\substack{h\\s}} = \frac{e^{-j\pi/4}\sin(\pi/n)}{n\sqrt{2\pi k}\sin\beta_0'} \times \left[\frac{1}{\cos(\pi/n) - \cos[(\phi - \phi')/n]} \pm \frac{1}{\cos(\pi/n) - \cos[(\phi + \phi')/n]}\right]$$

where D depends on ϕ, ϕ', and β_0. This formula holds if the field point is not close to the shadow or reflection boundary.

At points away from the edge,

$$D_{\substack{h\\s}} = \frac{-e^{-j\pi/4}}{2n\sqrt{2\pi k}\sin\beta_0'}$$

$$\times\left[\cot\frac{\pi + (\phi - \phi')}{2n} F[kla^{+}(\phi - \phi')] + \cot\frac{\pi - (\phi - \phi')}{2n} F[kla^{-}(\phi - \phi')]\right.$$

$$\left.\pm\left\{\cot\frac{\pi + (\phi + \phi')}{2n} F[kla^{+}(\phi + \phi')] + \cot\frac{\pi - (\phi + \phi')}{2n} F[kla^{-}(\phi + \phi')]\right\}\right]$$

where

$$F(X) = 2j\,|\sqrt{X}|\,e^{jX}\int_{\sqrt{X}}^{\infty} e^{-j\tau^2}\,d\tau,$$

with $X = kla^{\pm}(\phi \pm \phi')$, $a^{\pm}(\beta) = 2\cos^2\frac{1}{2}(2n\pi N^{\pm} - \beta)$, and $l = (1/\rho_e^i)s\rho_1^i\rho_2^i(\rho_e^i + s)(\rho_1^i + s)^{-1}(\rho_2^i + s)^{-1}\sin^2\beta_0'$. $N^{\pm}$ are integers most nearly satisfying the relations

$$2\pi nN^{+} - \beta = \pi, \qquad 2\pi nN^{-} - \beta = -\pi.$$

ρ_1^i and ρ_2^i are the principal radii of curvature of the wavefront at Q_E; ρ_e^i is the radius of curvature of the incident wavefront in the edge-fixed plane of incidence ($\rho = \rho_e^i$ for the wedge).

8-2-4 Ufimtsev's Physical Theory of Diffraction

Returning now to the MEC we seek expressions for J and K that are more general than those obtainable from the GTD by a stationary phase evaluation of Eq. (8-18). These have been obtained by Michaeli (1986a, 1986b) as follows:

From first principles the far-scattered field from a surface containing the edge C is

$$\mathbf{E}^s = jkZ\sum_{i=1}^{2}\iint_{S_i} \hat{s}\times[\hat{s}\times\mathbf{j}_i(\mathbf{r}')]g(\mathbf{r},\mathbf{r}')\,dS_i \tag{8-22}$$

where S_1 and S_2 are two surfaces with common boundary C and $\mathbf{j}_i(\mathbf{r}')$ is the induced current on S_i. If $\mathbf{r}$ is not on a caustic, then as $k\to\infty$ the integral (Eq. 8-22) can be reduced asymptotically to a sum of ray field contributions from (isolated) interior stationary phase points on S_1 and S_2 (if any) and a boundary contribution given by a line integral along C. So the MEC can be related to rigorous asymptotic radiation theory and, thereby, to Ufimtsev's physical theory of diffraction (PTD).

Let x_1 and x_2 be distances from C along the normals to C tangent to S_1 and S_2, respectively, and pointing to these surfaces. The MEC is related to the PTD by noting that edge diffraction is due to surface currents in narrow strips of S_1 and S_2 along C; let x_i and h be integration variables in these strips. Then the diffracted part of Eq. 8-22 becomes

$$\mathbf{E}^d \approx jkZ\int_C\left\{\sum_{i=1}^{2}\int_0 \hat{s}\times[\hat{s}\times\mathbf{j}_i(h,x_i)]g(\mathbf{r}'(h,x_i),\mathbf{r})W_i(h,x_i)\,dx_i\right\}dh \tag{8-23}$$

where $W_i(h, x_i)$ is the Jacobian of the transformation from local Cartesian coordinates to the coordinates h, x_i; and the absence of an upper limit for the x_i-integration indicates that only the asymptotic endpoint contribution at $x_i = 0$ is used. Then $W_i(h, 0) = 1$, $i = 1, 2$. For small x_i,

$$g(\mathbf{r}'(h,x_i),\mathbf{r}) \approx g(\mathbf{r}'(h,0),\mathbf{r})\exp(jkx_i\hat{x}_i\cdot\hat{s}). \tag{8-24}$$

Using these facts in Eq. 8-23 and comparing the result with Eq. 8-18 yields

$$K(h) = \sum_{i=1}^{2} K_i(h), \qquad J(h) = \sum_{i=1}^{2} J_i(h) \tag{8-25}$$

with

$$K_i(h) = Z \sin^{-2} \beta\, \hat{t} \cdot \hat{s} \times \boldsymbol{\kappa}_i(h), \tag{8-26a}$$

$$J_i(h) = \sin^{-2} \beta\, \hat{s} \cdot [(\hat{t} \times \hat{s}) \times \boldsymbol{\kappa}_i(h)], \tag{8-26b}$$

and

$$\boldsymbol{\kappa}_i(h) = \int_0 \mathbf{j}_i(h, x_i) \exp(jkx_i\hat{x}_i \cdot \hat{s})\, dx_i. \tag{8-27}$$

Since only a small neighborhood of $x_i = 0$ is used in evaluating Eq. 8-27, the actual surface current in Eq. 8-27 can be approximated by the PTD current appropriate to an infinite wedge.

8-2-5 Diffraction Coefficients Based on Canonical Wedge Results

Michaeli is now able to relate his endpoint integral evaluations to the canonical wedge problem. His results for J and K, which apply quite generally to an arbitrary bistatic situation, are

$$J = \frac{-2jE_{it}D_e}{kZ \sin^2 \beta'} - \frac{2jH_{it}D_{em}}{k \sin \beta'}, \tag{8-28}$$

$$K = \frac{2jZH_{it}D_m}{k \sin \beta \sin \beta'}, \tag{8-29}$$

where E_{it} and H_{it} are electric and magnetic incident field components along the edge element. The diffraction coefficients D_e, D_m, and D_{em} are

$$D_e = \frac{\frac{1}{n} \sin \frac{\phi'}{n}}{\cos \frac{1}{n}(\pi - \alpha_1) - \cos \frac{\phi'}{n}} + \frac{\frac{1}{n} \sin \frac{\phi'}{n}}{\cos \frac{1}{n}(\pi - \alpha_2) + \cos \frac{\phi'}{n}}, \tag{8.30}$$

$$D_m = \frac{\sin \phi}{\sin \alpha_1} \cdot \frac{\frac{1}{n} \sin \frac{1}{n}(\pi - \alpha_1)}{\cos \frac{1}{n}(\pi - \alpha_1) - \cos \frac{\phi'}{n}} + \frac{\frac{1}{n} \sin \frac{1}{n}(\pi - \alpha_2)}{\cos \frac{1}{n}(\pi - \alpha_2) + \cos \frac{\phi'}{n}} \cdot \frac{\sin(n\pi - \phi)}{\sin \alpha_2}, \tag{8-31}$$

$$D_{em} = \frac{Q}{\sin\beta'}\left\{\frac{\cos\phi}{\sin\alpha_1}\cdot\frac{\frac{1}{n}\sin\frac{1}{n}(\pi-\alpha_1)}{\cos\frac{1}{n}(\pi-\alpha_1)-\cos\frac{\phi'}{n}}\right.$$

$$\left.-\frac{\cos(n\pi-\phi)}{\sin\alpha_2}\cdot\frac{\frac{1}{n}\sin\frac{1}{n}(\pi-\alpha_2)}{\cos\frac{1}{n}(\pi-\alpha_2)+\cos\frac{\phi'}{n}}\right\}, \tag{8-32}$$

where $n\pi$ is the exterior wedge angle and

$$Q = \frac{2(1+\cos\beta\cos\beta')}{\sin\beta\sin\beta'}\sin\tfrac{1}{2}(\beta+\beta')\sin\tfrac{1}{2}(\beta-\beta'),$$

$$\sin\alpha_1 = \frac{1}{\sin\beta'}\sqrt{\sin^2\beta'-\sin^2\beta\cos^2\phi},$$

and

$$\sin\alpha_2 = \frac{1}{\sin\beta'}\sqrt{\sin^2\beta'-\sin^2\beta\cos^2(n\pi-\phi)}.$$

8-2-6 Monostatic Backscattering Case

It is worth noting that for the monostatic backscattering case where $\beta=\pi-\beta'$ and $\phi=\phi'$ (i.e., incidence and observation directions have opposite senses), J and K become:

$$J(\beta=\pi-\beta', \phi=\phi') = \frac{2jE^i_{z0}\sin\frac{\pi}{n}}{Znk\sin^2\beta'}$$

$$\times\left[\frac{1}{\cos\frac{\pi}{n}-1}-\frac{1}{\cos\frac{\pi}{n}-\cos\frac{2\phi'}{n}}\right]-\frac{4jH^i_{z0}\cot\beta'}{nk\sin\beta'}$$

$$\times\left\{\frac{\cot\phi'\sin\frac{1}{n}(\pi-\phi')}{\cos\frac{1}{n}(\pi-\phi')-\cos\frac{\phi'}{n}}-\frac{\cot(n\pi-\phi')\sin\frac{1}{n}(\pi+\phi')}{\cos\frac{1}{n}(\pi+\phi')-\cos\frac{\phi'}{n}}\right\}; \tag{8-33}$$

$$K(\beta=\pi-\beta', \phi=\phi') = \frac{2jZH_{z0}\sin\frac{1}{n}}{nk\sin^2\beta'}$$

$$\times\left[\frac{1}{\cos\frac{\pi}{n}-1}+\frac{1}{\cos\frac{\pi}{n}-\cos\frac{2\phi'}{n}}\right]; \tag{8-34}$$

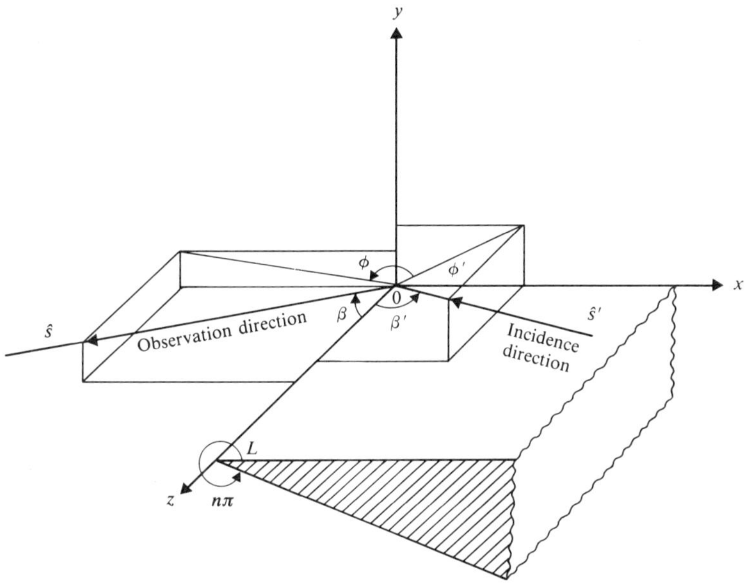

Figure 8-3 Incidence and observation directions for diffraction from a wedge of length L and exterior angle $n\pi$.

where E^i_{z0} and H^i_{z0} are z-components of the incident field at 0 (see Fig. 8-3, where it is assumed that the straight edge of the wedge is aligned along the z-axis).

8-2-7 Diffraction and PO Contributions to the Diffraction Integral

Michaeli's result includes a PO contribution from the wedge surface as well as a diffraction contribution from the wedge edge. If one wishes to obtain a contribution from the edge alone, then one must take measures to subtract out the PO result. Mitzner (1974) has performed such an analysis which he has called "incremental length diffraction coefficients." Recently, Knott (1985a) discussed this result and contrasted it with Michaeli's. We adapt Knott's discussion to our development here.

We use the Mitzner–Knott notation to compare the Michaeli and Mitzner diffracted fields for an *increment dc* of wedge edge. Set

$$\hat{e}^i_\perp = \frac{\hat{t} \times \hat{s}'}{\|\hat{t} \times \hat{s}'\|}, \qquad \hat{e}^i_\parallel = \hat{s}' \times \hat{e}^i_\perp; \tag{8-35}$$

$$\hat{e}^s_\perp = \frac{\hat{t} \times \hat{s}}{\|\hat{t} \times \hat{s}\|}, \qquad \hat{e}^s_\parallel = \hat{s} \times \hat{e}^s_\perp. \tag{8-36}$$

In addition, write the electric polarization of the incident wave as

$$\hat{p} = \hat{e}^i_\perp \cos\gamma + \hat{e}^i_\parallel \sin\gamma, \tag{8-37}$$

γ being the angle subtended by the incident electric field and the normal to the plane of incidence. Then E_{it} and H_{it} may be written in terms of γ and β', while $\hat{s} \times \hat{t}$ and $\hat{s} \times (\hat{s} \times \hat{t})$ can be given in terms of $\hat{e}^s_\perp$ and $\hat{e}^s_\parallel$. Now taking R_0 as distance from an origin near the scattering surface to the observation point, we may write

$$g \approx \frac{1}{4\pi R_0} \exp(-jkR_0) \exp(jk\mathbf{r}' \cdot \hat{s}') = g_0 \exp(jk\mathbf{r}' \cdot \hat{s}'). \qquad (8\text{-}38)$$

Discarding the relative phase $\exp(jk\mathbf{r}' \cdot \hat{s}')$ of the incident wave, the diffracted field obtained by Michaeli for an *increment* of edge becomes

$$\mathbf{E}^d = 2E_0 g_0 \, dc \left[D_m \hat{e}^s_\perp \cos\gamma - D_e \frac{\sin\beta}{\sin\beta'} \hat{e}^s_\parallel \sin\gamma - D_{em} \sin\beta \cos\gamma \hat{e}^s_\parallel \right]. \qquad (8\text{-}39)$$

Mitzner obtains the expression

$$\mathbf{E}^d = E_0 \frac{dc}{\sqrt{2\pi R}} \exp\left[-j\left(kR_0 - \frac{\pi}{4} \right) \right] \mathfrak{D} \cdot \hat{p}, \qquad (8\text{-}40)$$

where $\mathfrak{D}$ is a dyadic diffraction coefficient such that the dot product with the polarization vector is

$$\mathfrak{D} \cdot \hat{p} = (D_\perp - D'_\perp)\hat{e}^s_\perp \cos\gamma - (D_\parallel - D'_\parallel) \frac{\sin\beta}{\sin\beta'} \hat{e}^s_\parallel \sin\gamma$$
$$- (D_x - D'_x) \frac{\sin\beta}{\sin\beta'} \hat{e}^s_\parallel \cos\gamma. \qquad (8\text{-}41)$$

Comparing Eqs. 8-39 and 8-41 we see that they are identical if we remove the D'-terms from 8-41 and if we take

$$D_\perp = D_m, \qquad D_\parallel = D_e, \qquad D_x = D_{em} \sin\beta'. \qquad (8\text{-}42)$$

Thus, as Knott points out, the D'-terms represent the PO contribution that is subtracted out of the Mitzner expression to get the pure edge result. The D'-terms are

$$D'_\perp = U^+ \frac{\sin\phi}{\cos\alpha_1 + \cos\phi'} - U^- \frac{\sin(n\pi - \phi)}{\cos\alpha_2 + \cos(n\pi - \phi')} \qquad (8\text{-}43)$$

$$D'_\parallel = -U^+ \frac{\sin\phi'}{\cos\alpha_1 + \cos\phi'} - U^- \frac{\sin(n\pi - \phi')}{\cos\alpha_2 + \cos(n\pi - \phi')} \qquad (8\text{-}44)$$

$$D'_x = -U^+ \left[\frac{Q\cos\phi}{\cos\alpha_1 + \cos\phi'} - \cos\beta' \right]$$
$$+ U^- \left[\frac{Q\cos(n\pi - \phi)}{\cos\alpha_2 + \cos(n\pi - \phi')} - \cos\beta' \right] \qquad (8\text{-}45)$$

where the step functions U^+ and U^- depend on whether the upper (+) or lower (−) face of the wedge is illuminated:

$$U^+ = \begin{cases} 1, & \text{for the “+” face illuminated,} \\ 0, & \text{otherwise.} \end{cases} \tag{8-46}$$

$$U^- = \begin{cases} 1, & \text{for the “−” face illuminated,} \\ 0, & \text{otherwise.} \end{cases} \tag{8-47}$$

Knott makes these cogent observations: Michaeli's result extends the GTD to directions not on the Keller cone, while Mitzner extends the PTD. So the Keller-type of coefficients obtained by Michaeli will become singular along shadow and reflection boundaries. But the Ufimtsev-type of coefficients found by Mitzner remain finite on these boundaries. The D'-terms subtracted by Mitzner have a singularity that cancels the singularity in Michaeli's solution. As Knott (1985a, p. 114) puts it: "The price one pays for this highly desirable result is that two separate computations are required for any finite edged body. One is the edge contribution given by (Eqs. 8-40 and 8-41), and the other is a PO integral over the [body surface]."

8-2-8 An Example: Approximate Edge Diffraction RCS for a Straight-Edge Wedge

Direction of edge: $\hat{t} = \hat{z}$

Edgelength: L (from $z = 0$ to $z = L$)

Direction of observation: $\hat{s} = \hat{x} \sin\beta \cos\phi + \hat{y} \sin\beta \sin\phi + \hat{z} \cos\beta$

Direction of incidence: β', ϕ' with $\beta' = \pi - \beta$, $\phi' = \phi$

$$\hat{s} \times \hat{t} = \hat{x} \sin\beta \sin\phi - \hat{y} \sin\beta \cos\phi = \sin\beta(\hat{x} \sin\phi - \hat{y} \cos\phi)$$

$$\hat{s} \times (\hat{s} \times \hat{t}) = \sin\beta(\hat{x} \cos\beta \cos\phi + \hat{y} \cos\beta \sin\phi - \hat{z} \sin\beta)$$

$$s = \|\mathbf{r} - \mathbf{r}'\| \approx r - z(\hat{z} \cdot \hat{r}) = r - z \cos\beta$$

$$g(\mathbf{r}, \mathbf{r}') = \frac{1}{4\pi s} \exp(-jks) \approx \frac{1}{4\pi r} \exp(jkz \cos\beta).$$

The diffracted electric field is

$$\begin{aligned} \mathbf{E}^d &\approx jk[ZJ \sin\beta(\hat{x} \cos\beta \cos\phi + \hat{y} \cos\beta \sin\phi - \hat{z} \sin\beta) \\ &\quad + K \sin\beta(\hat{x} \sin\phi - \hat{y} \cos\phi)] \frac{1}{4\pi r} \exp(-jkr) \int_0^L \exp(jkz \cos\beta)\, dz \\ &= \frac{j}{2\pi r} \exp\left(-jkr + \frac{1}{2} jkL \cos\beta\right)[ZJ(\hat{x} \cos\beta \cos\phi \\ &\quad + \hat{y} \cos\beta \sin\phi - \hat{z} \sin\beta) \\ &\quad + K(\hat{x} \sin\phi - \hat{y} \cos\phi)] \sin\left(\frac{1}{2} kL \cos\beta\right). \end{aligned}$$

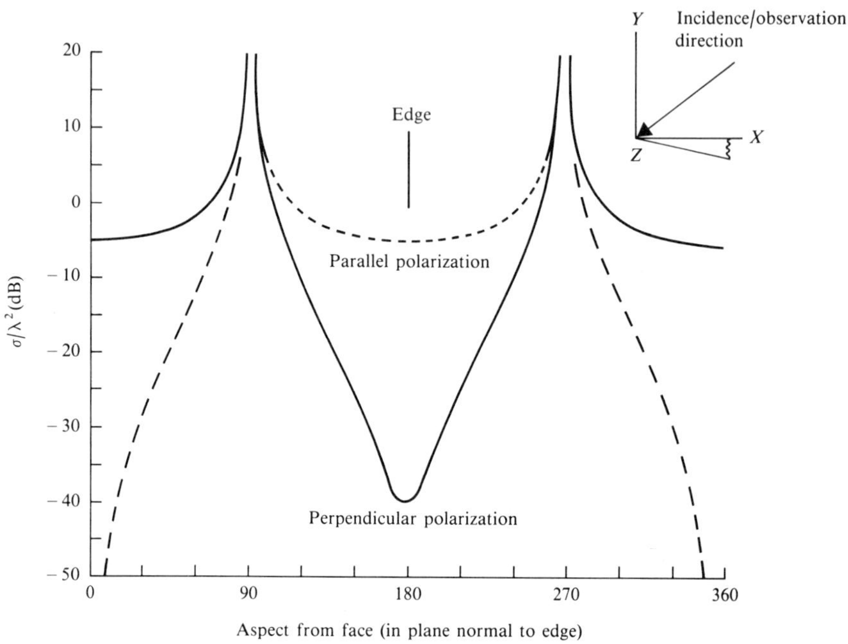

Figure 8-4 Wedge RCS for perpendicular and parallel polarizations (356° exterior angle).

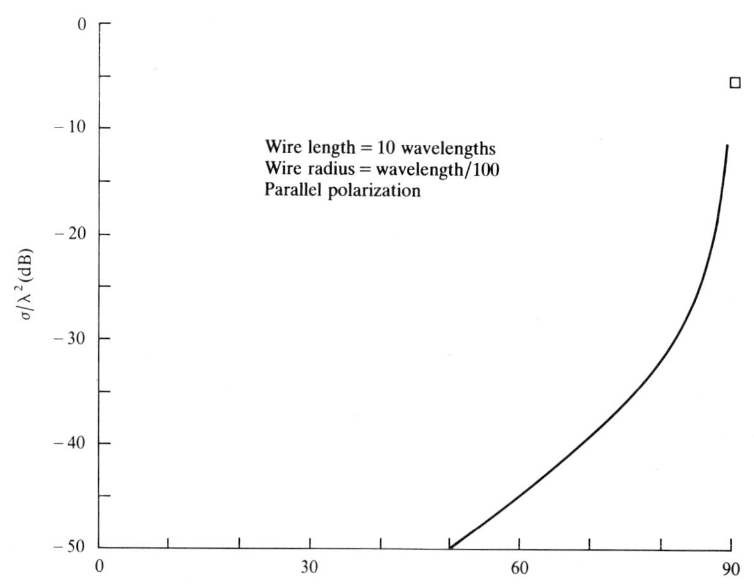

Figure 8-5 Straight-wire RCS.

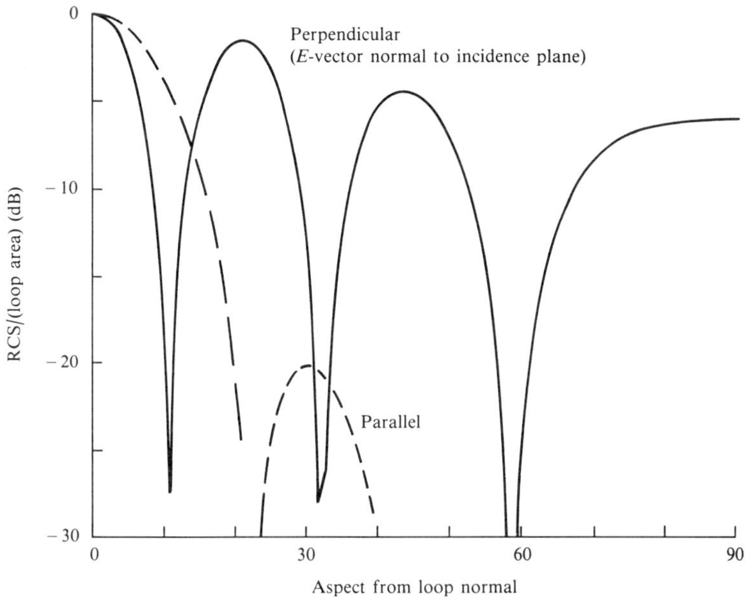

Figure 8-6 Wire loop RCS for perpendicular and parallel polarization. a = loop radius; $ka = 10$.

And the RCS is

$$\sigma^d = 4\pi r^2 \,\|\mathbf{E}^d\|^2 \approx \frac{1}{\pi}[Z^2 JJ^* + KK^*] \sin^2\left(\frac{1}{2} kL \cos\beta\right),$$

where we have assumed a unit incident field amplitude; $*$ means complex conjugate. J and K are given by Eqs. 8-28 and 8-29.

Figure 8-4 illustrates the scattering pattern (including the PO contribution) for an infinite wedge whose total interior angle is 4°. Figure 8-5 shows scattering from the edge only (PO contribution removed), which is analogous to wire scattering. Figure 8-6 shows the RCS pattern for a loop edge, or wire loop.

8-3 APPLICATION: A COMPOSITE MODEL FOR SEA SCATTERING

In the following sections we present an application of the diffraction results discussed above (Section 8-2) to modify an existing model so that it better describes the grazing angle scattering of microwave energy by a slightly rough sea.

Our composite model combines the Peake–Barrick (1967) model for scattering from a slightly rough sea surface with the Ryan–Peters (1970) [and later Knott and Senior (1974), Mitzner (1974), Michaeli (1984, 1986a, 1986b, 1987), and Knott (1985a, 1985b)] method of equivalent currents (MEC) for determination of edge diffraction.

Such edges are used to model the wedgelike wavelets, present on the surface of a slightly rough sea, which produce their most significant effect for shallow grazing angles, particularly for horizontal polarization. The curves derived from the composite model are compared with the measured data of Guinard and Daly (1970).

In Section 8-3-1 we review briefly the method of Peake and Barrick. We note in particular that the model predictions agree well with measured data for moderate-to-steep incidence angles, but not very well for small-to-moderate grazing angles. In Section 8-3-2 we outline the wedge, or edge, model as we intend to apply it to modify the Peake–Barrick model. Section 8-3-3 gives the composite result and we see that its predictions give greatly improved agreement with measured data in the small-to-moderate grazing angle region (Lyzenga, Maffett, and Shuchman, 1983).

8-3-1 Slightly Rough Surface RCS

The Peake–Barrick expression for the RCS of a slightly rough surface is (Peake and Barrick, 1967, p. 13)

$$\sigma_{0\gamma\delta}(\theta_s, \phi_s) = 4\pi k_0^4 \cos^2\theta_i \cos^2\theta_s\,|\alpha_{\gamma\delta}|^2 \times W[k_0(\sin\theta_s\cos\phi_s - \cos\theta_i),\ k_0\sin\theta_s\sin\phi_s] \quad (8\text{-}48)$$

where k_0 is the free space wave number, W is roughness height spectral density, and $\alpha_{\gamma\delta}$ are scattering matrix elements; each of the subscripts γ and δ may assume the polarization direction values h and v. The angles θ are incidence angles with subscripts i and s denoting illumination and scattering directions, respectively. The angle ϕ_s measures azimuth out of the bistatic plane (which contains the vertical direction and the direction of incidence) from the projection of the scattering direction on the ground plane. $\phi = 180°$ is backscattering.

Suppose surface roughness to be isotropic; then the height correlation function $R(\tau_x, \tau_y)$ is a function only of the separation r between surface points (x, y) and (x', y'); that is,

$$r = \sqrt{(x-x')^2 + (y-y')^2} = \sqrt{\tau_x^2 + \tau_y^2}.$$

So we may set

$$\rho(r) = R(\tau_x, \tau_y),$$

where $\rho(r)$ defines height correlation as a function of separation between any two points. If we write p and q as the arguments of W and set $t = \sqrt{p^2 + q^2}$, then we can express $W(p, q)$ as a function of the single variable t as

$$w(t) = W(p, q)$$

and express $w(t)$ as a Hankel transform of the roughness correlation function $\rho(r)$:

$$I = \frac{\pi^2}{h^2} w(t) = 2\pi \int_0^\infty r\rho(r)J_0(rt)\, dr, \tag{8-49}$$

where h^2 is surface mean square height and

$$t = k_0\sqrt{\cos^2\theta_i - 2\sin\theta_i \sin\theta_s \cos\phi_s + \sin^2\theta_s}.$$

For this situation the RCS becomes

$$\sigma_{0\gamma\delta}(\theta_s, \phi_s) = \frac{4}{\pi} k_0^4 h^2 \cos^2\theta_i \cos^2\theta_s\, |\alpha_{\gamma\delta}|^2\, I. \tag{8-50}$$

For backscattering, where $\theta_s = \theta_i$, $\phi_s = \pi$, Eq. 8-50 becomes

$$\sigma_{0\gamma\delta}(\theta_i, \pi) = \frac{4}{\pi} k_0^4 h^2 \cos^4\theta_i\, |\alpha_{\gamma\delta}|^2\, I, \tag{8-51}$$

with

$$t = k_0\sqrt{4\sin^2\theta_i} = 2k_0 \sin\theta_i, \tag{8-52}$$

and

$$I = 2\pi \int_0^\infty r\rho(r)J_0(2k_0 r \sin\theta_i)\, dr. \tag{8-53}$$

Note that no matter what form the roughness spectral density $W(p, q)$ takes, the dependence of σ_0 on incidence angle θ_i is basically of the form $\cos\theta_i$, although an angular dependence does enter into the computation of $\alpha_{\gamma\delta}$ and J_0. For application to a synthetic aperture radar (SAR) image of a slightly rough surface it would seem that θ_i must be determined from the average orientation of each resolution cell and not as the incidence angle with respect to an average surface level (or with respect to a flat earth).

Two types of roughness height correlation functions $\rho(r)$ produce especially interesting results because they permit the integration of I to be done exactly. They are:

1. Gaussian: $\rho(r) = e^{-r^2/s^2}$, where s is roughness correlation length, I becomes

$$I = \pi s^2 \exp[-(k_0 s \sin\theta_i)^2]$$

and so RCS is given by

$$\sigma_{0\gamma\delta} = 4(k_0 h)^2 (k_0 s)^2 \cos^4\theta_i\, |\alpha_{\gamma\delta}|^2 \exp[-(k_0 s \sin\theta_i)^2]. \tag{8-54}$$

2. Exponential: $\rho(r) = \exp[-|r|/s]$. Here

$$I = \frac{2\pi s^2}{[1 + (2k_0 s \sin \theta_i)^2]^{3/2}}$$

and from this

$$\sigma_{0\gamma\delta} = \frac{8(k_0 h)^2 (k_0 s)^2 \cos^4 \theta_i \, |\alpha_{\gamma\delta}|^2}{[1 + (2k_0 s \sin \theta_i)^2]^{3/2}} . \tag{8-55}$$

In both Eqs. 8-54 and 8-55 the scattering matrix entries are given by

$$\alpha_{hh} = \frac{\epsilon - 1}{(\cos \theta_i + \sqrt{\epsilon - \sin^2 \theta_i})^2} , \tag{8-56}$$

$$\alpha_{vv} = \frac{(\epsilon - 1)[(\epsilon - 1) \sin^2 \theta_i + \epsilon]}{(\epsilon \cos \theta_i + \sqrt{\epsilon - \sin^2 \theta_i})^2} , \tag{8-57}$$

$$\alpha_{hv} = \alpha_{vh} = 0, \tag{8-58}$$

where ϵ is the dielectric constant of the scattering surface.

For the exponential roughness height correlation function Fig. 8-7 shows plots of $\sigma_{0\gamma\delta}/(k_0 h)^2$ for $k_0 s = 6$ and 10 and $\epsilon = 48.3 - 34.9j$. This is the value of ϵ for seawater given by Guinard and Daly (1970). The shapes of these curves show a good resemblance to the measured data of Guinard and Daly provided that a suitable choice can be made for $(k_0 s)^2$.

Reasonable values for s are in the range of 1 to 10 cm with the most likely values being in the 3- to 5-cm range. So, for example, at a wavelength $\lambda_0 = 3$ cm, $k_0 s = 6$ is a reasonable value since it implies $s \approx \lambda_0 \approx 3$ cm. Heights h must be such that $h \leq s/7$; so for $k_0 s = 6$, $h \leq 0.4$ cm.

The value of h selected here certainly satisfies this restriction; it is $h = 0.024$ cm, so that $k_0 h = 0.05$. This value of $k_0 h$ makes Eq. 8-55 fit the Guinard and Daly measured data (see Fig. 8-8) for both polarizations in the

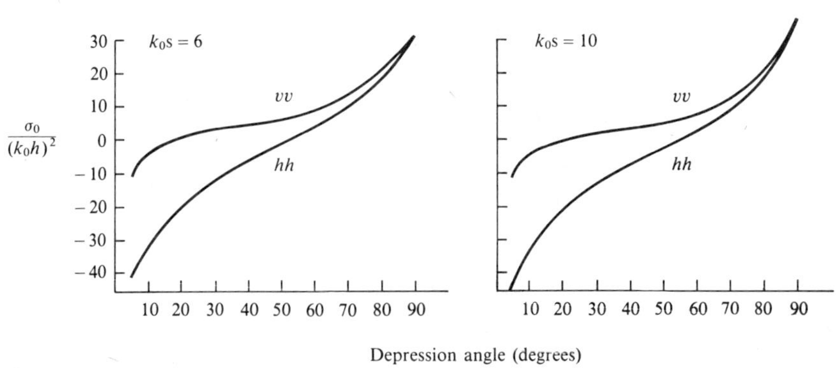

Figure 8-7 Normalized RCS for exponential roughness height correlation function.

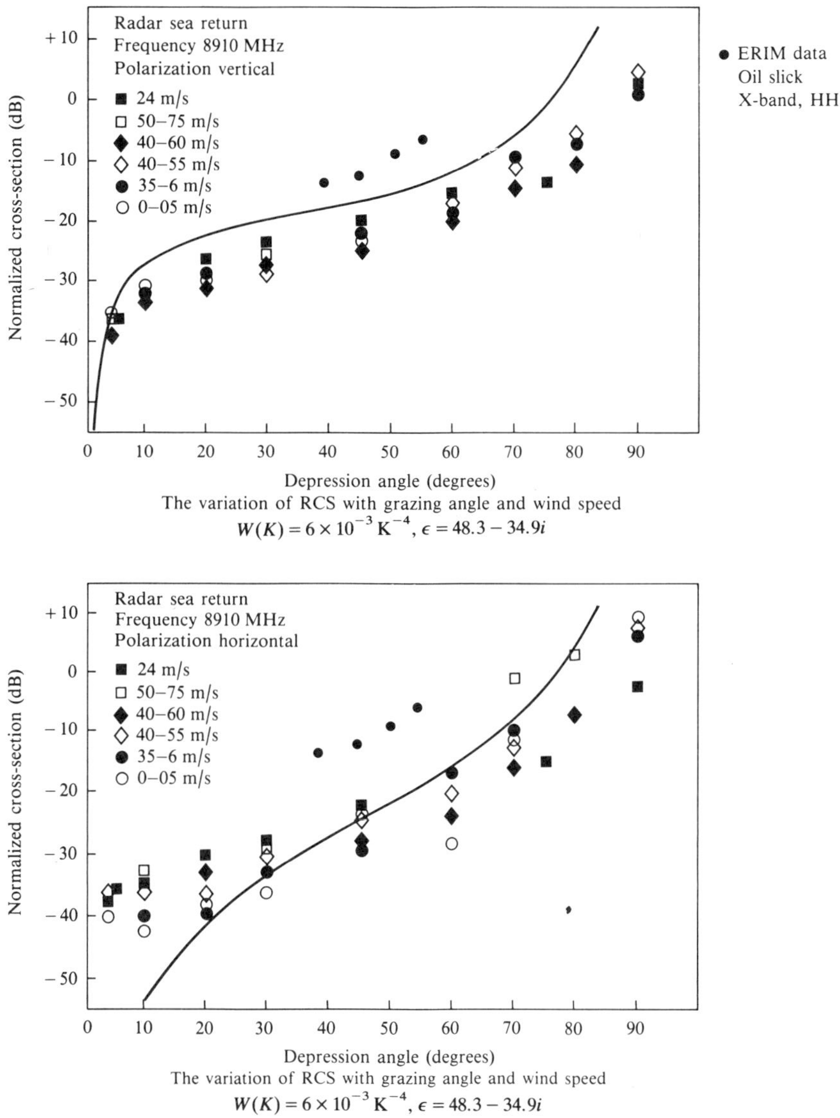

Figure 8-8 Measured sea clutter from Guinard and Daley (1970). © 1970 IEEE, reprinted with permission.

40 to 90° region of the depression angle. But this is really a *very small* value for surface roughness height, particularly in comparison to the radar wavelength: $h = 0.008\lambda_0$. And it may make one skeptical as to whether the next step can validly be taken.

When we examine the curves of Fig. 8-7 in connection with the Guinard and Daly data (see Fig. 8-8), we find that no value of k_0h can be selected to make the curves agree well with the data for both the shallow grazing angle

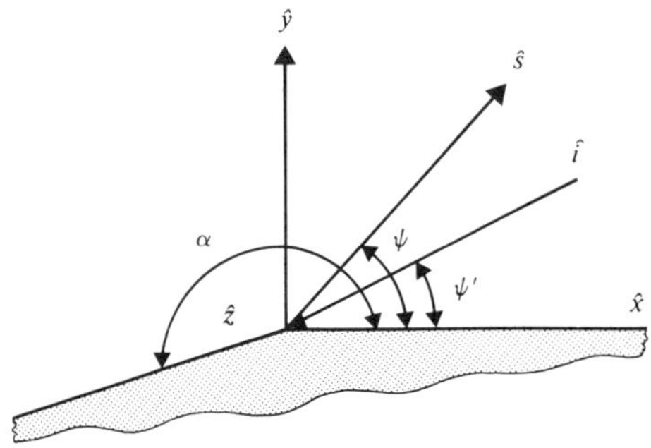

Figure 8-9 Incidence and scattering parameters for a wedge.

region and the steep depression angle region. In particular, if a value of k_0h is selected so as to make the curves fit the data nicely in the steep angle region, then they fall well below the data, particularly for horizontal polarization, for the shallow angle region. So the following tack is taken: it is assumed that the Peake–Barrick model does not account for all of the rough surface scattering effects at shallow grazing angles. In particular, it is assumed that the Peake–Barrick model does not account for scattering from the edges of small wedgelike wavelets. Such scattering would be significant, in comparison to the Peake–Barrick results, in the shallow angle region, especially for horizontal polarization; but in the steep angle region the edge scattering would not be of such significance as to destroy the good data fit of the Peake–Barrick model there. Wedgelike wavelets, which may be present on a slightly rough sea, have lengths L that are two to three times the peak value of the roughness correlation length s; so L could be 6–9 cm. The wedge interior angle must be at least 120°.

8-3-2 Wedge RCS

We have already met with the wedge and edge methods of Knott and Senior (1973, 1974), Mitzner (1974), Michaeli (1984, 1986a, 1986b, 1987), and (Knott (1985a, 1985b) in Section 8-2, for which we used the collective designation of method of equivalent currents (MEC). We propose to adapt this method of computing diffracted fields from edges and corners to compute the scattering return from the wedgelike wavelets. Consider a wedge with exterior angle α with edge direction $\hat{t}$ (in the $\hat{z}$-direction), together with the following incidence and scattering parameters (Fig. 8-9):

$\hat{i}$ = incidence direction, making angle ψ' with $\hat{x}$-axis;
$\hat{s}$ = scattering direction, making angle ψ with $\hat{x}$-axis;
$\mathbf{E}^i$, $\mathbf{H}^i$ = incident fields;
$\hat{e}_i$, $\hat{h}_i$ = unit incident fields;
$\hat{e}_r$, $\hat{h}_r$ = receiver polarization directions;
$\cos\beta_i = \hat{i}\cdot\hat{t}$, $\cos\beta_s = \hat{s}\cdot\hat{t}$, $e_{it} = \hat{e}_i\cdot\hat{t}$, $e_{rt} = \hat{e}_r\cdot\hat{t}$.

For backscattering, which is the only situation of interest here, $\hat{s} = -\hat{i}$, implying that $\psi = \psi'$ and $\beta_i = \beta_s = \beta$.

The scattering coefficient S for an edge defined by the curve C is given by

$$S = \frac{k_0}{2\pi} \int_C (f e_{it} e_{rt} + g h_{it} h_{rt}) \left(\frac{dc}{\sin \beta} \right) \exp(2jk_0 \mathbf{r} \cdot \hat{s}) \tag{8-59}$$

where

$$f = \begin{cases} (X - Y) - (X_1 - Y_1), & 0 \le \psi' \le \alpha - \pi \\ (X - Y) - (X_1 - Y_1) - (X_2 - Y_2), & \alpha - \pi \le \psi' \le \pi \\ (X - Y) - (X_2 - Y_2), & \pi \le \psi' \le \alpha; \end{cases} \tag{8-60}$$

$$g = \begin{cases} (X + Y) - (X_1 + Y_1), & 0 \le \psi' \le \alpha - \pi \\ (X + Y) - (X_1 + Y_1) - (X_2 + Y_2), & \alpha - \pi \le \psi' \le \pi \\ (X + Y) - (X_2 + Y_2), & \pi \le \psi' \le \alpha. \end{cases} \tag{8-61}$$

The three expressions of each set correspond, respectively, to face #1 lit, both faces lit, and face #2 lit.

The diffraction coefficients X and Y are

$$X = \frac{\dfrac{1}{n} \sin \dfrac{\pi}{n}}{\cos \dfrac{\pi}{n} - \cos \dfrac{\psi - \psi'}{n}} \tag{8-62}$$

$$Y = \frac{\dfrac{1}{n} \sin \dfrac{\pi}{n}}{\cos \dfrac{\pi}{n} - \cos \dfrac{\psi + \psi'}{n}} \tag{8-63}$$

where $\alpha = n\pi$. The quantities X_i, Y_i, $i = 1, 2$, are

$$X_1 = -\frac{1}{2} \frac{\sin \psi - \sin \psi'}{\cos \psi + \cos \psi'} \, [= 0 \text{ for } \psi = \psi'] \tag{8-64}$$

$$Y_1 = -\frac{1}{2} \frac{\sin \psi + \sin \psi'}{\cos \psi + \cos \psi'} \left[= -\frac{1}{2} \tan \psi' \text{ for } \psi = \psi' \right] \tag{8-65}$$

$$X_2 = -\frac{1}{2} \frac{\sin(\alpha - \psi) - \sin(\alpha - \psi')}{\cos(\alpha - \psi) + \cos(\alpha - \psi')} \, [= 0 \text{ for } \psi = \psi'], \tag{8-66}$$

$$Y_2 = -\frac{1}{2} \frac{\sin(\alpha - \psi) + \sin(\alpha - \psi')}{\cos(\alpha - \psi) + \cos(\alpha - \psi')}$$

$$\left[= \frac{1}{2} \tan(\alpha - \psi') \text{ for } \psi = \psi' \right]. \tag{8-67}$$

The scattering coefficient S is related to RCS by the equation

$$\sigma = \frac{4\pi}{k_0^2}|S|^2. \tag{8-68}$$

For the sea scattering application desired here the RCS σ must be divided by a suitable area to obtain the quantity σ_0. More on this matter below.

For polarization parallel to the wedge edge, the backscattering RCS is given by

$$\frac{\sigma_{\parallel}}{L^2} = \frac{f^2}{\pi}$$

$$= \frac{1}{\pi}\begin{cases} \left(X - Y - \frac{1}{2}\tan\psi'\right)^2, & 0 \le \psi' \le \alpha - \pi, \\ \left[X - Y - \frac{1}{2}\tan\psi' - \frac{1}{2}\tan(\alpha - \psi')\right]^2, & \alpha - \pi \le \psi' \le \pi, \\ \left[X - Y - \frac{1}{2}\tan(\alpha - \psi')\right]^2, & \pi \le \psi' \le \alpha. \end{cases} \tag{8-69}$$

When the polarization is perpendicular to the wedge edge, the RCS is

$$\frac{\sigma_{\perp}}{L^2} = \frac{g^2}{\pi}$$

$$= \frac{1}{\pi}\begin{cases} \left(X + Y + \frac{1}{2}\tan\psi'\right)^2, & 0 \le \psi' \le \alpha - \pi, \\ \left[X + Y + \frac{1}{2}\tan\psi' + \frac{1}{2}\tan(\alpha - \psi')\right]^2, & \alpha - \pi \le \psi' \le \pi, \\ \left[X + Y + \frac{1}{2}\tan(\alpha - \psi')\right]^2, & \pi \le \psi' \le \alpha. \end{cases} \tag{8-70}$$

The quantities f and g may be called modified diffraction coefficients; in their definitions (Eqs. 8-60 and 8-61; see also Knott and Senior, 1974a) the physical optics contribution has been subtracted from the total diffracted field of the wedge, leaving the contribution due to the edge alone. That is to say, if one were interested in the scattering coefficient S due to the total wedge diffracted field, one would have to replace f and g in Eq. 8-59 by $X - Y$ and $X + Y$, respectively. In the application here to sea scattering, no account has been taken of the reflection coefficient for the wedge surface, which should produce only a small effect on the wedge RCS (perhaps no more than 1 dB).

8-3-3 Combined Wedge Peake–Barrick RCS

If the expressions $\sigma_{\parallel}/L^2$ and $\sigma_{\perp}/L^2$ are computed for a set of wedgelike wavelets within an illuminated area A, then one obtains RCS per unit area by dividing by A. The resulting expressions may be called $\sigma_{0\parallel}/L^2$ and $\sigma_{0\perp}/L^2$ and may be combined with σ_{0hh} and σ_{0vv}, respectively, providing it is assumed that only those wedges with edges normal to the incidence direction yield a significant return. The combination is made by simply adding RCS values, that is, the assumption is made that the unknown phase between the two component scatterers is random (see Chapter 12, in particular, Section 12-1). The wedge return is insignificant for small incidence angles but is significant for small grazing angles.

Here are the assumptions: there exist 10 wedgelike wavelets per meter in range (along the surface, with edges approximately perpendicular to the range direction) and 3 per meter in azimuth. Of these wedgelike wavelets 10–50% may be oriented suitably to give a significant scattering return, which is then combined on a random phase basis (see Chapter 12) with the Peake–Barrick values, that is, the wedge and Peake–Barrick RCS returns are simply added.

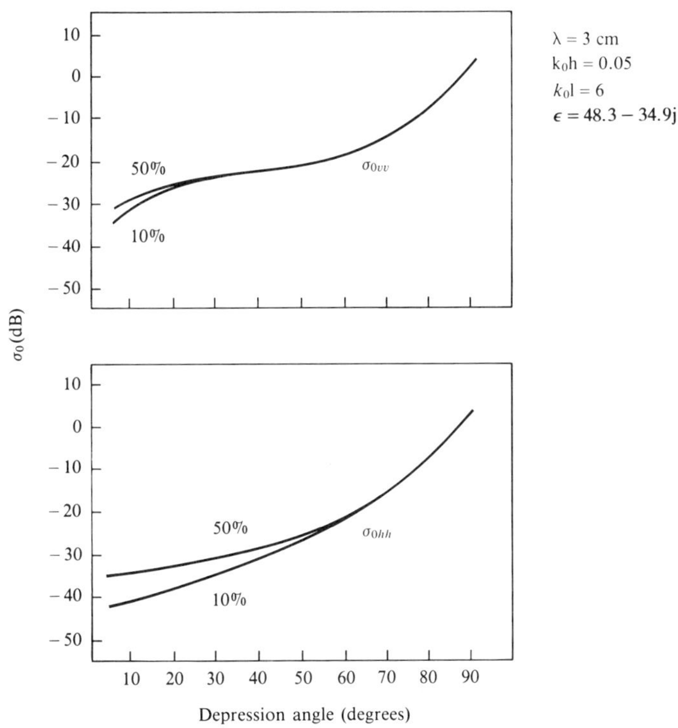

Figure 8-10 Combination of wedge and Peake–Barrick scattering for a slightly rough sea ($k_0s = 6$).

Figures 8-10 and 8-11 show the results. They should be compared with the Guinard–Daly data of Fig. 8-8. Here is a summary of conditions under which the curves of Figs. 8-10 and 8-11 were calculated:

1. $\lambda = 3$ cm.
2. Exponential roughness height correlation function with:
   (a) h, surface mean height, $= 0.024$ cm (0.008λ), or $k_0 h = 0.05$.
   (b) s, roughness correlation length, $= 3$ cm, or $k_0 s = 6$.
3. Number of wedgelike wavelets per meter in range, that is, approximately perpendicular to wedge edge, $= 10$.
4. Number of wedgelike wavelets per meter in azimuth $= 3$.
5. L, wedge length, ≈ 3 cm.
6. 10–50% of wedgelike wavelets in an illuminated area have edges approximately normal to the line of sight.
7. Wedge wavelet RCS contributions and exponential roughness contributions add with random (equally likely) phases.
8. ϵ, dielectric constant of the sea, $= 48.3 - 34.9j$.

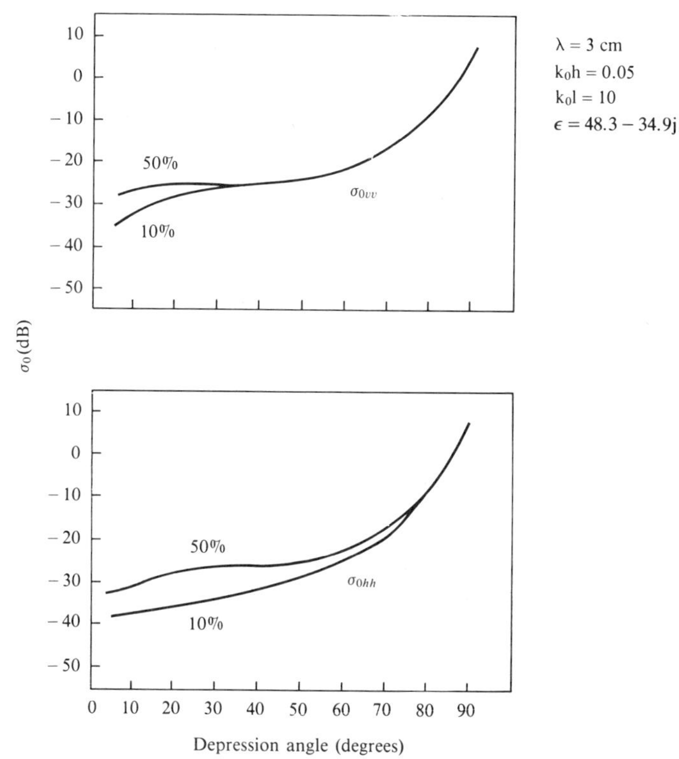

Figure 8-11 Combination of wedge and Peake–Barrick scattering for a slightly rough sea $(k_0 s = 10)$.

8-3-4 Discussion of Results

The computed curves for the composite theory described above are undeniably good fits (Fig. 8-10 is the better) to the Guinard–Daly data, and indeed rather good fits in the shallow grazing angle region where other theories do not provide good fits. However, there are caveats to be noted. The roughness height h is very small, perhaps too small to justifiably apply the equivalent current method; although it might be argued that one is really computing the diffraction from the edges only. In any case one would certainly feel better if h could be taken to be larger, or if the wavelength were somewhat smaller. The Gaussian roughness height correlation function has not been tested for a very large variety of parameters; this should be done to see whether larger values of h could be used.

The assumptions, particularly 2a, 2b, 3, 4, 5, and 6, should be carefully considered, both with respect to their effect on the validity of the equivalent current method and Peake–Barrick theories as used here and especially with respect to their accurate portrayal of reasonable descriptions of a slightly rough sea. Meanwhile, there remains the pragmatic view that the results do agree well with experimental data.

9

NUMERICAL APPROXIMATIONS AND NUMERICAL METHODS

When the illuminating wavelength is comparable to a major dimension of the target, the corresponding frequency region is called the *resonance region*. In this region there is no opportunity to employ the asymptotic methods that formed the basis of scattered field approximations in the Rayleigh and optics regions. This viewpoint may require some adjustment for a complex target, which may be viewed as a collection of simpler scattering components (or centers). In these circumstances, a resonance region classification for a complex target need not necessarily apply to each and every scattering component. So we shall eliminate such complex targets from the present discussion and return to them in Chapters 12 and 13.

But if asymptotic methods are denied to us in the resonance region, numerical methods take on added feasibility. Numerical methods can mean numerical evaluation of scattered field integrals, such as one might encounter in the use of PO; we begin with such a straightforward introduction to numerical approximations in Section 9-1 below. In even the simplest cases, however, it is necessary to have knowledge of the surface current of the target. Generally, however, no such knowledge is available, except in the form of an integral equation. Specifically, then, the problem we must treat by numerical methods is to solve the surface current integral equations for the surface current of the target (or else make an assumption about that surface current as we do in the case of PO). Once the surface current is in hand, the scattered field (see Eq. 4-8) and hence the RCS, is relatively easy to find. In all of these cases we must divide the scattering surface into increments, or patches, whose area is very small compared with a square wavelength. So the smaller the wavelength, the lengthier the numerical computation.

In Section 9-2 we restate the integral equations (including Green's functions) that yield surface current. In Section 9-3 we review the method of moments (MoM) which forms the basis for solving the surface current integral equations in Section 9-4. Section 9-5 shows that the PO method can be reintroduced to advantage in numerical methods. Finally, in Section 9-7 we comment briefly on historical aspects of early numerical work.

9-1 A NUMERICAL APPROXIMATION FOR PHYSICAL OPTICS

Refer to the physical optics integrals of Section 8.1 (Eqs. 8-8 and 8-9) which, for metallic targets, were written:

$$\mathbf{H}_{\mathrm{PO}} = \frac{jkH}{2\pi R} e^{-jkR}\{(\hat{r}\cdot\hat{h})\mathbf{f} - (\hat{r}\cdot\mathbf{f})\hat{h}\}, \tag{9-1}$$

where

$$\mathbf{f} = \int_{S_0} \hat{n}' e^{+jk(\hat{r}-\hat{k})\cdot\mathbf{r}'}\, dS. \tag{9-2}$$

We seek a numerical approximation for $\mathbf{f}$.

Divide S_0 into N small quadrilateral patches. Then split each quadrilateral patch into two triangular patches. Call the triangular patches† $\Delta_i S$, $i = 1, 2, \ldots, 2N$. Approximate $\mathbf{f}$ by Eq. 9-2:

$$\mathbf{f} = \sum_{i=1}^{2N} \hat{n}_i' e^{+jk(\hat{r}-\hat{k})\cdot\mathbf{r}_i}\, \Delta_i S.$$

Consider the ith patch. We may write

$$\hat{n}_i' \Delta_i S = \Delta_i A_{yz}\hat{x} + \Delta_i A_{xz}\hat{y} + \Delta_i A_{xy}\hat{z}, \tag{9-3}$$

where $\Delta_i A_{yz}$, $\Delta_i A_{xz}$, and $\Delta_i A_{xy}$ are, respectively, the *projections* of $\Delta_i S$ on the yz, xz, and xy-coordinate planes.

Each of the three corners of the ith triangular patch $\Delta_i S$ is located in space by a set of coordinates, say,

$$P_1^{(i)}(x_1^{(i)}, y_1^{(i)}, z_1^{(i)}),\ P_2^{(i)}(x_2^{(i)}, y_2^{(i)}, z_2^{(i)}),\ P_3^{(i)}(x_3^{(i)}, y_3^{(i)}, z_3^{(i)}).$$

Then we find for the projected areas of the ith patch $\Delta_i S$ (omitting the

† The longest leg of any triangular patch should not exceed $\lambda/10$.

superscripts i)

$$\Delta_i A_{yz} = \frac{1}{2} \det \begin{pmatrix} y_1 & z_1 & 1 \\ y_2 & z_2 & 1 \\ y_3 & z_3 & 1 \end{pmatrix}, \quad \Delta_i A_{xz} = \frac{1}{2} \det \begin{pmatrix} x_1 & z_1 & 1 \\ x_2 & z_2 & 1 \\ x_3 & z_3 & 1 \end{pmatrix},$$

$$\Delta_i A_{xy} = \frac{1}{2} \det \begin{pmatrix} x_1 & y_1 & 1 \\ x_2 & y_2 & 1 \\ x_3 & y_3 & 1 \end{pmatrix}, \tag{9-4}$$

where the absolute value of each determinant is to be taken.

The position vector $\mathbf{r}_i'$ may be taken as a vector from the coordinate origin to one of the vertices, say $P_1^{(i)}$, of the ith triangular patch (or to the centroid of a patch).

Specialize to the monostatic case, so that $-\hat{k} = \hat{r}$ and $\hat{r} \cdot \hat{h} = 0$. Taking $\hat{r} = +\hat{x} \sin\theta \cos\phi + \hat{y} \sin\theta \sin\phi + \hat{z} \cos\theta$, we may write the dot product $\hat{r} \cdot \mathbf{r}_i'$ as

$$\hat{r} \cdot \mathbf{r}_i' = +x_1^{(i)} \sin\theta \cos\phi + y_1^{(i)} \sin\theta \sin\phi + z_1^{(i)} \cos\theta. \tag{9-5}$$

Finally, using the above expressions, the components of $\mathbf{f}$ are:

$$f_x = \frac{1}{2} \sum_{i=1}^{2N} e^{j\psi_i} \det \begin{pmatrix} y_1^{(i)} & z_1^{(i)} & 1 \\ y_2^{(i)} & z_2^{(i)} & 1 \\ y_3^{(i)} & z_3^{(i)} & 1 \end{pmatrix}$$

$$f_y = \frac{1}{2} \sum_{i=1}^{2N} e^{j\psi_i} \det \begin{pmatrix} x_1^{(i)} & z_1^{(i)} & 1 \\ x_2^{(i)} & z_2^{(i)} & 1 \\ x_3^{(i)} & z_3^{(i)} & 1 \end{pmatrix} \tag{9-6}$$

$$f_z = \frac{1}{2} \sum_{i=1}^{2N} e^{j\psi_i} \det \begin{pmatrix} x_1^{(i)} & y_1^{(i)} & 1 \\ x_2^{(i)} & y_2^{(i)} & 1 \\ x_3^{(i)} & y_3^{(i)} & 1 \end{pmatrix},$$

where

$$\psi_i = 2k(x_1^{(i)} \sin\theta \cos\phi + y_1^{(i)} \sin\theta \sin\phi + z_1^{(i)} \cos\theta). \tag{9-7}$$

If one of the surface normal components n_x', n_y', or n_z' is zero, then the corresponding component f_x, f_y, or f_z of $\mathbf{f}$ is zero as well.

We compare, in Fig. 9-1, the approximate numerical PO bistatic RCS for a sphere of radius $a = 3.18\lambda$ ($ka = 20$), using the method described above, with an exact result given by Bowman, Senior, and Uslenghi (1969). A similar numerical approximation for monostatic RCS of a finite solid cylinder is compared with measured data in Fig 9-2.

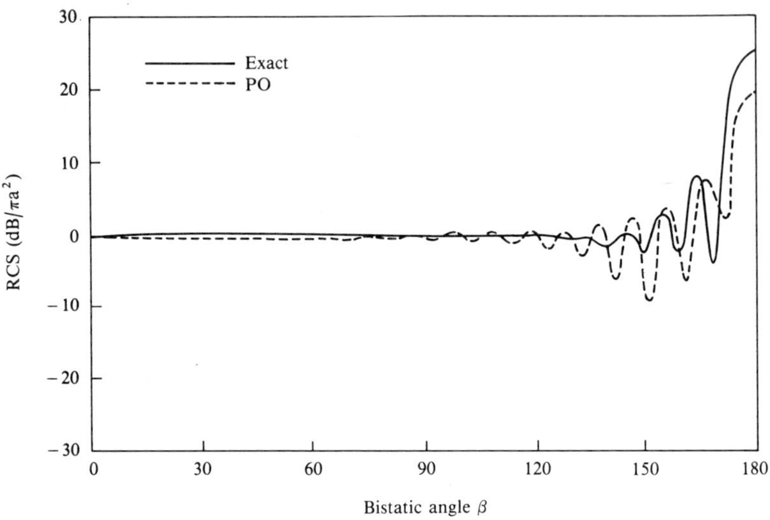

Figure 9-1 Bistatic RCS for a sphere with $ka = 20$.

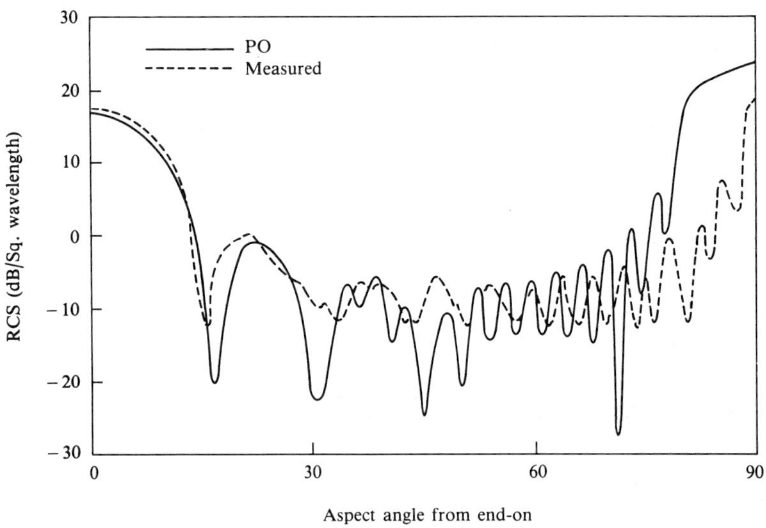

Figure 9-2 Monostatic RCS for a cylinder of radius 1.145λ and length 7.91λ.

9-2 INTEGRAL REPRESENTATIONS OF FIELDS†

Starting with the vector analog of Green's theorem (cf. Section 4-6):

$$\int_V dV(\mathbf{Q}\cdot\nabla\times\nabla\times\mathbf{P}-\mathbf{P}\cdot\nabla\times\nabla\times\mathbf{Q})=\int_{S'} dS(\mathbf{P}\times\nabla\times\mathbf{Q}-\mathbf{Q}\times\nabla\times\mathbf{P})\cdot\hat{n} \tag{9-8}$$

and setting $\mathbf{P}=\mathbf{E}$, $\mathbf{Q}=g\hat{a}$, where $\hat{a}$ is an arbitrary unit vector and $g=e^{-jkr}/r$, $r=\|\mathbf{r}-\mathbf{r}'\|$, is, as before, the free space Green's function, we find

$$\nabla\times\mathbf{Q}=\nabla g\times\hat{a}, \tag{9-9}$$

$$\nabla\times\nabla\times\mathbf{Q}=k^2g\hat{a}+\nabla(\hat{a}\cdot\nabla g),\quad \|\mathbf{r}-\mathbf{r}'\|\neq 0, \tag{9-10}$$

$$\nabla\times\nabla\times\mathbf{P}=k^2\mathbf{E}-j\omega\mu\mathbf{J}-\nabla\times\mathbf{K}.‡ \tag{9-11}$$

Since $\hat{a}$ is arbitrary and is common to all terms in Eq. (9-8), we find

$$\int_V dV\left[-j\omega\mu\mathbf{J}g-(\nabla'\times\mathbf{K})g+\frac{\rho}{\epsilon}\nabla'g\right]$$

$$=\int_{S'} dS[-j\omega\mu(\hat{n}'\times\mathbf{H})g+(\hat{n}'\times\mathbf{E})\times\nabla'g+(\hat{n}'\cdot\mathbf{E})\nabla'g-\hat{n}'\times\mathbf{K}g] \tag{9-12}$$

where the ∇' indicates an operation on the primed variables. The identity

$$\int_V dV(\nabla'\times\mathbf{K})g=\int_{S'}(\hat{n}'\times\mathbf{K}g)\,dS+\int_V dV(\mathbf{K}\times\nabla'g)$$

reduces Eq. (9-12) to

$$\int_V dV\left(-j\omega\mu\mathbf{J}g-\mathbf{K}\times\nabla'g+\frac{\rho}{\epsilon}\nabla'g\right)$$

$$=\int_{S'} dS[-j\omega\mu(\hat{n}'\times\mathbf{H})g+(\hat{n}'\times\mathbf{E})\times\nabla'g+(\hat{n}'\cdot\mathbf{E})\nabla'g]. \tag{9-13}$$

Now we must examine the singularity at $r=0$. Circumscribe a sphere of radius r_1 about $\mathbf{r}$ (or x, y, z) with its normal pointing *out* of V, that is, toward $\mathbf{r}$. Then

$$\nabla'g=\left(\frac{1}{r}-jk\right)\frac{\exp(-jkr)}{r}\hat{r}_o; \tag{9-14}$$

† Stratton (1941, p. 464); Mittra (1973, p. 162).

‡ We have occasion here to use (with $e^{j\omega t}$) the augmented form of Maxwell's equations (see Section 1-3):

$$\nabla\times\mathbf{E}+j\omega\mu\mathbf{H}=-\mathbf{K};\ \nabla\times\mathbf{H}-j\omega\epsilon\mathbf{E}=\mathbf{J};$$

$$\nabla\cdot\mathbf{H}=m/\mu;\ \nabla\cdot E=\rho/\epsilon.$$

and on the sphere, $\hat{r}_o = \hat{n}$. The area of the sphere vanishes with the radius as $4\pi r_1^2$. Thus, since

$$(\hat{n} \times \mathbf{E}) \times \hat{n} + (\hat{n} \cdot \mathbf{E})\hat{n} = \mathbf{E}, \tag{9-15}$$

the contribution of the spherical surface to the right-hand side of Eq. 9-13 is $4\pi\mathbf{E}(\mathbf{r})\exp(-jkr)$. So we may write, at any interior point of V, therefore, the electric field as

$$\begin{aligned}\mathbf{E}(\mathbf{r}) = &-\frac{T}{4\pi}\int_V dV\left(j\omega\mu\mathbf{J}g + \mathbf{K} \times \nabla' g - \frac{\rho}{\epsilon}\nabla' g\right)\\ &-\frac{T}{4\pi}\int_{S'} dS[j\omega\mu(\hat{n}' \times \mathbf{H})g - (\hat{n}' \times \mathbf{E}) \times \nabla' g - (\hat{n}' \cdot \mathbf{E})\nabla' g],\end{aligned} \tag{9-16}$$

where $T = 1/(1 - \Omega/4\pi)$ with $\Omega = 0$ for $\mathbf{r}$ on the surface, $= 2\pi$ otherwise. An application of the duality principle (Section 4-4) then gives the magnetic field

$$\begin{aligned}\mathbf{H}(\mathbf{r}) = &\frac{T}{4\pi}\int_V dV\left(-j\omega\epsilon\mathbf{K}g + \mathbf{J} \times \nabla' g + \frac{m}{\mu}\nabla' g\right)\\ &+\frac{T}{4\pi}\int_{S'} dS[j\omega\epsilon(\hat{n}' \times \mathbf{E})g + (\hat{n}' \times \mathbf{H}) \times \nabla' g + (\hat{n}' \cdot \mathbf{H})\nabla' g].\end{aligned} \tag{9-17}$$

In both integrals of Eq. 9-17 all vector fields have the argument $\mathbf{r}'$.

If charge and current are confined to a thin layer on S'; and if, as layer depth is decreased, densities are increased so that, in the limit, volume densities are replaced by surface densities (Chapter 3); and if there are no charge and no current sources inside or on S'; then the volume integral vanishes.

If S' is split into two pieces S and S_1 and the latter is allowed to recede to infinity, then the radiation condition requires that the contribution from the S_1 surface integral be due entirely to sources outside S_1. This contribution is then simply the incident field and can be split out of the above, giving:

$$\mathbf{H}(\mathbf{r}) = T\mathbf{H}^{\text{inc}}(\mathbf{r}) + \text{the volume integral} + \text{the surface integral over } S; \tag{9-18}$$

or, for the case referred to above where charge and current are confined to a thin layer on S,

$$\mathbf{H}(\mathbf{r}) = T\mathbf{H}^{inc}(\mathbf{r}) + \frac{T}{4\pi}\int_S dS[j\omega\epsilon(\hat{n}' \times \mathbf{E})g + (\hat{n}' \times \mathbf{H}) \times \nabla' g + (\hat{n}' \cdot \mathbf{H})\nabla' g]. \tag{9-19}$$

When the surface S is a perfectly conducting one, the boundary conditions at S require that (Chapter 3)

$$\hat{n}' \times \mathbf{E} = 0 \tag{9-20}$$

and

$$\hat{n}' \cdot \mathbf{H} = 0. \tag{9-21}$$

Using Eqs. 9-20 and 9-21 in Eq. 9-19 gives a much simpler expression for the magnetic field:

$$\mathbf{H}(\mathbf{r}) = T\mathbf{H}^{inc}(\mathbf{r}) + \frac{T}{4\pi}\int_S [\hat{n}(\mathbf{r}') \times \mathbf{H}(\mathbf{r}')] \times \nabla g(\mathbf{r}, \mathbf{r}')\, dS. \tag{9-22}$$

Indeed, if $T = 1$, Eq. 9-7 reduces to the expression for $\mathbf{H}^s(\mathbf{r})$ with which we introduced PO (Section 8-1). As we hinted there, the expression

$$\hat{n}(\mathbf{r}') \times \mathbf{H}(\mathbf{r}') = \mathbf{J}_S(\mathbf{r}') \tag{9-23}$$

may be called the surface current on S generated by the incident field $\mathbf{H}^{\text{inc}}$.

If $\mathbf{r}$ is allowed to approach the surface S, then Eq. 9-22 can be converted into the Maue (1949) equation:

$$2\hat{n}(\mathbf{r}) \times \mathbf{H}^{\text{inc}}(\mathbf{r}) = \mathbf{J}_S(\mathbf{r}) - \frac{1}{2\pi}\int_S \hat{n}(\mathbf{r}) \times [\mathbf{J}_S(\mathbf{r}') \times \nabla' g(\mathbf{r}, \mathbf{r}')]\, dS \tag{9-24}$$

which represents an integral equation in the surface current $\mathbf{J}_S$. The left-hand side of Eq. 9-24 is a known quantity. The unknown surface current $\mathbf{J}_S$, however, appears both outside and inside of the surface integral by virtue of the fact that the argument $\mathbf{r}$ of the current expression $\mathbf{J}_S(\mathbf{r})$ appearing outside the integral is not dependent on the variable of integration $\mathbf{r}'$ which is the argument of the current expression $\mathbf{J}_S(\mathbf{r})$ appearing inside the integral. The quantity g is the free-space three-dimensional† Green's function encountered in the discussion of PO. It is

$$g(\mathbf{r}, \mathbf{r}') = \frac{e^{-jk\,\|\mathbf{r}-\mathbf{r}'\|}}{\|\mathbf{r}-\mathbf{r}'\|} = \frac{e^{-jkR}}{R}, \quad R = \|\mathbf{r}-\mathbf{r}'\|. \tag{9-25}$$

Notice that there is a good possibility for difficulty in Eq. 9-24; when the observation point $\mathbf{r}$ is placed on S, the variation of $\mathbf{r}'$ over S will permit $\mathbf{r}' = \mathbf{r}$ with a consequent singularity in $g(\mathbf{r}, \mathbf{r}')$. It is, however, possible to deal with such problems by using a Cauchy principal value treatment for the integral in the neighborhood of $\mathbf{r} = \mathbf{r}'$.

Before going on to an introduction of the method of moments (Section 9-3), with which we expect to be able to solve integral equations like Eq. 9-24, we make a brief detour to look again at Green's functions and their role in the theory of linear space and operators.

9-2-1 Green's and δ-functions (from a different viewpoint)

The dyadic Green's function, which is a solution of the vector Helmholtz equation, has already been reviewed in Chapter 4. Since it represents the

† In two dimensions the Hankel function plays the role of the free-space Green's function. See Magnus and Oberhettinger (1949).

effect of a single source point (Eq. 4-2), it may be combined with the surface current distribution of a scattering body to obtain the fields generated by that surface current (Eq. 9-1). In the following discussion we examine this concept from a somewhat simpler viewpoint (cf. van der Pol and Bremmer, 1950).

The influence of a force or action concentrated in a single point (that is, a δ-function†) is described by a Green's function. To keep matters simple, consider a second order ordinary differential equation of the Sturm–Liouville type:

$$\frac{d}{dx}\left[f(x)\frac{d}{dx}y(x)\right]-u(x)y(x)=-\phi(x), \text{ or } Ly(x)=-\phi(x)$$

which has to be satisfied in some interval $a<x<b$ together with homogeneous boundary conditions at the endpoints $x=a$, b [such as, for example, $y(a)=y(b)=0$].

A Green's function $g(x, \xi)$ solves this problem, and it is uniquely determined by the following two statements:

1. For $x\neq\xi$, g is a continuous function of x and satisfies both the differential equation and the boundary conditions.
2. For $x=\xi$, both $\partial g/\partial x$ and $\partial^2 g/\partial x^2$ are continuous; at $x=\xi$, the first derivative has a jump:

$$\left.\frac{\partial g(x, \xi)}{\partial x}\right|_{x=\xi-0}^{x=\xi+0}=-\frac{1}{f(\xi)}.$$

† There are many possible forms for the δ-function. For example, taking a unit function in the form

$$U(t, \lambda)=\frac{1}{2}+\frac{1}{\pi}\arctan(\lambda t),\ |\arctan|\leq\frac{\pi}{2},$$

the derivative is

$$\delta(t, \lambda)=\frac{d}{dt}U(t, \lambda)=\frac{\lambda}{\pi(\lambda^2t^2+1)}.$$

In the limit $\lambda\to\infty$ we have the unit and δ-functions.

$$\delta(t)=\lim_{\lambda\to\infty}\frac{\lambda}{\sqrt{\pi}}e^{-\lambda^2t^2}$$

also serves the purpose. The δ-function was introduced by Dirac but Cauchy and Hermite used the idea, as well as Poisson, Kirchhoff, and Kelvin. Hermite's form is

$$\lim_{\lambda\to\infty}\int_{\alpha}^{\beta}\frac{\lambda\,dt}{\lambda^2t^2+1}$$

with α and β on opposite sides of $t=0$.

These conditions can both be satisfied if g is a solution of

$$Lg(x, \xi) = -\delta(x-\xi),$$

where $\delta(x-\xi)$ is a δ- or impulse, function with impulse at the point $x=\xi$. Then

$$y(x) = \int_a^b g(x, \xi)\phi(\xi)\, d\xi$$

(Lighthill, 1959; Schwartz, 1961 (in French); and Morse and Feshbach, 1953).

Example 1. Consider

$$\frac{d^2y}{dx^2} = -(1+4x^2), \qquad y(0) = y(1) = 0. \tag{9-26}$$

Solve

$$\frac{d^2}{dx^2} g(x, \xi) = -\delta(x-\xi), \qquad g(0, \xi) = g(1, \xi) = 0 \tag{9-27}$$

by observing that at $x=0$ a solution of the form x satisfies the boundary condition, while at $x=1$ a solution of the form $1-x$ satisfies the boundary condition. However, then

$$g = \begin{cases} x, & x < \xi \\ 1-x, & x > \xi \end{cases} \tag{9-28}$$

would not be continuous at $x=\xi$. So to remedy this unfortunate circumstance, we multiply the first equation by $(1-\xi)$ and the second by ξ. Moreover, we see that

$$\left.\frac{\partial g}{\partial x}\right|_{x=\xi+0} - \left.\frac{\partial g}{\partial x}\right|_{x=\xi-0} = -1. \tag{9-29}$$

Thus

$$g(x, \xi) = \begin{cases} x(1-\xi), \; x < \xi \\ (1-x)\xi, \; x > \xi \end{cases} \text{ and } g(0, \xi) = g(1, \xi) = 0; \tag{9-30}$$

and

$$\begin{aligned} y(x) &= \int_x^1 x(1-\xi)(1+4\xi^2)\, d\xi + \int_0^x (1-x)\xi(1+4\xi^2)\, d\xi \\ &= \frac{5x}{6} - \frac{x^2}{2} - \frac{x^4}{3} \end{aligned} \tag{9-31}$$

which could, of course, have been obtained by direct integration.

9-2-2 Linear Spaces and Operators

Consider the inhomogeneous differential equation

$$Lf = h \tag{9-32}$$

To solve for f we need the idea of inner product $\langle f, h \rangle$ which satisfies the rules

1. $\langle f, h \rangle = \langle h, f \rangle$;
2. $\langle \alpha f + \beta h, q \rangle = \alpha \langle f, q \rangle + \beta \langle h, q \rangle$, α, β scalars;
3. $\langle f^*, f \rangle > 0$ if $f \neq 0$, $= 0$ if $f = 0$,

with asterisk meaning complex conjugate as usual.

We also define an adjoint operator L^a such that

4. $\langle Lf, h \rangle = \langle f, L^a h \rangle$ for all f in the domain of L. The operator is *self-adjoint* if $L^a = L$. If the solution of $Lf = h$ exists and is unique for all h, then L^{-1} exists and $f = L^{-1}h$.

For instance, the Sturm–Liouville problem could be stated as

$$Ly(x) = -\phi(x),\ L = \frac{d}{dx}\left[f(x)\frac{d}{dx}\right] - u(x) \tag{9-33}$$

Then as an inverse operator L^{-1} we have, in terms of the Green's function,

$$y(x) = L^{-1}\phi = \int_a^b g(x, \xi)\phi(\xi)\, d\xi. \tag{9-34}$$

9-3 METHOD OF MOMENTS (MoM)

We still consider linear equations of the form $Lf = h$, L being a linear operator and h a given function. Expand f in a series of functions $f_1, f_2, f_3, \ldots$ in the domain of L:

$$f = \sum_n \alpha_n f_n,\ \alpha_n \text{ constants}; \tag{9-35}$$

then since L is linear,

$$Lf = L\Big(\sum_n \alpha_n f_n\Big) = \sum_n \alpha_n L f_n = h. \tag{9-36}$$

Next, assume that a set of weighting, or testing, functions w_1, w_2, $w_3, \ldots$ has been chosen and that a suitable dot product is available, that is, an operation satisfying rules 1 through 3 above. Form a dot product of each w_m with the above:

$$\sum \alpha_n \langle w_m, Lf_n \rangle = \langle w_m, h \rangle, \tag{9-37}$$

which can be written in the form of a matrix equation

$$[t][\alpha] = [h], \tag{9-38}$$

with $[t]$ a matrix of m rows by n columns given by

$$[t_{mn}] = \begin{pmatrix} \langle w_1, Lf_1\rangle \langle w_1, Lf_2\rangle \ldots \\ \langle w_2, Lf_1\rangle \langle w_2, Lf_2\rangle \ldots \\ \vdots \qquad\qquad \ldots \end{pmatrix} = [t]. \tag{9-39}$$

$[\alpha]$ is a column matrix of n rows $\alpha_1, \ldots, \alpha_n$ and $[h]$ is also a column matrix, but of m rows $\langle w_1, h\rangle, \ldots, \langle w_m, h\rangle$. If $[t]^{-1}$ exists, then

$$[\alpha] = [t]^{-1}[h] \tag{9-40}$$

is our (perhaps approximate) solution. If $w_n = f_n$, then the procedure is known as *Galerkin's method.*

Example 2. Solve $-d^2f/dx^2 = 1 + 4x^2$, $f(0) = f(1) = 0$, the boundary value problem we have considered above in a Green's function context. Take

$$f_n = x - x^{n+1}, \quad n = 1, 2, \ldots, N \tag{9-42}$$

noting that x is included in the expression so that each f_n satisfies $f_n(0) = f_n(1) = 0$. Take also

$$w_n = f_n. \tag{9-43}$$

The domain of the w_n must be the domain of L^a. But here $L^a = L$ and so the domains of w_n and f_n are the same. Now

$$\langle w_m, Lf_n\rangle = \int_0^1 (x - x^{m+1})[L(x - x^{n-1})]\,dx = n(n+1)\int_0^1 (x - x^{m+1})x^{n-1}\,dx$$

$$= n(n+1)\left[\frac{x^{n+1}}{n+1} - \frac{x^{m+n+1}}{m+n+1}\right]_0^1$$

$$= n(n+1)\left[\frac{1}{n+1} - \frac{1}{m+n+1}\right]$$

$$= \frac{mn}{m+n+1} = t_{mn} \tag{9-44}$$

$$\langle w_m, h\rangle = \int_0^1 (x - x^{m+1})(1 + 4x^2)\,dx = \int_0^1 (x - x^{m+1} + 4x^3 - 4x^{m+3})\,dx$$

$$+\left[\frac{x^2}{2} - \frac{x^{m+2}}{m+2} + x^4 - 4\frac{x^{m+4}}{m+4}\right]_0^1$$

$$= \frac{3}{2} - \frac{1}{m+2} - \frac{4}{m+4} = \frac{m(3m+8)}{2(m+2)(m+4)}. \tag{9-45}$$

CASE 1. $N=1$, $t_{11}=\frac{1}{3}$, $h_1=\frac{11}{30}$. So $\frac{1}{3}\alpha=\frac{11}{30}$, whence $\alpha_1=\frac{11}{10}$; and our solution is

$$f=\frac{11}{10}(x-x^2). \tag{9-46}$$

CASE 2. $N=2$, $t_{11}=\frac{1}{3}$, $t_{12}=\frac{2}{4}=t_{21}$, $t_{22}=\frac{4}{5}$, $h_1=\frac{11}{30}$, $h_2=\frac{28}{48}$.

$$\text{So}\quad \begin{pmatrix}1/3 & 1/2\\ 1/2 & 4/5\end{pmatrix}\begin{pmatrix}\alpha_1\\ \alpha_2\end{pmatrix}=\begin{pmatrix}11/30\\ 7/12\end{pmatrix},\qquad \text{whence}\qquad \begin{pmatrix}\alpha_1\\ \alpha_2\end{pmatrix}=\begin{pmatrix}1/10\\ 2/3\end{pmatrix}.$$

Our solution is

$$f=\frac{1}{10}(x-x^2)+\frac{2}{3}(x-x^3)=\frac{23}{30}x-\frac{1}{10}x^2-\frac{2}{3}x^3. \tag{9-47}$$

CASE 3. $N=3$. $[t_{mn}]=\begin{pmatrix}1/3 & 1/2 & 3/5\\ 1/2 & 4/5 & 1\\ 3/5 & 1 & 9/7\end{pmatrix}$, $[h_m]=\begin{pmatrix}11/30\\ 7/12\\ 51/70\end{pmatrix}$. Therefore,

$[\alpha]=\begin{pmatrix}1/2\\ 0\\ 1/3\end{pmatrix}$. So our solution is

$$f=\frac{1}{2}(x-x^2)+\frac{1}{3}(x-x^4)=\frac{5}{6}x-\frac{1}{2}x^2-\frac{1}{3}x^4, \tag{9-48}$$

which, we see, is *exact*.

For $N\geq 4$, the exact solution continues to be the result.

Example 3.† Consider the step, or pulse, functions (Fig 9-3a) defined by

$$P(x)=\begin{cases}1, & |x|<\dfrac{1}{2(N+1)}\\[2ex] 0, & |x|>\dfrac{1}{2(N+1)}.\end{cases} \tag{9-49}$$

If we replace 5 points (6 spaces) between 0 and 1, then at x_2

$$P(x-x_2)=\begin{cases}1, & |x-x_2|<\dfrac{1}{2(N+1)}=\dfrac{1}{12},\\[2ex] 0, & |x-x_2|>\dfrac{1}{2(N+1)}=\dfrac{1}{12}.\end{cases} \tag{9-50}$$

$f_n=P(x-x_n)$ gives a step function approximation for f. But for $L=-d^2/dx^2$, LP is *not* in the domain of L.

† Adapted from Harrington (1968), with permission of Robert E. Krieger Publishing Co.

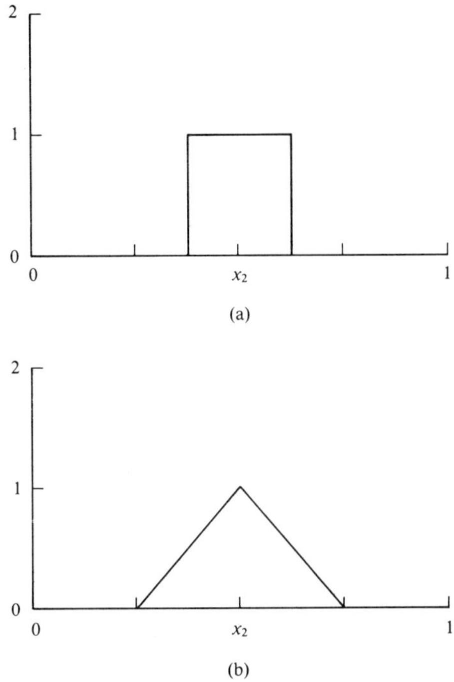

Figure 9-3. (a) A square pulse; (b) a triangular pulse.

However, this situation suggests the idea of using triangle functions whose first derivatives are step functions and whose second derivatives are δ-functions. Take

$$T(x) = \begin{cases} 1 - |x|\,(N+1), & |x| < \dfrac{1}{N+1} \\ 0, & \text{otherwise.} \end{cases} \tag{9-51}$$

For $N = 5$, the triangle for the second point (Fig. 9-3b) is, as above,

$$T(x - x_2) = \begin{cases} 1 - 6(x - x_2), & |x - x_2| < 1/6 \\ 0, & \text{otherwise.} \end{cases} \tag{9-52}$$

With

$$f = \sum_{n=1}^{N} \alpha_n T(x - x_n) \tag{9-53}$$

we have

$$Lf = (N+1)[-\delta(x - x_{n-1}) + 2\delta(x - x_n) - \delta(x - x_{n+1})]. \tag{9-54}$$

We obtain, using testing functions $w_m = P(x - x_m)$

$$t_{mn} = \begin{cases} 2(N+1), & m = n \\ -(N+1), & m - n = 1 \\ 0, & m - n > 1 \end{cases} \tag{9-55}$$

and

$$h_m = \frac{1}{N+1}\left[1 + \frac{4m^2 + (1/3)}{(N+1)^2}\right]. \tag{9-56}$$

For additional details as well as other appropriate approaches to problems like the ones suggested above, we refer the reader to Harrington (1968).

9-4 APPLICATION OF THE MoM TO SCATTERING

The MoM has been used to considerable advantage in the treatment of electromagnetic scattering problems, particularly by Harrington (in Mittra, 1975) and by Poggio and Miller (in Mittra, 1973), as well as others who will be mentioned subsequently. We synthesize from these, particularly the latter, to give a general view of the approach, which is based largely on integral equations. The time dependence is $e^{j\omega t}$.

Following the operator format introduced above, we write the magnetic field integral equation (MFIE) and the electric field integral equation (EFIE) as

$$L_H[\mathbf{J}_S(\mathbf{r}'), \mathbf{K}_S(\mathbf{r}')] = 2\hat{n} \times \mathbf{H}^{\text{inc}}(\mathbf{r}) \tag{9-57}$$

and

$$L_E[\mathbf{J}_S(\mathbf{r}'), \mathbf{K}_S(\mathbf{r}')] = 2\hat{n} \times \mathbf{E}^{\text{inc}}(\mathbf{r}), \tag{9-58}$$

respectively, where

$$L_H[\ ] = \mathbf{J}_S(\mathbf{r}) - \frac{1}{2\pi}\hat{n} \times \int_S (-j\omega\epsilon\mathbf{K}_S g + \mathbf{J}_s \times \nabla' g + (m_S/\mu)\nabla' g)\, dS' \tag{9-59}$$

and

$$L_E[\ ] = -\mathbf{K}_S(\mathbf{r}) + \frac{1}{2\pi}\hat{n} \times \int_S (j\omega\mu\mathbf{J}_S g + \mathbf{K}_S \times \nabla' g - (\rho_S/\epsilon)\nabla' g)\, dS'. \tag{9-60}$$

In the above equations we have used the expressions $\mathbf{K}_S$ and $\mathbf{J}_S$ for equivalent surface magnetic and electric currents, respectively:

$$\mathbf{K}_S = -\hat{n} \times \mathbf{E}, \quad \mathbf{J}_S = \hat{n} \times \mathbf{H}; \tag{9-61}$$

m_S and ρ_S are the magnetic and electric surface charges, respectively. As usual,

$$g(\mathbf{r}, \mathbf{r}') = \frac{e^{-jk\,\|\mathbf{r}-\mathbf{r}'\|}}{\|\mathbf{r} - \mathbf{r}'\|} \tag{9-62}$$

is the free space Green's functions with $\mathbf{r}'$ and $\mathbf{r}$ denoting source (or integration) and observation points, respectively; $\nabla' g$ indicates that the gradient operation is to be performed on the primed variable of g. Either of the above equations may be represented symbolically as

$$L[\mathbf{A}(\mathbf{r}')] = \mathbf{B}(\mathbf{r}), \tag{9-63}$$

where $\mathbf{A}$ is the unknown response to the source $\mathbf{B}$. With this format we can introduce the MoM. Let

$$\mathbf{A}(\mathbf{r}) \approx \sum_{n=1}^{N} a_n \mathbf{f}_n(\mathbf{r}), \tag{9-64}$$

where, as above, the a_n are to be determined. When $\mathbf{A}$ is introduced into the operator equation, a residual error can be expressed as

$$\boldsymbol{\varepsilon}(\mathbf{r}) = \sum_{n=1}^{N} a_n L[\mathbf{f}_n(\mathbf{r}')] - \mathbf{B}(\mathbf{r}). \tag{9-65}$$

We define our inner (dot) product, over S, of two vectors $\mathbf{P}$ and $\mathbf{Q}$ tangential to S as

$$\langle \mathbf{P}, \mathbf{Q} \rangle = \int_S \mathbf{P} \cdot \mathbf{Q} \, dS \tag{9-66}$$

Then, choosing weighting functions $\{\mathbf{W}_m\}$ and employing the inner product, we have, for $m = 1, 2, \ldots, M$,

$$\langle \mathbf{W}_m, \boldsymbol{\varepsilon} \rangle = \sum_{n=1}^{N} a_n \langle \mathbf{W}_m, L[\mathbf{f}_n(\mathbf{r}')] \rangle - \langle \mathbf{W}_m, \mathbf{B} \rangle. \tag{9-67}$$

Of course, the name of the game is to make $\langle \mathbf{W}_m, \boldsymbol{\varepsilon} \rangle = 0$. The result is a set of linear equations for the a_n:

$$[Z][A] = [E], \tag{9-68}$$

where $[A]$ is the column of a_n, $n = 1, 2, \ldots, N$; $[E]$ is the column of $\langle \mathbf{W}_m, \mathbf{B} \rangle$, $m = 1, 2, \ldots, M$; and $[Z]$ is the $m \times n$ matrix of elements:

$$\langle \mathbf{W}_m, L[\mathbf{f}_n(\mathbf{r}')] \rangle, \; m, \; n \text{ as above.} \tag{9-69}$$

If $[Z]$, called the impedance matrix, is invertible, then the column $[A]$ can be found, since $[E]$ is known.

The basic functions can be defined over subdomains $\Delta\mathbf{r}_n$ of L, which is equivalent to replacing the integration by a Riemann sum. We can write

$$\mathbf{A}(\mathbf{r}) = U(\mathbf{r}_n) a_n \mathbf{f}_n(\mathbf{r}), \qquad \mathbf{r} \text{ in } \Delta\mathbf{r}_n, \tag{9-70}$$

using the unit function

$$U(\mathbf{r}_n) = \begin{cases} 1, & \mathbf{r} \text{ in } \Delta\mathbf{r}_n, \\ 0, & \text{otherwise.} \end{cases} \tag{9-71}$$

The trick is to choose the $\mathbf{W}_m$ suitably. We should note that the matrix elements Z_{mn} involve three integrations: two for the surface integral and one for the inner product, which, in general, defy analytic execution. The expansion referred to above was used, with δ-function weights, by Oshiro (1965) and by Andreason (1965). Ideally, the functions $\mathbf{f}_n$ and $\mathbf{W}_m$ must be chosen to meet the conflicting requirements of solution accuracy and cost of

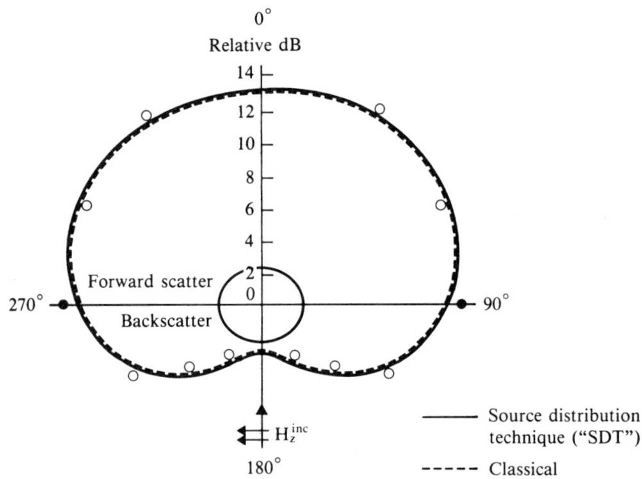

(a) Bistatic cross-section

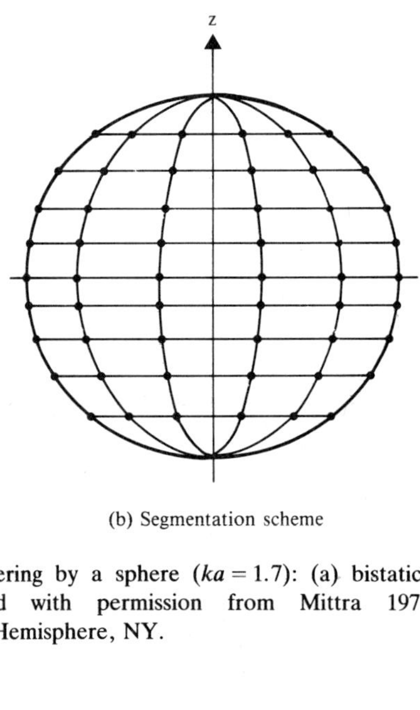

(b) Segmentation scheme

Figure 9-4 Scattering by a sphere ($ka = 1.7$): (a) bistatic RCS; (b) segmentation of the sphere. Reprinted with permission from Mittra 1973, *Computer Techniques for Electromagnetics,* Hemisphere, NY.

computation. Fenlon (1969) records weighting functions, of which the most useful are the Galerkin, already mentioned (following Eq. 9-40 above), and the delta function weighting, which eliminates the need for a double (surface) integration. For example, Oshiro (1969), using MFIE together with delta function weighting, accomplished his integration using a simple

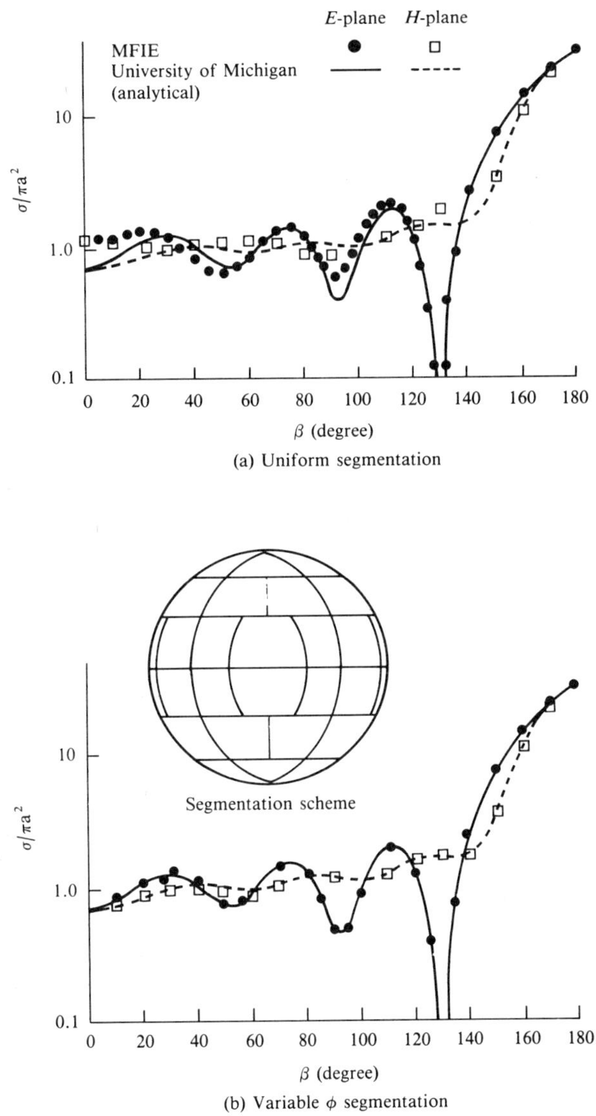

Figure 9-5 Scattering by a sphere ($ka = 5.3$): (a) uniform segmentation; (b) variable segmentation. Reprinted with permission from Mittra, 1973, *Computer Techniques for Electromagnetics*, Hemisphere, NY.

rectangular rule wherein the integration over each subsectional patch was carried out by using the integrand value at $\mathbf{r}_m$ multiplied by patch area. Moreover, the principal value of the integral was obtained by simply overlooking the patch with the singularity, that is, where $\mathbf{r}_n = \mathbf{r}_m$. Even with this type of approximation, Oshiro obtained good results.

Figures 9-4 and 9-5 illustrate the bistatic RCS for a sphere using two different segmentation schemes. Figure 9-4 uses equal area patches, once again excluding the "self-patch." Figure 9-5 uses equal polar angle segmentation for each of 13 bands: 4, 8, 12, 16, 20, 24, 24, 24, 20, 16, 12, 8, 4. These patches maintain an aspect ratio (length-to-width) more nearly equal to unity over the sphere. The latter seems clearly better than the former.

9-5 PO COMBO

From earlier sections we know

$$\mathbf{H}^s_{PO} = \frac{1}{2\pi}\int_I (\hat{n}' \times \mathbf{H}^{\text{inc}}) \times \boldsymbol{\nabla}' g \, dS', \tag{9-72}$$

where I is the illuminated portion of a scattering surface S and g is the free-space Green's function. Since PO makes its greatest error near the shadow boundary, we can use the integral equation idea to find a better expression for the current in this region, still using the PO current approximation $2\hat{n}' \times \mathbf{H}^{inc}$ for the illuminated region. The result is

$$\mathbf{J}_S = \begin{cases} 2\hat{n} \times \mathbf{H} & \text{on the illuminated surface,} \\ 2\hat{n} \times \mathbf{H}^{\text{inc}} + \dfrac{1}{2\pi}\hat{n} \times \left[\displaystyle\int_U \mathbf{J}_S \times \boldsymbol{\nabla}' g \, dS' + \int_I (2\hat{n}' \times \mathbf{H}^{\text{inc}}) \times \boldsymbol{\nabla}' g \, dS'\right] & \\ \qquad \text{on the shadowed surface;} & \end{cases} \tag{9-73}$$

here the U-integration is taken over the shadowed surface; the I-integration is over the illuminated surface. The surface over which the integral equation is defined is now smaller to the extent by which the shadowed surface is smaller than the total surface.

This idea has been used by Oshiro and Mitzner (1967) to find the H-plane bistatic RCS of a sphere for $ka = 4.1$ and 10 (a being sphere radius and k being wave number as usual). Their results are shown in Fig. 9-6.

Wire grid structures have also been used to model solid surface targets (Figs. 9-7 to 9-9). Again the method is usually one of collocation with constant current samples.

Bistatic computations for a finite cylinder of diameter 2.158λ and length 18.667λ have been carried out at three different large bistatic angles by Paddison, Follin, Mitchell, and Maffett (1985a). Their results are shown in Figs. 9-10 through 9-12.

It is important to observe here that the PO approximation introduces a

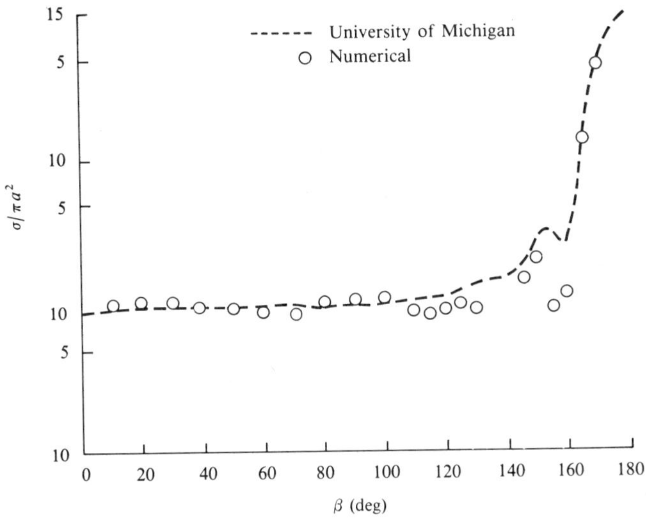

Figure 9-6a Physical optics H-plane bistatic RCS of a sphere ($ka = 10$). Figure 9-6a and b reprinted with permission from Mittra, 1973, *Computer Techniques for Electromagnetics*, Hemisphere, NY.

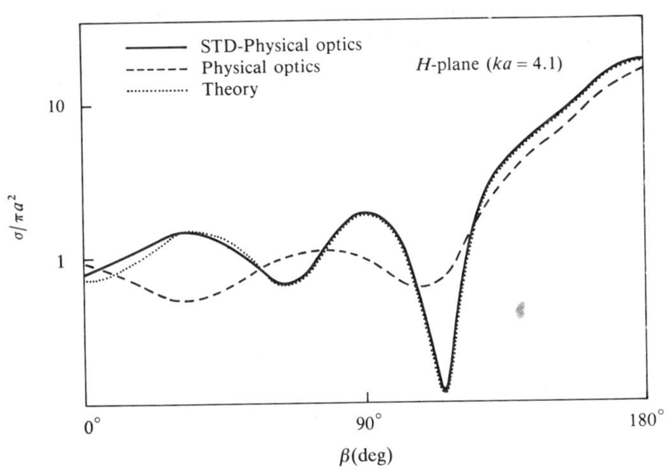

Figure 9-6b SDT-physical optics H-plane bistatic RCS of a sphere ($ka = 4.1$).

dependence on the incident wave direction (in the shadow region). This fact is in contrast to a non-PO case where the coefficient matrix $[Z]$ is independent of the source direction.

Along these lines, one might remark on the number and kind of computer operations required to invert $[Z]$, or to solve the system of

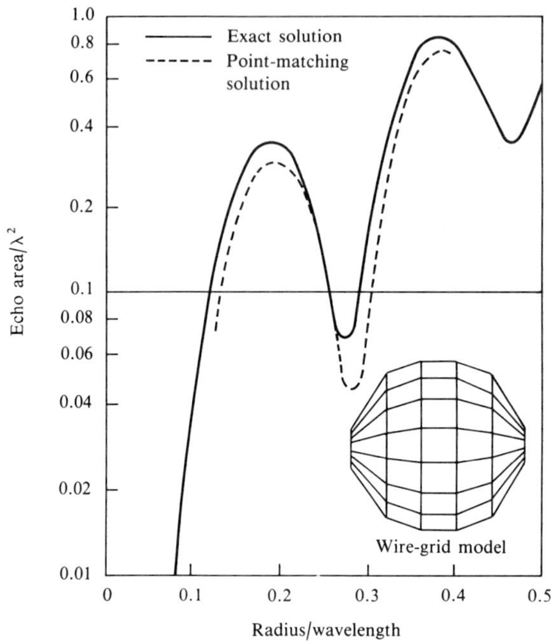

Figure 9-7 Backscatter echo area of a wire-grid sphere. Reprinted with permission from Mittra, 1973, *Computer Techniques for Electromagnetics,* Hemisphere, NY.

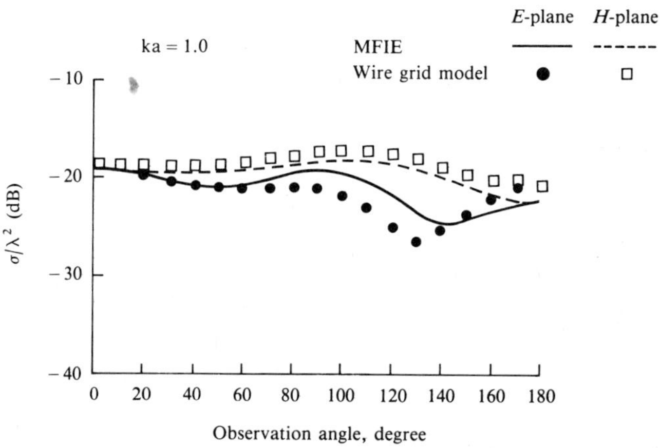

Figure 9-8 Cone-sphere bistatic scattering pattern for tip incidence (30° included cone angle) as computed using a wire-grid model with the magnetic field integral equation. Reprinted with permission from Mittra, 1973, *Computer Techniques for Electromagnetics,* Hemisphere, NY.

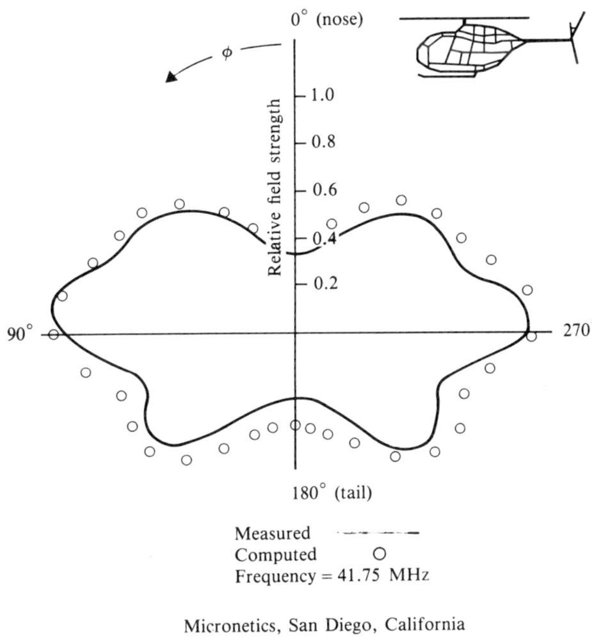

Figure 9-9 Comparison of numerical and experimental scattered field patterns for an OH–6A helicopter wire-grid model. Reprinted with permission from Mittra, 1973, *Computer Techniques for Electromagnetics,* Hemisphere, NY.

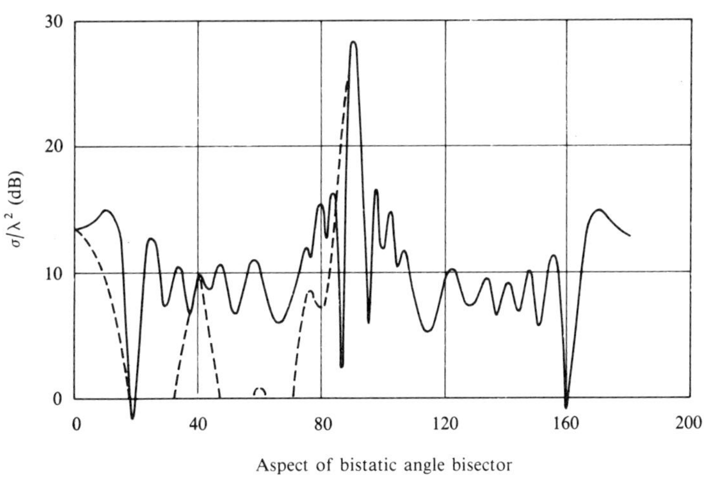

Figure 9-10 Comparison of computed and measured bistatic RCS for a circular cylinder of diameter 1.85 cm and length 16 cm at 35 GHz at a bistatic angle of 140°. Solid curve: measured by Delco; dashed curve: computed by JHU/APL using NEC.

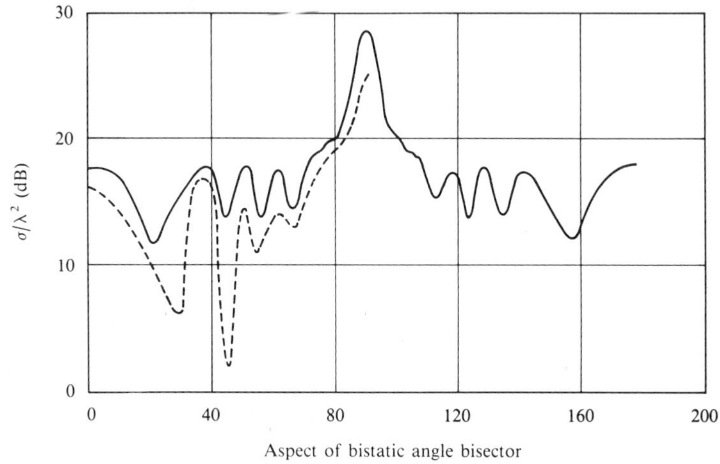

Figure 9-11 Comparison of computed and measured bistatic RCS for a circular cylinder of diameter 1.85 cm and length 16 cm at 35 GHz at a bistatic angle of 160°. Solid curve: measured by Delco; dashed curve: computed by JHU/APL using NEC.

equations. One such is an iteration technique that requires only $O(N^2)$ computations; however, this procedure must be repeated every time the source direction changes. The other two, more standard, methods invert $[Z]$ and require, in general, on the order of N^3 operations. But once $[Z]^{-1}$ is in hand, the unknown vector $[A]=[Z]^{-1}[E]$ requires on the order of N^2 operations only, for each matrix $[E]$ (See Mittra, 1973, p. 191].

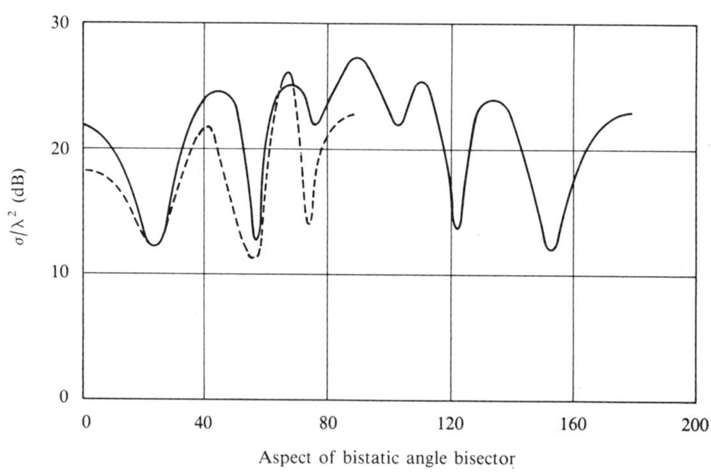

Figure 9-12 Comparison of computed and measured bistatic RCS for a circular cylinder of diameter 1.85 cm and length 16 cm at 35 GHz at a bistatic angle of 170°. Solid curve: measured by Delco; dashed curve: computed by JHU/APL using NEC.

9-6 TWO-DIMENSIONAL SCATTERERS

We define two-dimensional scatterers to mean closed cylindrical bodies of infinite extent along the cylinder generator. Call the axial direction $\hat{z}$. The advantage in considering this type of scatterer is that with the assumption of a uniform current distribution in the $\hat{z}$-direction, all currents and fields have only two components. So the pertinent equations for surface currents and their resultant scattered fields become much simpler to find and deal with. Indeed, even though surface currents must be obtained from integral equations, the solution methods are straightforward enough to avoid such treatments as the MoM discussed above. Moreover, integrals of the volume and surface types encountered earlier reduce here to contour integrals along a curve C formed by the intersection of the cylinder and the incidence plane normal to $\hat{z}$.

Equations for surface currents and scattered fields can be formulated in terms of a complex scalar surface impedance η (η is zero for a perfectly conducting surface, unity for a perfectly absorbing surface). That is, if the scattering body has a coating of some kind, all effects occurring within the coating can be expressed as a single effect η on the surface of the coating. Such a condition is called an impedance boundary condition (IBC). It states that on the surface the fields must obey the relation:

$$\mathbf{E} - (\hat{n} \cdot \mathbf{E})\hat{n} = \eta Z_0 \hat{n} \times \mathbf{H}, \tag{9-74}$$

which can also be written as

$$\hat{n} \times \mathbf{E} = -\eta Z_0 \hat{n} \times (\hat{n} \times \mathbf{H}). \tag{9-75}$$

If we interpret $\hat{n} \times \mathbf{E}$ as a magnetic surface current $\mathbf{K}_S$ (Section 4-6) and $\hat{n} \times \mathbf{H}$ as an electric surface current $\mathbf{J}_S$, Eq. (9-75) becomes

$$\mathbf{K}_S = -\eta Z_0 \hat{n} \times \mathbf{J}_S. \tag{9-76}$$

The IBC can also be expressed in the form

$$\mathbf{H} - (\hat{n} \cdot \mathbf{H})\hat{n} = -\left(\frac{Y_0}{\eta}\right)\hat{n} \times \mathbf{E} \tag{9-77}$$

under the dual transformations $\mathbf{E} \rightarrow \mathbf{H}$, $\mathbf{H} \rightarrow -\mathbf{E}$, $Z_0 \rightarrow Y_0$, $\eta \rightarrow 1/\eta$ (Senior, 1962). Corresponding to Eq. 9-76, then, we find

$$\mathbf{J}_S = \left(\frac{Y_0}{\eta}\right)\hat{n} \times \mathbf{K}_S. \tag{9-78}$$

For the present discussion, we shall retain the factor η where appropriate but shall reserve a detailed discussion of the IBC for Chapter 16.

We have the opportunity here to revisit and give an application for the Maxwell and wave equations which make use of a magnetic current term (Section 1-3, Eqs. 1-42, 1-43, and 1-47). Formal solutions for Eqs. 1-42 and

1-43 can be written as

$$\mathbf{H} = \nabla \times \mathbf{A} - j\omega\epsilon\mathbf{F} + \frac{1}{j\omega\mu}\nabla(\nabla \cdot \mathbf{F}) \tag{9-79}$$

and

$$\mathbf{E} = -\nabla \times \mathbf{F} - j\omega\mu\mathbf{A} + \frac{1}{j\omega\epsilon}\nabla(\nabla \cdot \mathbf{A}). \tag{9-80}$$

Using the expression (9-80) for **E** in the Faraday equation:

$$\nabla \times \mathbf{E} = -j\omega\mu\mathbf{H} - \mathbf{K}$$

we find that **F** must satisfy a wave equation

$$(\nabla^2 + k^2)\mathbf{F} = -\mathbf{K} \tag{9-81}$$

and so has solution

$$\mathbf{F} = \int_V \mathbf{K} \cdot \mathfrak{I} g \, dV. \tag{9-82}$$

On the other hand, using the expression (9-79) for **H** in the Ampère equation:

$$\nabla \times \mathbf{H} - j\omega\epsilon\mathbf{E} = \mathbf{J},$$

we find that **A** also satisfies a wave equation

$$(\nabla^2 + k^2)\mathbf{A} = -\mathbf{J} \tag{9-83}$$

and so has solution

$$\mathbf{A} = \int_V \mathbf{J} \cdot \mathfrak{I} g \, dV; \tag{9-84}$$

where $\mathfrak{I}$ is a unit dyadic and g is a free-space scalar Green's function (Chapter 4, Eq. 4-4) which, for the two-dimensional case, is given by the Hankel function $H_n^{(2)}(kr)$ of the second kind with the orders n of 0 and 1 both being required; $\mathbf{r} = \boldsymbol{\rho} - \boldsymbol{\rho}'$ is a vector from source (or integration) point to observation point. The volume V is enclosed by the surface over which the electric and magnetic currents are described. [Time dependence $\exp(j\omega t)$ is used here; if the time dependence $\exp(-i\omega t)$ were used, then the conjugate Hankel function of the first kind would serve as the Green's function.]

Consider now a plane wave with incidence direction normal to the $\hat{z}$-axis, with $\mathbf{H}^i = \hat{z}H_z^i$ (called H-polarization). We then have $\mathbf{H} = \hat{z}H_z$ and $\mathbf{J} = \hat{s}J_s$, $\hat{s}$ being the tangent vector at the point of interest on C; $\hat{s}$, the unit normal $\hat{n}$, and $\hat{z}$ form a right-hand system, and s measures distance along C. The impedance boundary condition (9-84) requires $E_s = \eta Z_0 J_s$. Then

$$H_z^i(s) = \frac{1}{2} J_s(s) + \frac{1}{4} \int_C \eta(s') J_s(s') H_0^{(2)}(kr) d(ks')$$
$$+ \frac{j}{4} \fint_C J_s(s') H_1^{(2)}(kr)(\hat{n}' \cdot \hat{r}) d(ks') \quad (9\text{-}85)$$

The slashed integral means that the Cauchy principal value (Churchill and Brown, 1984) must be used.

The E-polarization case can be obtained from the H-polarization case by making the dual substitutions $H_z \to -E_z$, $E_s \to H_s$, $J_s \to -\eta Z_0 J_z$, $Y_0 \to Z_0$, and $\eta \to 1/\eta$. The result is

$$Y_0 E_z^i(s) = \frac{1}{2} \eta(s) J_z(s) + \frac{1}{4} \int_C J_z(s') H_0^{(2)}(kr) d(ks')$$
$$+ \frac{j}{4} \fint_C \eta(s') J_z(s')(\hat{n}' \cdot \hat{r}) H_1^{(2)}(kr) d(ks'). \quad (9\text{-}86)$$

The curves C in Eqs. 9-85 and 9-86 must be approximated by a set of segments such that the length of the longest segment does not exceed 0.2λ. And for each segment the integrals can be approximated by finite sums. The result is a set of equations in the unknown current distribution (J_s-component or J_z-component) along C. The radar cross-width (Section 5-2) is then easily obtained from the current distributions. Note that Eqs. 9-85 and 9-86 become much simplified for a perfect conductor ($\eta = 0$). For a perfect absorber ($\eta = 1$), they can, with the help of the duality substitutions, be made identical. In Section 16-6, we shall discuss further those cases for which $\eta \neq 0$.

In using Eqs. 9-85 and 9-86, we must observe some caveats having to do with Cauchy principal values of the integrals where discontinuities may occur and with very thin cylinders. But by and large the method is remarkably trouble-free and can be used to study various scattering features of cylindrical shapes (refer to Section 11-9).

To illustrate the type of result that can be obtained from Eqs. 9-85 and 9-86, Fig. 9-13 shows a plot of RCW (actually RCW normalized to λ and called $\sigma(\lambda)$ for a cylinder whose section is a wedge capped by a circle. The wedge included angle is 25° and the circle radius a is such that $ka = 3$.

Both E- and H-polarization returns are shown in the figure. For the H-polarization case, notice the peaks at angles 28° and 139° from the wedge axis, zero degrees being toward the wedge edge. Both of these peaks appear to be due to what are called traveling wave returns; we address this subject and this example again in Section 11-9 in an attempt to discern the nature of the surface currents which cause these peaks. For the E-polarization case, we take note of the higher return (almost 10 dB) at nose-on which appears to arise from the wedge edge. Finally, if the cylinder length of the

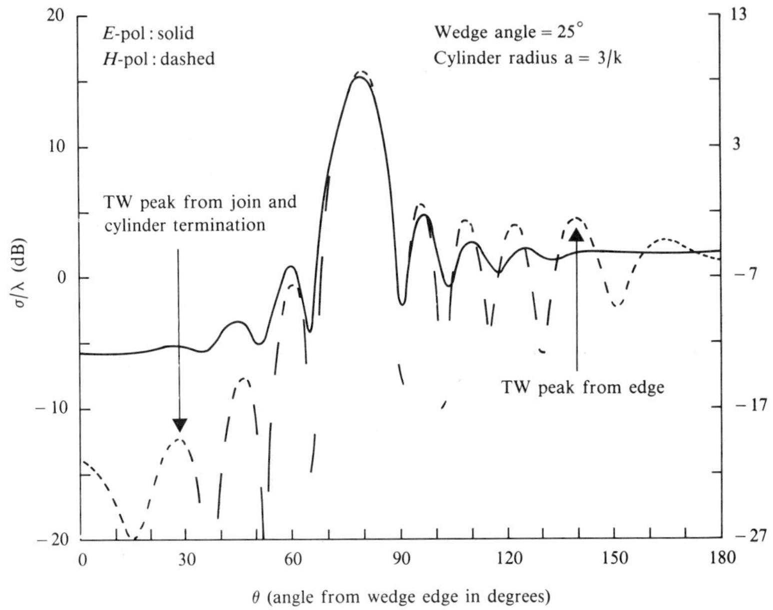

Figure 9-13 RCS of a wedge cylinder for E- and H-polarizations.

wedge-circle-cylinder is assumed to be 12 in. we can use the formula 5-8 from Section 5-2 to convert the two-dimensional RCW (normalized to λ) to standard RCS in dBsm; the right-hand ordinate depicts the latter.

9-7 HISTORICAL PERSPECTIVE

It is of some interest to take notice of early RCS results using numerical methods. One of the pioneering treatments was Oshiro's SDT (source distribution technique) (1965) with which he found the bistatic RCS of a sphere with $ka = 1.7$ (cf. Fig. 9-14a). Following quickly thereafter were results for a prolate spheroid (Fig. 9-14b) (Oshiro, 1965) and for a cone-sphere (Fig. 9-15) (Mautz and Harrington, 1969). The MFIE has also been applied to the same cone-sphere in both bistatic and monostatic modes to compare with the Mautz and Harrington results (Fig. 9-16).

The flat-back cone (Fig. 9-17) and cylinder (Fig. 9-18) have also been studied; and the sphere-cylinder as well (Fig. 9-19).

An interesting modeling approach for the numerical work has already been mentioned above; it is the wire grid method. Richmond (1966) used it to model a sphere (Fig. 9-7). Poggio and Miller also used it with MFIE to examine the bistatic RCS of the cone-sphere (Fig. 9-8); and even to find the monostatic RCS for a helicopter (Fig. 9-9).

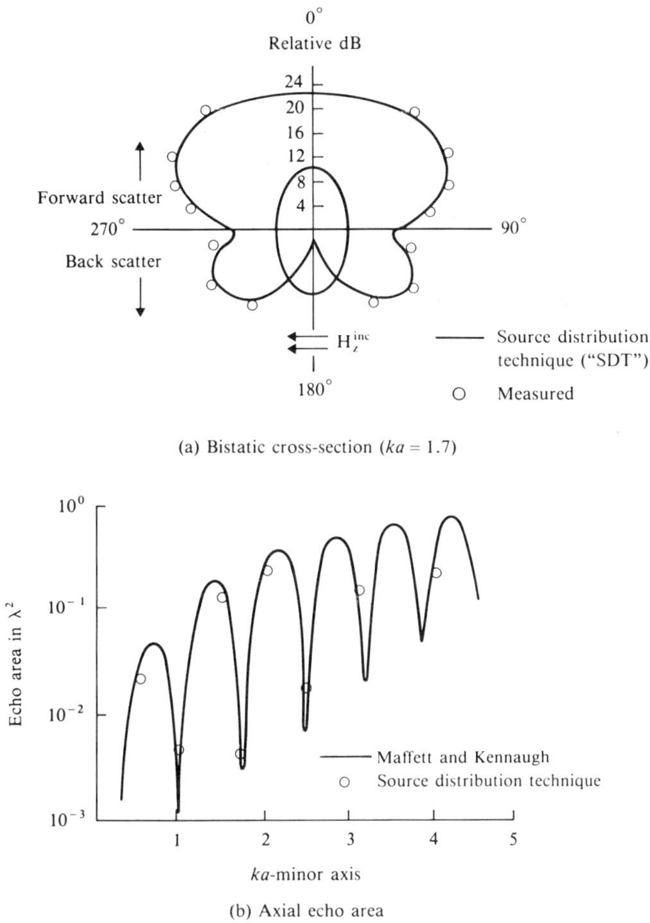

Figure 9-14 Scattering by a spheroid: (a) bistatic RCS, $ka = 1.7$; (b) axial RCS. Reprinted with permission from Mittra, 1973, *Computer Techniques for Electromagnetics*, Hemisphere, NY.

All figures illustrating the above remarks are reproduced from Mittra (1973; chapter by Poggio and Miller).†

This historical perspective would be incomplete without emphasizing the two-dimensional work of Knott and Senior (1974b) and of Liepa, Knott, and Senior, (1974), which has been reviewed in Section 9-6 for perfectly conducting surfaces. The importance of their work lies in the fact that the integral equations from which they find the fields scattered by a surface can incorporate a nonzero surface impedance. Thus, the RCS of coated bodies

† With permission, Hemisphere Publishing Co., New York.

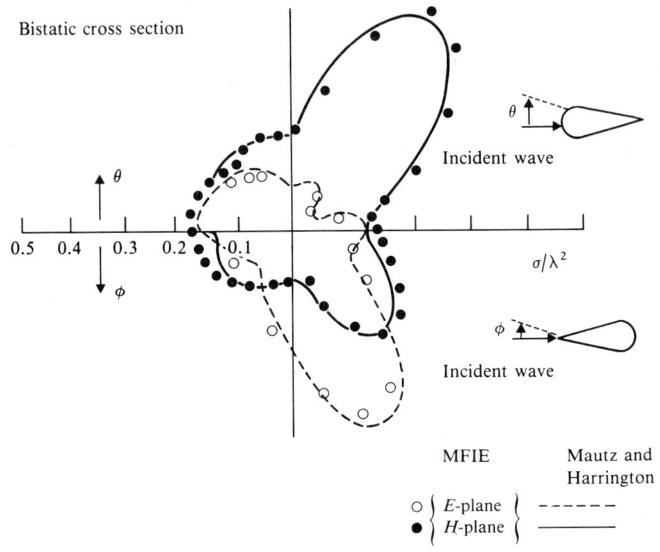

Figure 9-15 Cone-sphere scattering. $ka = 1.26$, half-cone angle $= 10°$. Reprinted with permission from Mittra, 1973, *Computer Techniques for Electromagnetics,* Hemisphere, NY.

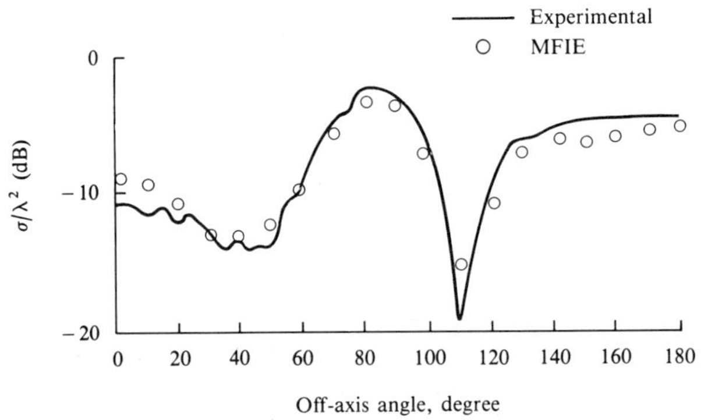

Figure 9-16 Comparison of measured and calculated H-plane monostatic RCS of a cone-sphere. $ka = 1.7$, half-cone angle $= 15°$. Reprinted with permission from Mittra, 1973, *Computer Techniques for Electromagnetics,* Hemisphere, NY.

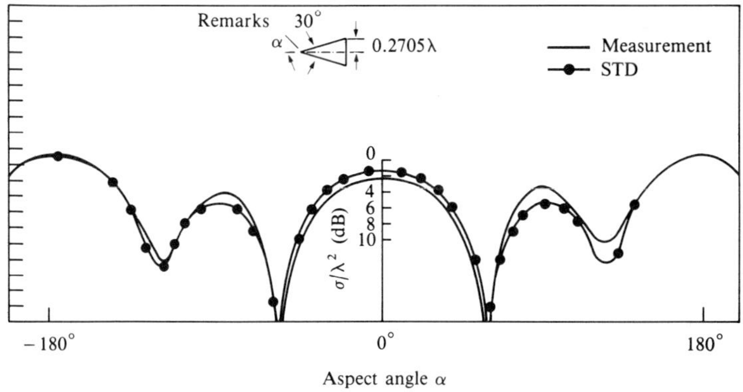

Figure 9-17 *H*-plane monostatic RCS of a flat-back cone. $ka = 1.7$. Reprinted with permission from Mittra, 1973, *Computer Techniques for Electromagnetics,* Hemisphere, NY.

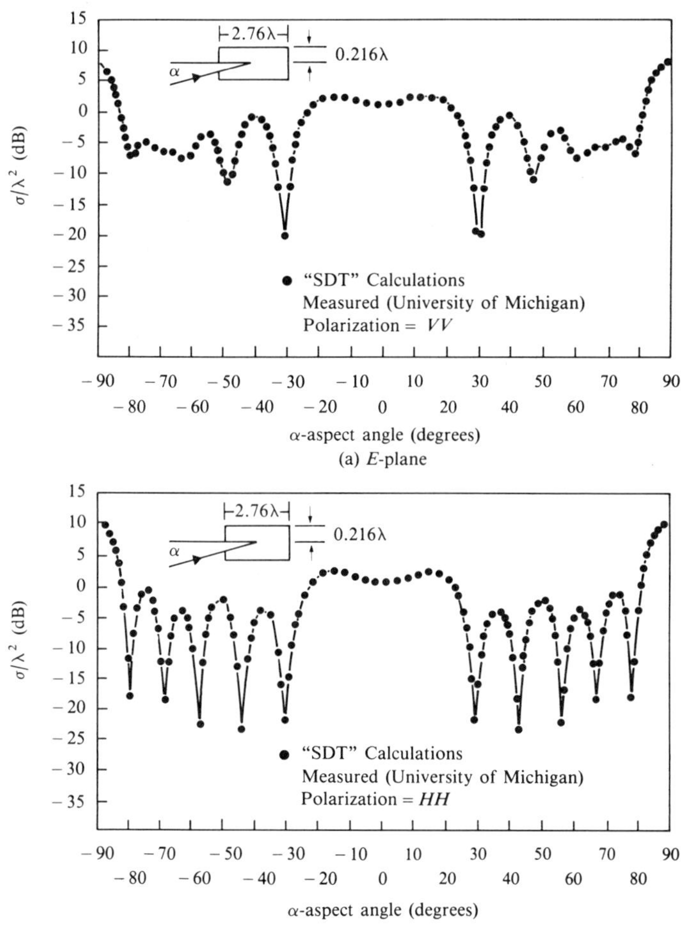

Figure 9-18 Monostatic RCS of a right circular cylinder: (a) *E*-plane; (b) *H*-plane. Reprinted with permission from Mittra, 1973, *Computer Techniques for Electromagnetics,* Hemisphere, NY.

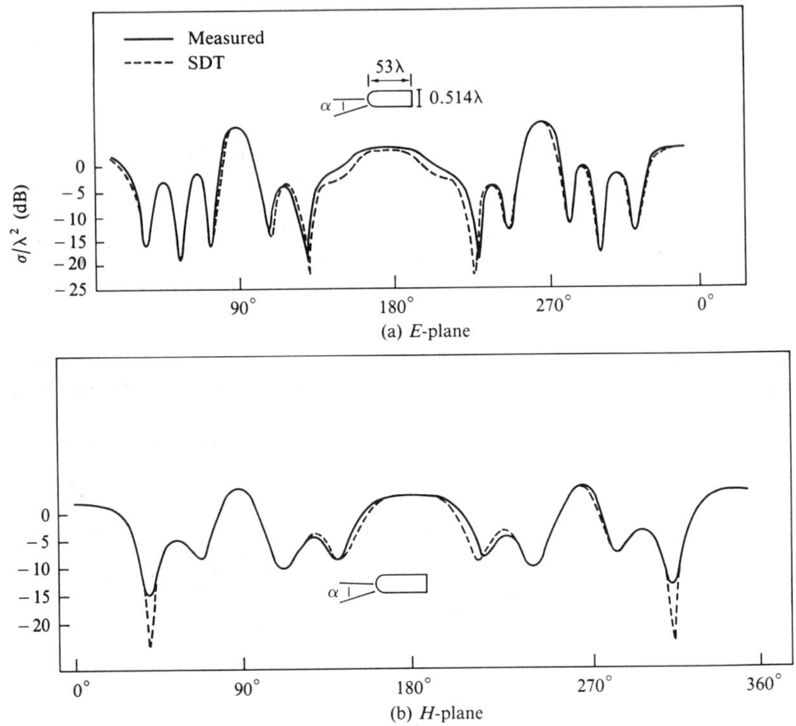

Figure 9-19 Monostatic RCS of a sphere-cylinder. Reprinted with permission from Mittra, 1973, *Computer Techniques for Electromagnetics*, Hemisphere, NY.

of certain types can be investigated. We consider this topic more carefully in Chapter 16.

The two-dimensional integral equation method is additionally important in that it can be used to illustrate the nature of the current distributions which generate such "simple" phenomena as edge, traveling wave, and creeping wave scattering. These subjects are specifically addressed in Section 11-9.

10

RCS MEASUREMENTS

RCS approximation methods can be evaluated and validated by two methods: exact solutions and experimental measurements. There are few of the former; none at all for complex targets. The latter have certain criteria that must be met if measured data are to be regarded as reliable RCS representations.

At the foundation of any measurement procedure is the notion of *calibration.* In order to assign a RCS value to power (or to amplitude and phase) measured for a candidate target, conversion of received power to RCS must be established from some standard whose RCS is known exactly or to a good approximation. The sphere satisfies the exactly-known-RCS criterion. Its RCS can be calculated from the exact Mie series solution (Born and wolf, 1959) to any desired degree of precision for any ka value (a being sphere radius and k wave number as usual). [Of course, such a computation becomes increasingly expensive as ka grows larger because the series solution becomes more slowly convergent and so requires more terms for a desired accuracy. However, the GO approximation (Fig. 7-1) becomes increasingly acceptable as ka increases.] Secondary standards, such as the circular cylinder, dihedral cylinder, and corner reflector, are often used. There exist no exact scattering solutions for these standards but their approximate ones are acceptable for large ratios of dimension to wavelength.

Circular polarization measurements require additional care with the calibration process, a necessity occasioned by the manner in which circular polarization must be generated. A circularly polarized wave is synthesized by splitting a coherent signal, with half of the power transmitted as a linear horizontally polarized wave in phase quadrature with the other half of the power which is transmitted as a linear vertically polarized wave. To accomplish such a transmission a balance between the two linearly polarized waves must be achieved and held. So the two components must be orthogonal, matched in amplitude, and held in precise phase quadrature.

These requirements can be realized with a combination of an orthomode antenna feed,† a power splitter (a quadrature hybrid is best), and an appropriate combination of phase shifter and attenuator to be used for fine tuning.‡ The tuning, or calibration, is done by illuminating and observing a flat plate at normal incidence that returns maximum RCS for RL (transmit right circular, receive left circular) or LR and minimum RCS for RR or LL; the balance is adjusted to achieve these minimum and maximum values. The ratio of minimum to maximum is called the cancellation ratio. Additional steps must then be taken to equalize RL and LR signal paths by using thin-wire and dihedral targets. The ratio of amplitudes of the linearly polarized components is called the axial ratio. A ratio of unity (0 dB; in actual practice 1/2 to 1 dB) indicates perfect circularity; a ratio other than unity indicates elliptic polarization (refer to the discussion of the ellipticity angle χ in Section 1-6). In practice the cancellation ratio is easier to measure and tune than the axial ratio.

Once a balance is achieved it must be held long enough to make a measurement. The only way to be sure this requirement is met is to perform a post calibration.§

Beyond the fundamental calibration principle, this chapter looks into the nature of a set of criteria that ensure measurement conditions conducive to acceptable RCS data. The following statements should be considered to be a minimum set of rules for RCS measurements. (See Mack, 1981).

10-1 BASIC PRINCIPLES OF ELECTROMAGNETIC SCATTERING MEASUREMENTS

1. Measurements of scattered fields must be made in the *far field* of the scattering target; and the incident field at the target must be a good approximation to a *plane wave.*
2. *Background signals* (wall reflections, leakage, ground clutter, etc.) must be reduced to levels that do not induce unacceptable errors into the *lowest RCS* to be measured.

† An orthomode feed is a square feed horn that has a pair of orthogonal feed probes.

‡ The phase shifter and attenuator are placed in either the horizontal probe feed line or the vertical probe feed line, depending on the amplitude and phase adjustments required. An attenuator in each of these two lines is an added and desirable convenience.

§ Information on circular polarization calibration, together with permission to use Figs. 10-17b and 10-21 and quote the range capabilities of Section 10-12, supplied through the courtesy and kindness, and with the permission, of William D. Fortner, Mission Gorge Range, San Diego, California.

3. Power relations in electromagnetic scattering are governed by the radar range equation (Chapter 5). The *smallest RCS* that can be measured with a specific error, for a given experimental setup, must be defined by this equation.
4. Linear passive electromagnetic systems can be *scaled* up or down in frequency; the scaling must be governed by the relations

$$\frac{L_2}{\lambda_2} = \frac{L_1}{\lambda_1} \qquad \text{and} \qquad \frac{\sigma_2}{\lambda_2^2} = \frac{\sigma_1}{\lambda_1^2},$$

respectively, between full-scale and fractional-scale targets, where L denotes some representative length of the target, λ is wavelength, and σ is RCS. The subscripts 1 and 2 might refer, respectively, to the full- and fractional-scale targets.

These are the principles that must govern the design or choice of the following *subsystems* which comprise a measurement facility (except perhaps for the last two items of the list, which are not so much electromagnetic concerns as they are state-of-the-art and budgetary constraints):

1. Range or chamber,
2. Model mounts,
3. Transmit/receive antennas,
4. RF transmitting sources,
5. RF plumbing,
6. Receiving equipment;
7. Data-handling equipment.

In the following we discuss the above principles in terms of many ideas already in hand from preceding chapters. To begin, we use the integral for $\mathbf{H}^s$ to derive the far field criterion.

10-2 FAR FIELD CRITERION

The need for a far field criterion is generated by both theoretical and practical considerations. We emphasize in both earlier and later chapters plane wave illumination of targets and scattering solutions that behave like plane waves for the observer. Both of these situations require the target to be far (in wavelengths) from the illuminator/observer. Interest in such solutions has been promoted by the large distances at which radars must illuminate and observe practical targets such as aircraft, missiles, and satellites. So we must seek some practical definition for "far" in terms of such parameters as wavelength, target size (and perhaps the sizes of illuminating and receiving antennas), and distance from illuminator/observer to target.

We can achieve this goal by considering the role of these parameters in the expression for the scattered magnetic field (from Chapter 8):

$$\mathbf{H}^s = \frac{1}{4\pi} \int_S (\hat{n} \times \mathbf{H}(\mathbf{r}')) \times \nabla g \, dS, \tag{10-1}$$

where $\mathbf{H}(\mathbf{r}')$ is the total field on the surface, and the Green's function is

$$g = \frac{1}{R} e^{-jkR}, \tag{10-2}$$

R being distance from transmitter/receiver to surface integration point. Carrying out the gradient differentiation yields

$$\mathbf{H}^s = \frac{-1}{4\pi} \int_S [(\hat{n} \times \mathbf{H}) \times \nabla R] \left[jk + \frac{1}{R} \right] e^{-jkR} \frac{dS}{R}. \tag{10-3}$$

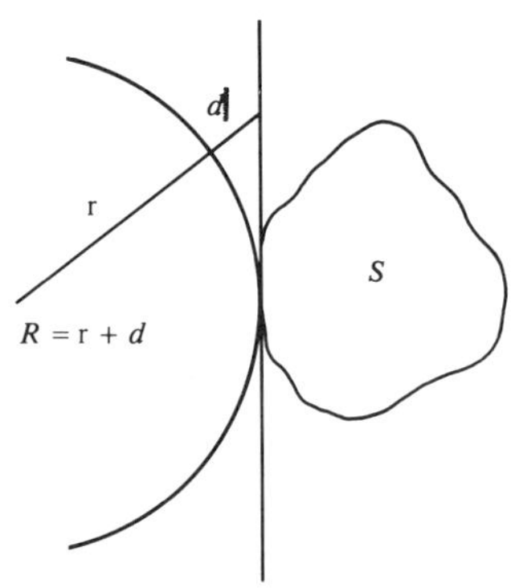

Figure 10-1a Spherical and plane wavefronts.

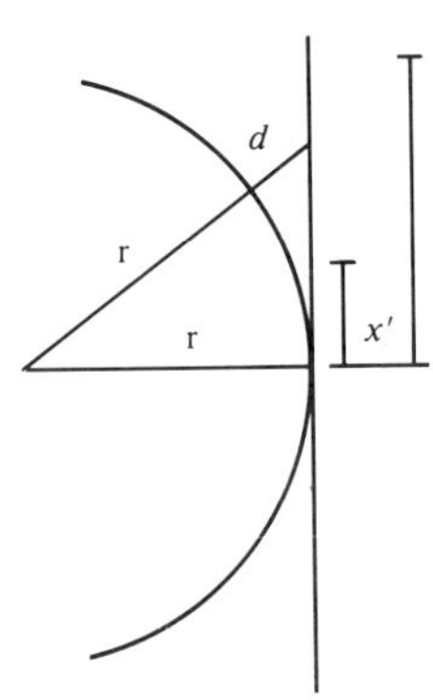

Figure 10-1b Parameters for a far field criterion.

$\mathbf{H}^s$ has two components, one of $O(1/R)$, the other of $O(1/R^2)$, differing in phase by 90°; and having a magnitude ratio of $\lambda/2\pi R$. If $R \gg \lambda$, then the latter component can be discarded with negligible error.

Let $R = r + d$ as shown in Fig. 10-1; d is the amount by which the spherical wave front lacks being a plane wave. Assume $r \gg d$. Then we may replace R by r in the denominator of $\mathbf{H}^s$. But in the exponential, we set

$$R = r + d = (r^2 + x'^2)^{1/2} = r\left[1 + \frac{x'^2}{2r^2}\right] = r + x'^2/2r. \qquad (10\text{-}4)$$

At the extremities, $x' = L/2$, we find that

$$d = \frac{L^2}{8r}. \qquad (10\text{-}5)$$

By standard usage, $d = \lambda/16$ is the accepted deviation of the phase front from planarity at the extremities of the target; so

$$r = \frac{2L^2}{\lambda} \qquad (10\text{-}6)$$

prescribes the *conventional far field criterion.*

There exist, of course, other criteria based on different choices of d, or based on the size of the illuminator/receiver, as well as on target size (Kouyoumjian and Peters, 1965). If a larger value for d is acceptable, then a

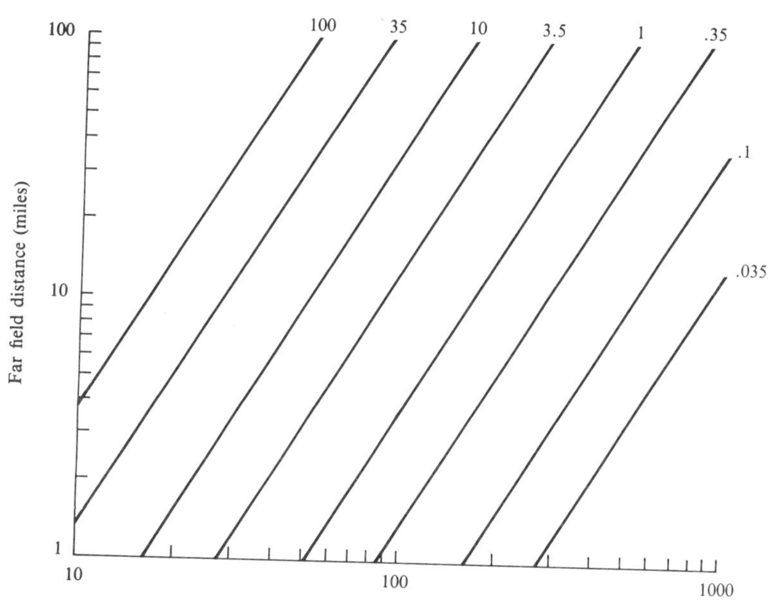

Figure 10-2 Far field range criterion for indicated frequencies in GHz.

penalty is exacted in the form of degraded peaks and nulls in the RCS pattern. Figure 10-2 shows far field distance as a function of target size and frequency when the conventional criterion is used.

10-3 FIELD VARIATION

Not only does the phase of the field vary over the target at a finite distance from the illuminator/observer, but its amplitude may vary as well. We consider the nature of longitudinal and transverse field strength variation.

10-3-1 Longitudinal Field Strength Variation

From Fig. 10-3 we observe that field strength variation δE_L in the longitudinal direction is bounded below by

$$\delta E_L \geq \frac{E_L(r - L/2) - E_L(r + L/2)}{E_L(r)} \tag{10-7}$$

Then if we assume that E_L varies like $1/r$, we may write the above as

$$\delta E_L \geq \frac{\dfrac{1}{r - L/2} - \dfrac{1}{r + L/2}}{1/r} = \frac{L/r}{1 - (L/2r)^2} \approx L/r \tag{10-8}$$

10-3-2 Transverse Field Strength Variation

The field strength variation in the transverse direction is bounded by

$$\delta E_T \geq E_T(0) - E_T(\theta), \ \theta = \arctan((L/2)/r), \tag{10-9}$$

the transverse field being specified by the antenna pattern of the transmit/receive antenna. So to be more explicit about δE_T we must assume some knowledge of the antenna. Suppose the antenna aperture to be

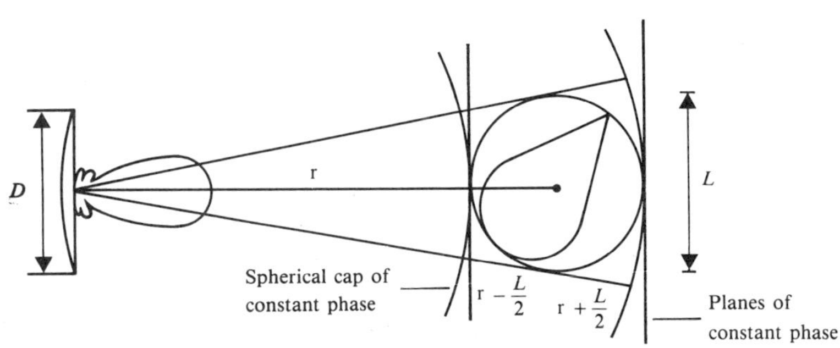

Figure 10-3 Field strength variation over a target.

circular with uniform illumination. Then, if the antenna radius is a,

$$E_{TC} = \frac{2J_1(ka \sin \theta)}{ka \sin \theta} \tag{10-10}$$

where $J_1(\cdot)$ is the first order Bessel function: thus, from Eq. 10-9,

$$\begin{aligned} \delta E_{TC} \geq 1 - E_{TC} &= \frac{1 - 2J_1(kLa/2r)}{kLa/2r} \\ &= \frac{1}{8}\left(\frac{kaL}{2r}\right)^2 \\ &\approx \frac{1}{8}\left(\frac{\pi a}{2L}\right)^2, \text{ if } r = 2L^2/\lambda. \end{aligned} \tag{10-11}$$

In the last approximation, we assume $2a/L \ll 1$ and replace the first order Bessel function by the first two terms of its power series.

Example 1. Suppose a 12-in. model is to be measured with a 12-in. diameter antenna at 10 GHz ($\lambda = 1.18$-in.). Then

$$r = \frac{2L^2}{\lambda} = 20.3'.$$

$$\delta E_L = \frac{L}{r} = \lambda/2L = 0.049$$

$$\delta E_{TC} = \frac{1 - 2J_1(\pi/4)}{\pi/4} = 0.08$$

Example 2. As in example 1, but with a 24-in. antenna: Then

$$r = \frac{2(2a)^2}{\lambda} = 81.28'.$$

$$\delta E_L = \frac{L}{r} = L\lambda/2(2a)^2 = 0.012$$

$$\delta E_{TC} = 0.019$$

Example 3. As in example 2, but using $r = L^2/2\lambda$:

$$r = (2a)^2/2\lambda = 20.3'.$$

$$\delta E_L = \frac{L}{r} = 2\lambda L/(2a)^2 = 0.049.$$

$$\delta E_{TC} = 1 - \frac{J_1(k2L\lambda/2(2a)^2)}{k2L\lambda/2(2a)^2} = \frac{1 - J_1(\pi/2)}{\pi/2} = 0.308.$$

The latter of course represents a rather considerable transverse field variation.

Finally, the 3-dB (power) beamwidths at a distance R from the antenna cover a lateral extent s of

$$s = 2R \tan\frac{\phi}{2} \approx \frac{2R\phi}{2} \approx \frac{R\lambda}{D},$$

where $D = 2a$ and ϕ is the angle between 3-dB points, $\approx\lambda/D$. If $R = 2D^2/\lambda$, then

$$s = \frac{2D^2\lambda}{\lambda D} = 2D;$$

that is, at $R = 2D^2/\lambda$ the lateral distance between 3-dB points is twice the antenna diameter D.

10-4 GROUND PLANE EFFECTS

One item that affects field variation in the target space is the proximity of the transmitting/receiving antenna to the ground. Indeed, some ranges, such as RATSCAT (see Fig. 10-17) at Holloman AFB in New Mexico, exploit the ground plane effect. Such exploitation requires, of course, that the ground plane by kept in a very smooth condition (asphalt, concrete, or careful grading). Figure 10-4 illustrates the four paths available to energy

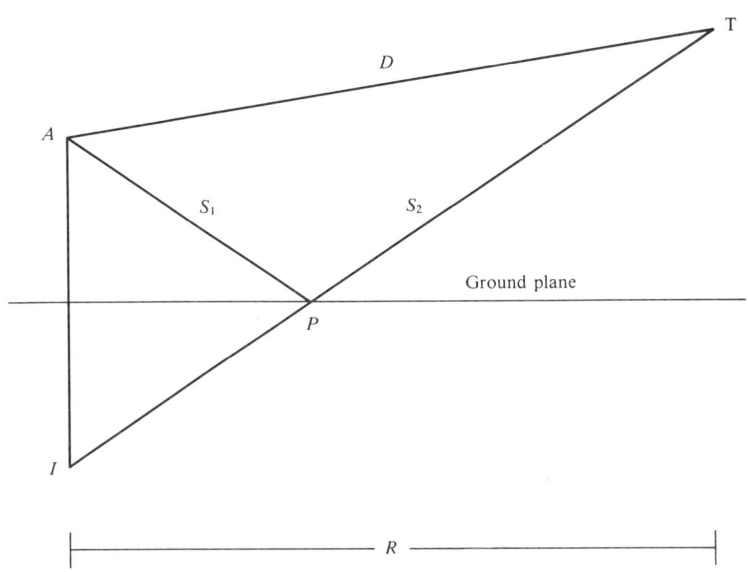

Figure 10-4 Ground plane geometry.

between the antenna at A and the target at T:

1. ATA: direct-direct;
2. APTPA: image-image;
3. ATPA: direct-image; and
4. APTA: image-direct.

Paths 3 and 4 are bistatic ones but generally the bistatic angles ATP and PTA are sufficiently small to make target return insensitive to small differences in arrival angle between direct and indirect paths, that is, the target is viewed at essentially the same aspect along these paths.† Paths 2, 3, and 4 all involve the ground plane reflection coefficient ρ. Altogether, then, the RCS can be written

$$\sigma = \sigma_f \,| e^{2jkD} + \rho^2 e^{2jkI} + 2\rho e^{jk(D+I)} |^2, \tag{10-12}$$

where

σ_f = free-space target RCS;

ρ = effective reflection coefficient of ground plane at P;

D = direct pathlength, AT;

I = indirect pathlength, APT, $= S1 + S2$; and

k = free-space wave number, as usual.

It can be observed that the above expression is a perfect square and so can be written as

$$\sigma = \sigma_f \,|e^{jkD} + \rho e^{jkI} |^4. \tag{10-13}$$

If the ground plane is perfectly conducting so that $\rho = -1$, then

$$\sigma = 16\sigma_f \sin^4\left[\frac{1}{2} k(D - I)\right]. \tag{10-14}$$

Thus, one sees that if antenna height h_A and target height h_T are suitably chosen so as to make Eq. 10-14 assume a peak value, then received power will be 12 dB ($10 \log_{10} 16$) higher than if the target were measured in free space.

Example 4. Equation 10-14 above has maxima when the sine argument is an odd integer multiple m of $\pi/2$; that is

$$k(D - I) = m\pi.$$

From Fig. 10-4, then, the path difference $\delta = D - I$ is

$$\delta = D - I = \frac{m\lambda}{2},$$

† These bistatic angles are small because the antenna height A and target height T are $\ll$ than the range R.

or

$$D = \sqrt{(h_T - h_A)^2 + R^2} \approx R + \frac{(h_T - h_A)^2}{2R}$$

and

$$I = \sqrt{(h_T + h_A)^2 + R^2} \approx R + \frac{(h_T + h_A)^2}{2R}$$

So the path difference δ is

$$\delta = D - I \approx \frac{2h_A h_T}{R},$$

whence

$$h_A h_T \approx \frac{mR\lambda}{4}.$$

This relationship is illustrated in Fig. 10-5 using the first lobe ($m = 1$; the value of m can be chosen to suit the range parameters R, h_A, h_T, and λ) at a frequency of 3 GHz for a range of 1400′.

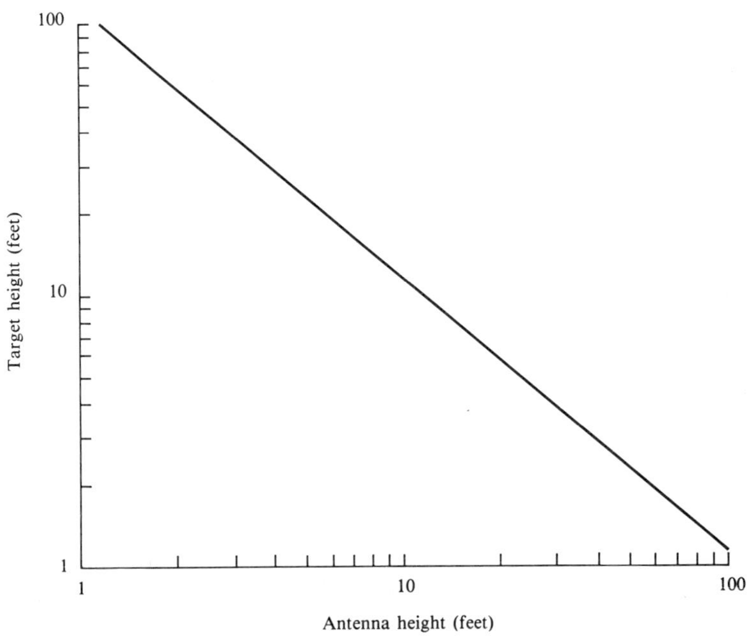

Figure 10-5 Target height versus antenna height at 3 GHz for an ideal ground plane; $R = 1400$ feet.

10-4-1 Reflection Coefficient

If the ground plane is composed, say, of soil, then its reflection coefficient is not -1. For higher microwave frequencies, say 1.5–15 GHz, one may find $\rho \approx 0.9 \exp(j\,170°)$. And the reflection coefficient may also depend strongly on angle of incidence, especially grazing angles of incidence. We find

$$\rho_{\perp} = \frac{\sin \psi - (\epsilon - \cos^2 \psi)^{1/2}}{\sin \psi + (\epsilon - \cos^2 \psi)^{1/2}} \tag{10-15}$$

$$\rho_{\parallel} = \frac{\epsilon \sin \psi - (\epsilon - \cos^2 \psi)^{1/2}}{\epsilon \sin \psi + (\epsilon - \cos^2 \psi)^{1/2}} \tag{10-16}$$

where ψ denotes grazing angle (1/2 of angle API in Fig. 10-4) and ϵ is relative permittivity. The subscripts $\perp$ and $\parallel$ denote E-vector perpendicular to and parallel to, respectively, the incidence plane (formed by the incidence direction and normal to the reflecting plane). When ψ takes the Brewster angle value, ρ can be zero and so energy is propagated into the ground.

Surface roughness, which depends on surface height variation, may also enter into the composition of the reflection coefficient. One would then write an effective reflection coefficient as

$$\rho_e = \rho_s \rho \tag{10-17}$$

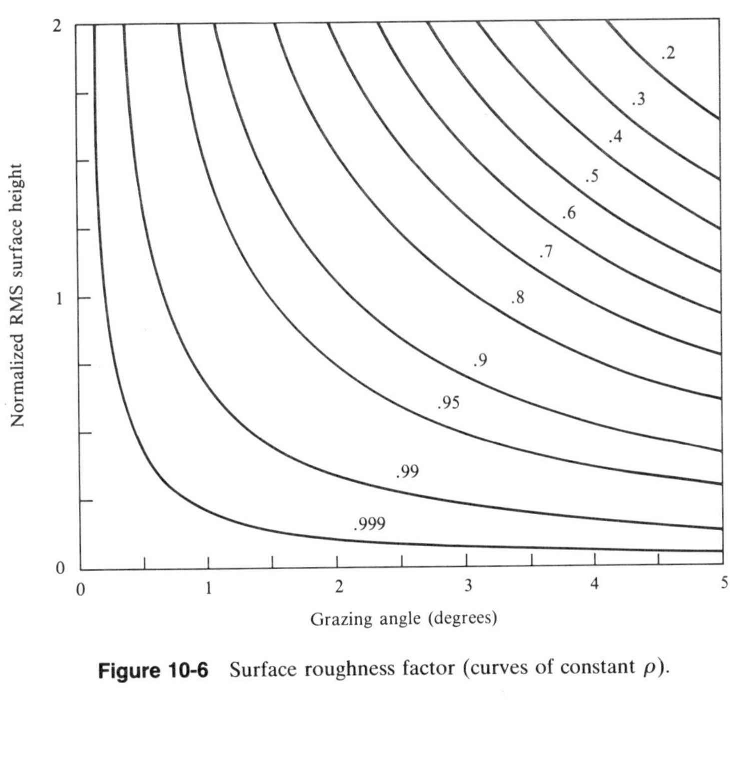

Figure 10-6 Surface roughness factor (curves of constant ρ).

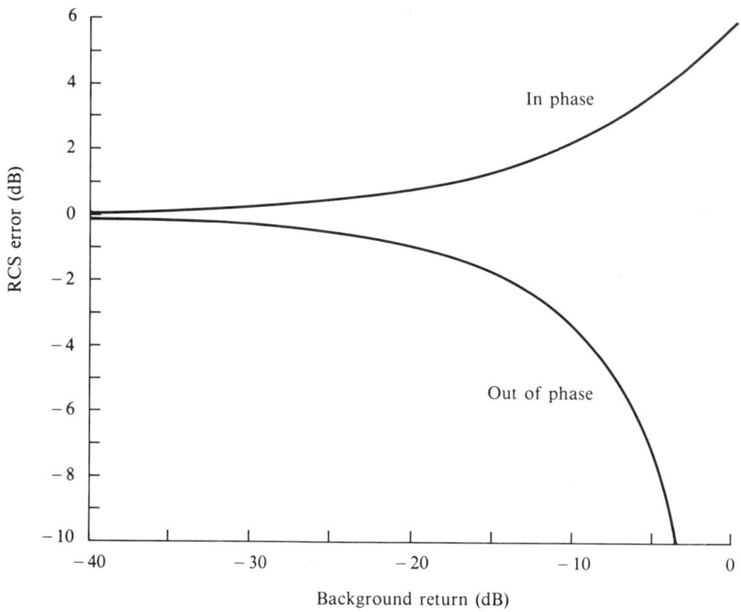

Figure 10-7 RCS error from background clutter.

where the roughness factor ρ_s is given by (compare Section 8-3-1):

$$\rho_s = \exp[-2(k\sigma_h \sin \psi)^2], \tag{10-18}$$

with σ_h being rms surface height variation. Figure 10-6 shows σ_h/λ versus grazing angle up to 5° (since typical outdoor operation seldom calls for greater angles) for a range of the roughness factor ρ_s. For example, suppose one wishes to have $\sigma_h = 3\lambda/2$, then at a grazing angle of 2° the surface roughness must be approximately 0.81. So for millimeter wave measurements it may be a serious problem indeed to suppress surface roughness clutter.

10-5 BACKGROUND

The background is any unwanted, or error, or clutter, signal that gets mixed with the true signal to be measured. We first examine its nature, then its sources, and finally, what can be done to eliminate or deal with it.

Let

$$e_t = \text{true signal}, \tag{10-19}$$

$$e_c = \text{clutter, or background, signal, and} \tag{10-20}$$

$$e_m = \text{signal actually measured.} \tag{10-21}$$

We have

$$e_m = e_t + e_c = e_t[1 + s_c e^{j\phi}], \tag{10-22}$$

where

$$s_c e^{j\phi} = \frac{e_c}{e_t}. \tag{10-23}$$

Then RCS is

$$\sigma = e_m e_m^* = e_t e_t^*[1 + 2s_c \cos\phi + s_c^2]. \tag{10-24}$$

The bracketed factor of Eq. 10-24 is plotted in Fig. 10-7 for inphase ($\phi = 0$) and out-of–phase ($\phi = \pi$) conditions. (The abscissa represents s_c^2 in dB, where s_c is of course a real number.)

10-5-1 Columns

One of the chief sources of background return is the target support column. Consider the following simplified analysis of a cylindrical column of height h and diameter d, which assumes the front and back surfaces to be planar. We then have

$$\sigma(0) = \frac{1}{2} kdh^2 \, |1 + \exp(2jkd\sqrt{\epsilon})|^2 \, |R(0)|^2 \tag{10-25}$$

where $R(0)$ is the voltage reflection coefficient at normal incidence:

$$\begin{aligned} R(0) &= \frac{\sqrt{\epsilon} - \sqrt{\mu}}{\sqrt{\epsilon} + \sqrt{\mu}} \\ &\approx \frac{1}{4}(\epsilon - 1) \qquad \text{for} \qquad \mu \approx 1 \text{ and } \epsilon - 1 \ll 1. \end{aligned} \tag{10-26}$$

Then expansion of the absolute value in Eq. 10-25 yields

$$\sigma(0) = \frac{kdh^2}{16}(\epsilon - 1)^2(1 + \cos(2kd\sqrt{\epsilon})), \tag{10-27}$$

which

$$= \text{maximum} = \frac{1}{8} kdh^2(\epsilon - 1)^2 \tag{10-28}$$

when $2kd\sqrt{\epsilon} = 2n\pi$, n = a positive integer, or when

$$\begin{aligned} f &= \frac{15n}{d\sqrt{\epsilon}} \text{ (in GHz);} \\ &= \text{minimum} = 0, \end{aligned} \tag{10-29}$$

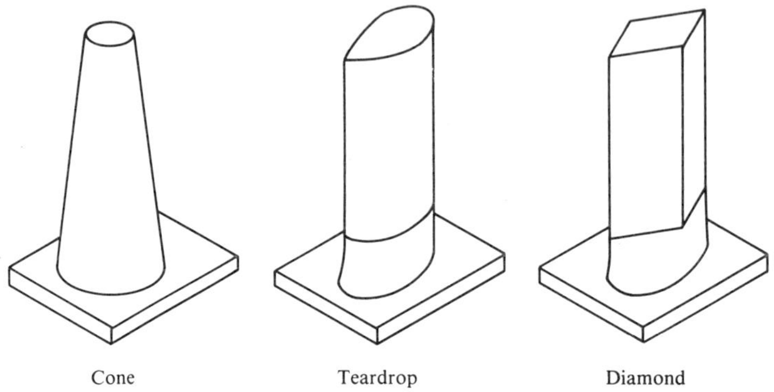

Figure 10-8 Support column shapes.

when $2kd\sqrt{\epsilon} = \frac{1}{2}\pi(4n - 1)$, or when

$$f = \frac{30(2n-1)}{4d\sqrt{\epsilon}} \text{ (in GHz).} \tag{10-30}$$

These results can be assumed to hold approximately over a small angular interval near the normal to the column.

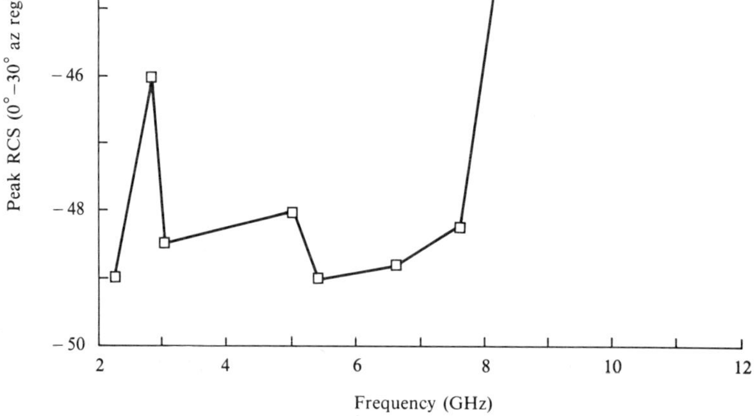

Figure 10-9 RCS of diamond column.

Example 5. Let

$$h = 8', \ \epsilon = 1.05, \ d = 14''.$$

Then we find (when $n = 25$, for example) that

$$f_{\max} = 10.3 \text{ GHz}$$

where $\sigma_{\max} \approx 0.14\, m^2$ ($= -8.5$ dBsm);

$$f_{\min} = 10.1 \text{ and } 10.5 \text{ GHz}$$

when $n = 25$ and 26, respectively. Clearly, we can "tune out" a column by choosing an appropriate measurement frequency.

In addition to tuning, one can also exploit shaping to reduce column RCS. Figure 10-8 illustrates several types and Fig. 10-9 shows the general level of RCS to be expected from a diamond column.

10-5-2 Multipath

Multipath returns may arise from, in addition to column supports, other nearby objects, such as buildings, poles, and trees. Surfaces of equal time delay are ellipsoids (actually prolate spheroids); Fig. 10-10 shows a slice of such a spheroid (symmetric about its long axis). Let s be the distance between radar and target and $s_1 + s_2$ be an indirect distance by way of the

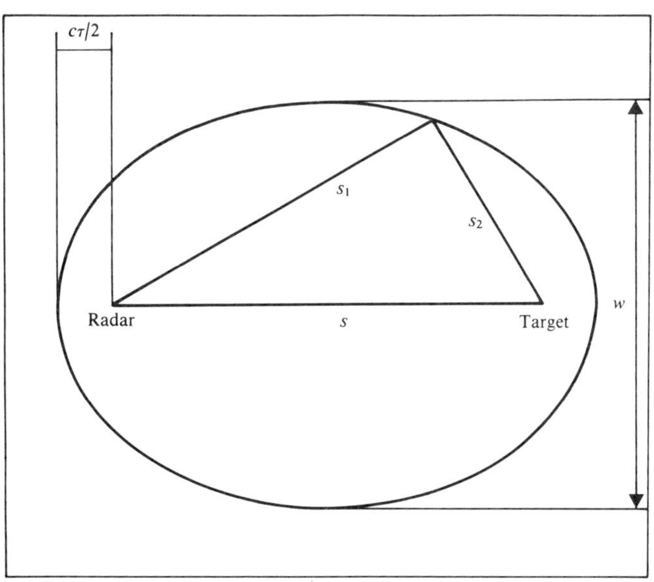

Figure 10-10 Ellipse of constant error.

spheroid. If L is the spheroid length and τ the smaller of pulse and range gate width, then

$$L = s + c\tau,$$

where c is the speed of light. The spheroid width is

$$w = c\tau\left(1 + \frac{2s}{c\tau}\right)^{1/2}.$$

Example 6. Let $\tau = 0.2\,\mu$ sec, so that $c\tau = 197$ feet. If the target range $s = 1400$ feet, then $L = 1577$ feet, $w = 768$ feet. So the length of this range should be twice its width.

Clutter fences have been considered as a measure to reduce undesired returns. But a study by the Physical Science Laboratory of New Mexico State University indicates that they offer no distinct advantages; fence edges diffract energy toward the target and these edge effects are very difficult to suppress. One way to deal with this problem might be to make the antenna beam smaller; but again according to the above-mentioned study, a smaller antenna beam reduces fence-diffracted energy but also reduces some of the unwanted returns. So, in effect, the fences are unnecessary.

10-5-3 Column Forward Scattering

If the range is an indoor one, then an added source of column problems is their possible interaction with the chamber back wall. Consider the reflection coefficients at the interface between two media, say air (medium 1) and polyfoam (medium 2):

$$\Gamma_{\perp} = \frac{\cos\theta - \sqrt{(\epsilon_2/\epsilon_1) - \sin^2\theta}}{\cos\theta + \sqrt{(\epsilon_2/\epsilon_1) - \sin^2\theta}};$$

$$\Gamma_{\parallel} = \frac{(\epsilon_2/\epsilon_1)\cos\theta - \sqrt{(\epsilon_2/\epsilon_1) - \sin^2\theta}}{(\epsilon_2/\epsilon_1)\cos\theta + \sqrt{(\epsilon_2/\epsilon_1) - \sin^2\theta}}.$$

If the incidence angle is grazing ($\theta \to \pi/2$), then

$$\Gamma_{\perp} = \Gamma_{\parallel} \to -1$$

and so the column might just as well be a metallic one insofar as forward scattering toward the back wall is concerned. This means that the back wall would have to absorb the enhanced forward-scattering RCS of the column support. The increase over backscattering RCS for a cylindrical column could be as much as 20 dB.

An example is instructive. We use (for the sake of easy computations) a sphere of radius a, which places the sphere in or near the optics region. We have

$$\sigma_{\text{back}} = \pi a^2,$$

$$\sigma_{\text{fwd}} = (ka)^2(\pi a)^2.$$

The ratio $\sigma_{\text{fwd}}/\sigma_{\text{back}}$ increases rapidly with ka:

σ(fwd/back)	ka
9.5 dB	3.0
14.0	5.0
17.5	7.5
20.0	10.0
23.5	15.0
26.0	20.0
28.0	25.0
29.5	30.0

The remedies for the forward-scattering problem are a suitable design for the back wall, both in terms of shaping and radar absorbing material (RAM) installation; and suitable choices of column shapes. Consider the RAM problem on the back wall:

Without a scatterer we have the following condition governing transmitted power P_t and received power P_r:

$$\frac{P_r}{P_t} = \frac{G^2\lambda^2}{16(4\pi)^2(r_1 + r_2)^2}\,\theta_{el}\phi_{az}\Gamma_w$$

where θ_{el} and ϕ_{az} are elevation and azimuth beamwidths of the antenna whose gain is G; r_1 is the distance from antenna to the center of the model space, r_2 is the distance from the center of the model space to the wall, and Γ_w is the (power) reflection coefficient of the wall. We observe that the illuminated (elliptical) area of the wall is

$$\frac{\pi}{4}(r_1 + r_2)^2\theta_{\text{el}}\phi_{az}.$$

The important role of Γ_w is demonstrated by a further example.

Example 7. Suppose that the antenna diameter D is chosen in accordance with the far field criterion, that is,

$$\frac{2D^2}{\lambda} = r_1 + r_2.$$

Then using this diameter to compute gain G, with the additional assumption of 60% efficiency, gives

$$G = \frac{0.3\pi^2(r_1 + r_2)}{\lambda}.$$

Take $f=3\,\text{GHz}$ ($\lambda=0.1\,\text{m}$), and suppose that elevation and azimuth beamwidths, θ_{el} and ϕ_{az}, are each λ/D. Then, with $r_1=8\,\text{m}$ and $r_2=2\,\text{m}$,

$$P_r/P_t=6.9\times10^{-5}\Gamma_w=\begin{cases}6.9\times10^{-9} & \text{for 40-dB RAM,}\\ 6.9\times10^{-7} & \text{for 20-dB RAM.}\end{cases}$$

Now consider a spherical target with $ka=20$ at $r_1=8\,\text{m}$. Then the power returned from the sphere is

$$\frac{P_r}{P_t}=3.4\times10^{-5}.$$

Hence, with 40-dB RAM, no significant error arises from the back wall. *But,* with 20-dB RAM, at least 10 dB of cancellation would be required to ensure ±1-dB error in the measured RCS σ_m (cf. Fig. 10-7 above).

10-5-4 Pyramidal RAM

To control the background clutter it is usually necessary, especially for an indoor range, to employ high-grade RAM. The high grade of absorptivity is usually achieved with large thicknesses and pyramidally shaped carbon-loaded foams. Figure 10-11 gives absorptivity performance as a function of thickness at a representative frequency of 1 GHz (although, generally speaking, such RAM is broadband in absorptivity).

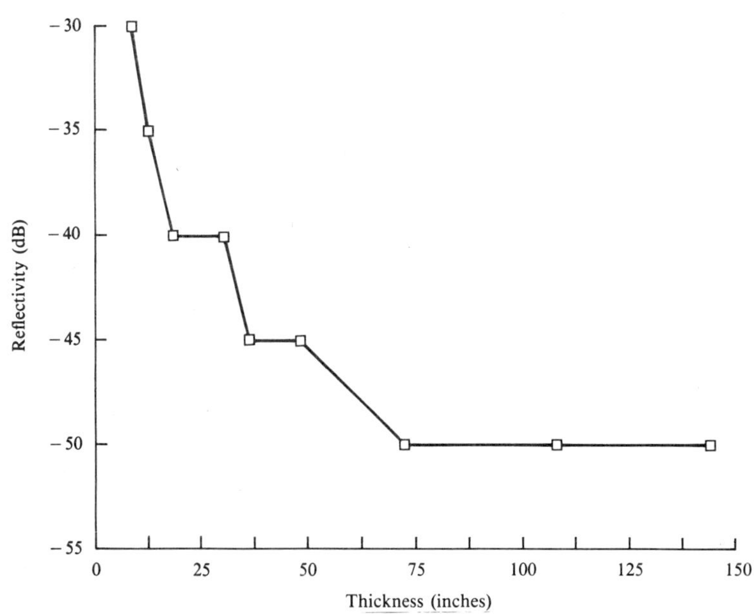

Figure 10-11 Thickness characteristics for pyramidal RAM.

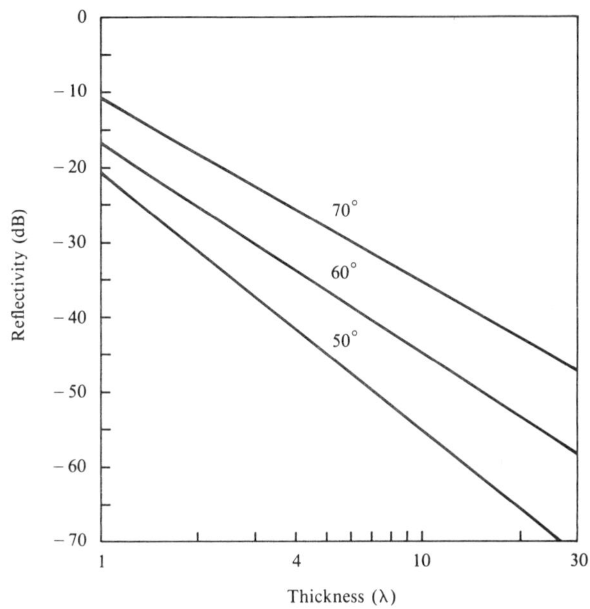

Figure 10-12 Wide-angle reflectivity for pyramidal RAM.

Even with large thicknesses, however, the performance of pyramidal RAM falls off with incidence angle. Figure 10-12 illustrates that, for example, at a thickness of 10 in there is a 20-dB fall-off in performance when the incidence angle is increased from 50 to 70°.

10-6 POWER RELATIONS

The minimum measurable received power establishes the smallest RCS, $\sigma_{\min}$, measurable with a given set of equipment parameters. From the radar range equation

$$\sigma = \frac{P_r}{P_t}\frac{(4\pi)^3 r^4}{\alpha_r \alpha_t G^2 \lambda^2} \tag{10-31}$$

where α is the attenuation of the transmission (subscript t) or receiving (subscript r) line path and G is antenna gain (with a single antenna used for transmitting and receiving). Writing the equation above in decibel notation yields

$$\sigma_{\mathrm{dBsm}} = 33 + (P_{r\mathrm{dBm}} - P_{t\mathrm{dBm}}) - \alpha_{T\mathrm{dB}} - 2G_{\mathrm{dB}} + 40 \log r_m - 20 \log \lambda, \tag{10-32}$$

where

$$\alpha_{T\mathrm{dB}} = \alpha_{t\mathrm{dB}} + \alpha_{r\mathrm{dB}}. \tag{10-33}$$

If we wish to specify the measurement of σ, admitting an error of ± 1 dB, then we must add 20 dB to the right-hand side in Eq. (10-32) [cf. the σ_e graph (Fig. 10-7) for maximum and minimum ϕ]:

$$\sigma_{\text{dBsm}} = 53 + (P_{r\text{dBm}} - P_{t\text{dBm}}) - \alpha_{T\text{dB}} - 2G_{\text{dB}} + 40 \log r_m - 20 \log \lambda, \qquad (10\text{-}34)$$

where r_m is the range at which the minimum power is required and dBm means decibels relative to a milliwatt. Equation 10-34 is called the *system sensitivity equation* for ± 1-dB error.

Consider the following example.

Example 8. What is $\sigma_{\min}$ for a system whose parameters are:

$$f = 10\ \text{GHz}\ (\lambda = 1.18\ \text{in.} = 0.02997\ \text{m})$$
$$r = 10\ \text{m}$$
$$G = 30\ \text{dB}$$
$$\alpha_T = -8\ \text{dB}$$
$$P_t = 10\ \text{dBm},\ P_r = -100\ \text{dBm}.$$

SOLUTION.

$$\sigma_{\min} = 53 - 100 - 10 + 8 - 60 + 40 \log 10 - 20 \log (0.03) = -38.5\ \text{dBsm}\ (=1.4 \times 10^{-4}\ \text{m}^2). \qquad (10\text{-}35)$$

For outdoor ranges low RCS levels may impose much more severe requirements on measurement parameters because of the much greater ranges required to accommodate larger targets. For example, ranges of as much as 3000 m may be required. If the 10-m range used in Eq. 10-35 is replaced by 3000 m, we see at once the slightly ridiculous results with which we must be content with a minimum RCS level of slightly more than 60 dBsm. So a critical reexamination must be given to transmitted and received power capabilities, as well as antenna gains and system losses.

10-7 ANTENNAS

The choice of antennas for a measurement range can be phrased in terms of antenna gain G. For a 60% efficient reflector type of circular antenna of diameter D, with a uniform current distribution, G is given by

$$G = 0.6\left(4\pi \frac{A}{\lambda^2}\right) = 0.6\pi^2\left(\frac{D}{\lambda}\right)^2. \qquad (10\text{-}36)$$

Since the conventional far field of an antenna is analogous to that of a scatterer, we have from $r = 2D^2/\lambda$,

$$D = \sqrt{\lambda r/2}; \qquad (10\text{-}37)$$

and so

$$G = 0.6\pi^2 \frac{r\lambda}{2\lambda^2} = 0.3\pi^2 \frac{r}{\lambda}. \tag{10-38}$$

Thus, we can choose the required antenna to suit the frequency of interest. But the number of antennas needed to cover a reasonably wide frequency band would obviously create an economic strain. So it is customary to compromise with several antennas which can provide an acceptable approximation to the idealized curves of antenna parameters versus frequency.

Figures 10-13 and 10-14 illustrate the idealized curves of aperture D and half-power distance at the target space versus frequency for a range having $r = 32'$. Assuming the availability of four antennas, say dishes of 6″, 12″, 18″, and 36″, Figs. 10-13 and 10-14 show how the idealized curves are approximated by the performance curves of these four dish antennas. The approximations offer a reasonable coverage of the specified frequency band.

The lateral distance s between half-power beamwidths for a circular aperture is

$$s = \frac{rc}{fD} \quad \text{or} \quad s = \frac{32c}{fD} \text{ if } r = 32', \tag{10-39}$$

c being the speed of light. The optimum is $s = 2D$, as was pointed out

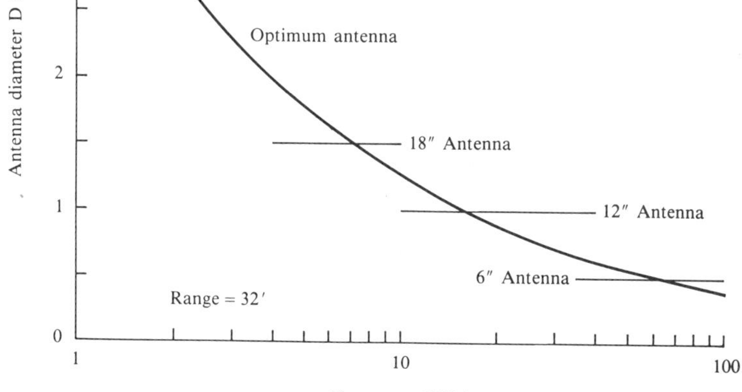

Figure 10-13 Four-antenna (diameter) practical approximation to optimum for frequency coverage of 1–100 GHz.

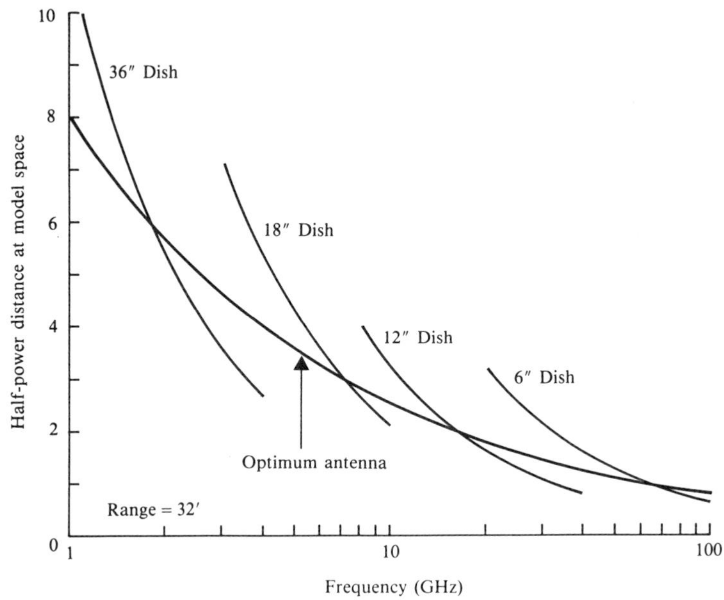

Figure 10-14 Four-antenna (half-power) practical approximation to optimum for frequency coverage of 1–100 GHz.

earlier. Lateral half-power distances are charted in Fig. 10-14 for four antennas of 36, 18, 12, and 6″ diameters.

10-8 COORDINATE SYSTEMS AND TRANSFORMATIONS

There is scarcely anything that can cause as many difficulties as coordinate systems and transformations. The choice of a standard system seems unlikely, since some disciplines even employ left-handed systems. For our discussions here we choose to follow the RATSCAT definitions, overlaid by some of our own nomenclature.

Referring to Fig. 10-15 we choose the axes of our coordinate system as indicated there. Rotations about these axes are referred to by the same names as the axes. Measurements of the rotation angles are made as follows: Assume a pointer to be fixed at the 0° marks. When the target is rotated about an axis, it carries its scale with it. The pointer then indicates on the scale the value of the rotation angle.

In order to introduce a mathematical formulation to these ideas we make the following identifications:

Yaw axis $= \hat{x}$-axis, positive upward;
Pitch axis $= \hat{y}$-axis, positive starboard; and
Roll axis $= \hat{z}$-axis, positive forward.

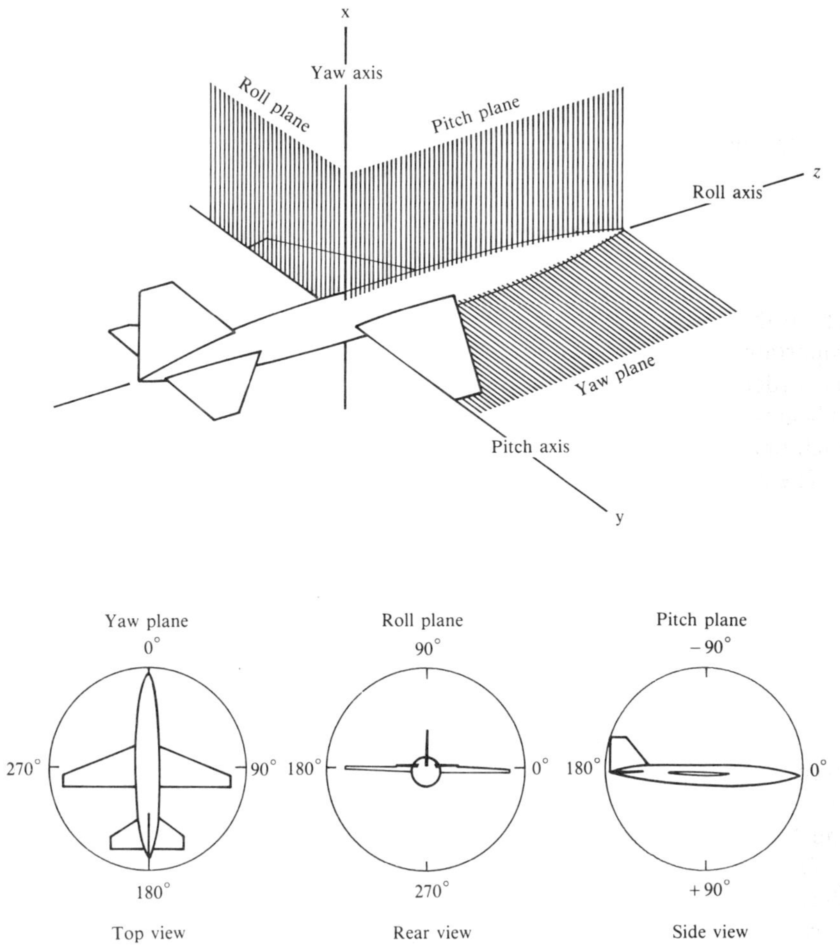

Figure 10.15 Coordinate system pitch, roll, and yaw planes.

Imagine these axes to be fixed to the target initially, then to be rotatable to new positions, called $\hat{x}'$, $\hat{y}'$, $\hat{z}'$, through angles y, p, and r which correspond to yaw, pitch, and roll angles, respectively. To be precise, we have

$$\text{Yaw: } \mathfrak{Y} = \begin{pmatrix} 1 & 0 & 0 \\ 0 & \cos y & \sin y \\ 0 & -\sin y & \cos y \end{pmatrix} \begin{pmatrix} \text{about } x\text{-axis,} \\ \text{positive for } +y \\ \text{toward } +z \end{pmatrix} \tag{10-40}$$

$$\text{Pitch: } \mathfrak{P} = \begin{pmatrix} \cos p & 0 & -\sin p \\ 0 & 1 & 0 \\ \sin p & 0 & \cos p \end{pmatrix} \begin{pmatrix} \text{about } y\text{-axis,} \\ \text{positive for } +z \\ \text{toward } +x \end{pmatrix} \tag{10-41}$$

$$\text{Roll: } \mathfrak{R} = \begin{pmatrix} \cos r & \sin r & 0 \\ -\sin r & \cos r & 0 \\ 0 & 0 & 1 \end{pmatrix} \begin{pmatrix} \text{about } z\text{-axis,} \\ \text{positive for } +x \\ \text{toward } +y \end{pmatrix} \tag{10-42}$$

For any one, $\mathfrak{A}$, of these transformations the axes are given by

$$\begin{pmatrix} \hat{x}' \\ \hat{y}' \\ \hat{z}' \end{pmatrix} = \mathfrak{A} \begin{pmatrix} \hat{x} \\ \hat{y} \\ \hat{z} \end{pmatrix}, \tag{10-43}$$

where the caret denotes unit vector. Note that order of transformations is important, that is, $\mathfrak{Y}$, $\mathfrak{P}$, and $\mathfrak{R}$ do not, in general, commute. Note also that the order of application of transformation is right to left, that is, $\mathfrak{P}\mathfrak{Y}\mathfrak{R}$ indicates roll first through an angle r, yaw next through an angle y, and pitch finally through an angle p.

The combination $\mathfrak{P}\mathfrak{R}\mathfrak{Y}$, say, represents a most general rotation

$$\mathfrak{P}\mathfrak{R}\mathfrak{Y} = \begin{pmatrix} \cos p \cos r & \cos p \sin r \cos y + \sin p \sin y & \cos p \sin r \sin y - \sin p \cos y \\ -\sin r & \cos r \cos y & \cos r \sin y \\ \sin p \cos r & \sin p \sin r \cos y - \cos p \sin y & \sin p \sin r \sin y + \cos p \cos y \end{pmatrix} \tag{10-44}$$

since this is an orthonormal transformation and can therefore be achieved by a set of direction angles. Compare these remarks about PITCH, ROLL, and YAW angles with those about Euler angles in Chapter 19.

One can also overlay polar systems on the rectangular systems described above. For example, let THETA (θ) be the polar angle, measured from the $\hat{z}$-axis, $0 \leq \theta \leq \pi$, and ϕ the azimuth angle, measured in the $\hat{x}\hat{y}$-plane from $\hat{x}$ toward $\hat{y}$, $-\pi < \phi \leq \pi$. Then ϕ fixed, θ varying from 0 to π yields a *planar cut*† containing the longitudinal axis ($\hat{z}$). And θ fixed, ϕ varying from $-\pi$ to π yields a *conical cut* about the longitudinal axis.

With the help of the PITCH, ROLL, and YAW transformations, these planar (or great circle, as they are often called) and conical cuts can be referred to an arbitrary axis. For example, suppose we wish to refer our cuts to a vertical axis rather than a longitudinal axis. Then we PITCH the *axes* (not the target) $\pi/2$ and follow this by a ROLL of π, with no YAW transformation. Then ϕ fixed, θ varying from 0 to π yields a planar cut† containing the vertical axis. And θ fixed, ϕ varying from $-\pi$ to π yields a conical cut about the vertical axis. Of course, arbitrary conical and planar (also referred to as great circle) cuts are only theoretically possible. For

† Note that this is a "one-sided," or half-plane, cut. For example, $\phi = -\pi/2$, $0 \leq \theta \leq \pi$ views the *port* side of the target; while $\phi = \pi/2$, $0 \leq \theta \leq \pi$ views the *starboard* side.

measurements, target support is obviously the paramount constraint. Yet many practical conical and great circle cuts are available, particularly with modern support capabilities (see Fig. 10-17*a* below). Great circle cuts are achieved by rotating the column or target support platform (or saddle) about an axis perpendicular to the ground plane. Cuts are varied by changing target orientation (with pitch, roll, and yaw transformations) on its platform. Conical cuts are achieved by tilting the rotation axis away from vertical (usually toward the radar). Target orientation can be varied here as well to produce a variety of conical aspects. Suitable target orientations can usually be combined with achievable great circle and conical cuts to produce satisfactorily dense coverage of an aspect region of interest.

10-9 RECORDING OF DATA

The rotation of the target during a measurement defines the appearance of data from a rectilinear recorder. This statement is true for both analog and digital recording, although with today's processing capabilities, it is possible to display data in a fairly arbitrary format. Nevertheless, it is important to know how target movement and data records are related.

When analog recording was common, most facilities used the standard format shown in Fig. 10-16. For example, suppose the target to be supported in the horizontal plane. From a look-down viewpoint, let the measurement begin with the radar viewing the target tail. Then let the target be rotated clockwise so that the radar views first the starboard side, then the nose, then the port side, and finally the tail again. The analog recording chart would show the data with port side on the left (recorded last) and the starboard side on the right (recorded first).

With digital recording, the data can be "managed" as recorded, so that various statistics can be produced. For example, one popular practice (particularly for high frequencies) is to "medianize" the data, that is, medians are recorded in a sliding window of some chosen size. The result is a smoothed RCS curve.

Other percentiles and averages are also useful in the study of RCS data. We give some definitions here to be sure that these statistics are distinguished from each other as well as from the statistics to be presented in Chapters 12 and 13.

Let $\sigma_1, \sigma_2, \ldots, \sigma_n$ be a finite number (n) of RCS values (for example, n would equal 181 if data were recorded at 1° intervals over half a great circle cut). Two means, arithmetic and geometric, are in popular usage and either can be weighted with weights w_i, $i = 1, \ldots, n$, $w_i \geq 0$. The arithmetic mean is defined as

$$m_a = \left(1 \Big/ \sum_{i=1}^{n} w_i\right) \sum_{i=1}^{n} w_i \sigma_i . \tag{10-45}$$

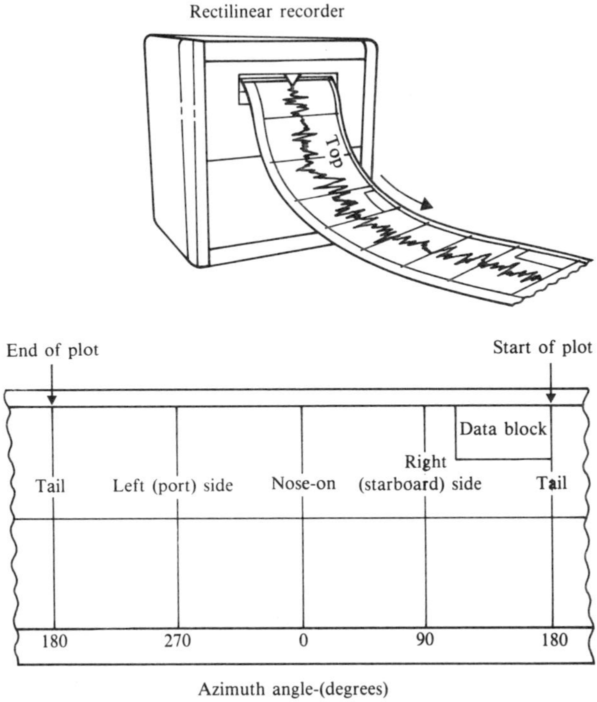

Figure 10-16 Format for rectilinear plots.

The geometric mean is defined as

$$m_g = \left(\prod_{i=1}^{n} \sigma_i \right)^{(w_i / \sum_{m=1}^{n} w_m)}; \tag{10-46}$$

or, taking the logarithm of both sides of Eq. 10-46,

$$\log m_g = \left(1 \Big/ \sum_{j=1}^{n} w_j \right) \sum_{i=1}^{n} w_i \log \sigma_i. \tag{10-47}$$

If we multiply both sides of Eq. 10-47 by 10, we see at once that the geometric mean in dBsm is the arithmetic mean of the RCS data, where each datum is expressed in dBsm. It can be proven in general that (Hardy, Littlewood, and Polya, 1959)

$$m_g \le m_a. \tag{10-48}$$

The pth percentile of a set of RCS data may be defined as the RCS level for which $p\%$ of the RCS values falls below the corresponding abscissa value (of RCS). The median is therefore the 50th percentile. A knowledge of all of the percentiles furnishes the distribution of RCS values over the interval of interest. There are no general theorems relating any percentile (other than 0 and 100), and especially the median, to either the arithmetic mean or the geometric mean (as they were related to each other). However, experience has shown that the median almost always lies in the vicinity of the geometric-arithmetic means interval.

10-10 SCALING

We have already taken some notice above of the background problems which can arise from target supports. Such problems may be aggravated by the size and weight of a target. Figure 10-17† illustrates a heavy and bulky target of the sort which may confront a measurement facility. Such very large and/or very heavy targets introduce the added problems of safety constraints which in turn may interfere unduly with the RCS level to be measured.

To attempt to alleviate these problems one must examine the merits of scaling. The principles of scaling are simple ones. Because of the linearity of the equations governing scattering, one must arrange to change model sizes and measurement frequencies in such a way as to keep the ratio of target dimensions to wavelength fixed. Then the ratio of RCS to squared wavelength will remain fixed as well.

But all quantities appearing in the scattering equations must be scaled similarly. And this is where the first problem arises, for the constitutive parameters cannot (in general) be scaled. In fact, for other than perfectly conducting bodies, the materials required to scale ϵ and μ, particularly their frequency dependence, simply do not exist. However, one can often get round this problem for other than perfectly conducting bodies, and for theory but not measurements, by employing only the effect of the reflection coefficient. But there may still be some conductivity difficulties if one scales to very small millimeter wavelengths. Table 10-1 (Blacksmith, Hiatt, and Mack, 1965) enumerates all pertinent quantities and how they scale.

Detail can be difficult and expensive to achieve for very small models. This fact is particularly true for targets having inlet ducts, engines, and exhausts. However, the effects of such scattering components can be mitigated if one is interested chiefly in RCS statistics because inlet, engine, and exhaust detail affect, in general, only the detail of RCS patterns and have little effect on first order statistics such as mean RCS and standard deviation of RCS.

Clearly, one of the chief advantages of scaling is to be found in the far field criterion, $r = 2L^2/\lambda$. Small targets require much less real estate and, indeed, will usually permit the use of a chamber, with all of its attendant advantages.

† With the permission and courtesy of 6585th Test Group (RATSCAT) Holloman Air Force Base, NM, and the kind assistance of Colonel Richard W. Denny and Lieutenant Raymond P. Lentz III.

Figure 10-17a A target model on an Ogival pylon support.

Figure 10.17b A target model on a polystyrene column.

TABLE 10-1 Scaling by a Factor *p*

Quantity	Full-Scale System	Fractional Scale System
Length	L	$L' = L/p$
Time	t	$t' = t/p$
Frequency	f	$f' = fp$
Wavelength	λ	$\lambda' = \lambda/p$
Wave number	k	$k' = kp$
Conductivity	s	$s' = sp$
Permittivity	ϵ	$\epsilon' = \epsilon$
Permeability	μ	$\mu' = \mu$
Impedance	Z	$Z' = Z$
Antenna gain	G	$G' = G$
RCS	σ	$\sigma' = \sigma/p^2$

After Blacksmith, Hiatt, and Mack (1965), © 1965 IEEE, reprinted with permission.

In fact, there is now in fairly popular use a type of range that permits larger targets at shorter ranges, and so lends itself to indoor operation. It is the compact range.

10-11 THE COMPACT RANGE

The driving reason for the far field criterion is of course the need to achieve, with as little error as possible, a plane wave illumination at the target. It now appears practical to modify the illumination source so as to achieve this goal for moderately large targets with a smaller than usual amount of real estate. The result is the so-called compact range. The subject is discussed by Young and Walton (1984) and by Tuley in Currie's collection of measurement methods (1984). In particular, we refer the reader to the latter author for a discussion of such matters as source feeds, power requirements, and other systems subjects that are outside of the scope of our endeavors here.

There are essentially two ways to achieve a plane wave at the target: with a dielectric lens or with a reflector. Two kinds of lenses that might be appropriate: conventional and reversed. But the difficulties, such as dielectric nonuniformities, connected with the fabrication of suitable lenses, as well as the problems arising from internal reflections reduce the practicability of this approach. Indeed, this writer knows of no operational compact range that uses lenses.

But the parabolic reflector can be and is employed successfully in a compact range. Looking at Fig. 10-18, we can imagine a large parabolic reflector arranged so as to bathe a target in plane wave illumination. Only a portion of the reflector can be used if interference with the source is to be

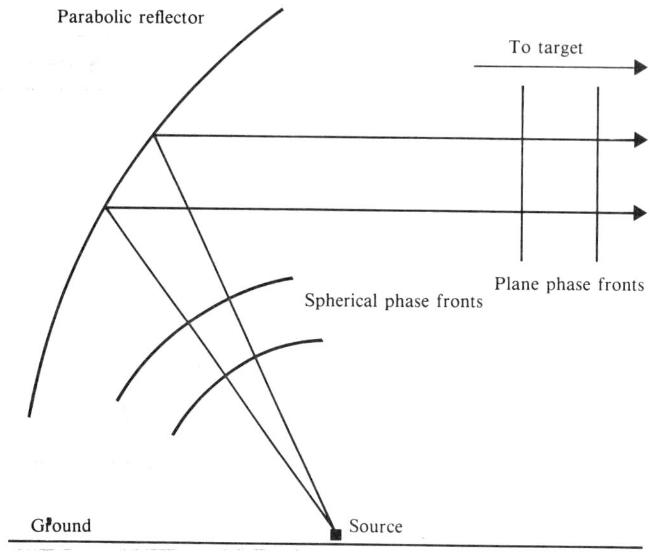

Figure 10-18 Section of a parabolic reflector for a compact radar range.

avoided. The figure shows only the upper portion of such a reflector which, by the properties of a paraboloid, provides equal pathlengths from source to plane wave front, and so converts a spherical wave from the source to a plane wave at the target. To avoid diffraction effects from the edge of a circular paraboloid, the outer edge is made quite jagged in practice. Young and Walton (1984) describe their very successful use of the parabolic reflector design in the creation of a compact range at the Ohio State University ElectroScience Laboratory.

10-12 NEAR FIELD MEASUREMENTS

The last five to 10 years has seen a continuing increase in the ideas of near field measurements that can then be converted analytically into far field results. The main impetus has been from antenna measurements but the ideas work as well for scattering as for radiation. Actually, the idea was first put forward in 1950 by Booker and Clemmow and somewhat later on more extensively by Clemmow (1966). We have already met up with the idea in our examination of diffraction by a perfectly absorbing half-plane; it is called a plane wave spectrum. Basically, the idea is this (see Joy et al., 1978): Let

$$\mathbf{E} = \mathbf{A}(\mathbf{k}) \exp(-j\mathbf{k} \cdot \mathbf{r}) \tag{10-49}$$

be a solution of the Helmholtz equation

$$\nabla^2\mathbf{E} + k^2\mathbf{E} = 0, \tag{10-50}$$

where

$$\mathbf{k} = k_x\hat{x} + k_y\hat{y} + k_z\hat{z} \tag{10-51}$$

and

$$\mathbf{k}\cdot\mathbf{k} = k_x^2 + k_y^2 + k_z^2 = k^2 = \omega^2\epsilon\mu. \tag{10-52}$$

At a fixed frequency only two of the components of $\mathbf{k}$ can be specified. Let these be k_y and k_z. Then

$$k_x = \begin{cases} \sqrt{k^2 - k_y^2 - k_z^2} & \text{if } k_y^2 + k_z^2 < k^2, \\ -j\sqrt{k_y^2 + k_z^2 - k^2} & \text{otherwise.} \end{cases} \tag{10-53}$$

If there is no free charge in the region of interest, $\nabla\cdot\mathbf{E} = 0$, which yields, when substituted into the expression for $\mathbf{E}$ (Eq. 10-49),

$$\mathbf{k}\cdot\mathbf{A}(\mathbf{k}) = 0. \tag{10-54}$$

So only two components of $\mathbf{A}(\mathbf{k})$ can be chosen independently. Let these be A_y and A_z:

$$A_x(\mathbf{k}) = -(k_x^{-1})(k_yA_y + k_zA_z). \tag{10-55}$$

We note that

$$\mathbf{H} = -\frac{1}{j\omega\mu}\nabla\times\mathbf{E} = \frac{1}{\omega\mu}\mathbf{k}\times\mathbf{A}(\mathbf{k})e^{-j\mathbf{k}\cdot\mathbf{r}} \tag{10-56}$$

Since the field equations are linear, and since the medium is linear, the general solution for $\mathbf{E}$ and $\mathbf{H}$ can be constructed as linear combinations over all k_y and k_z:

$$\mathbf{E}(\mathbf{r}) = \iint_{-\infty}^{\infty} \mathbf{A}(\mathbf{k})e^{-j\mathbf{k}\cdot\mathbf{r}}\,dk_y\,dk_z \tag{10-57}$$

$$\mathbf{H}(\mathbf{r}) = \frac{1}{\omega\mu}\iint_{-\infty}^{\infty} \mathbf{k}\times\mathbf{A}(\mathbf{k})e^{-j\mathbf{k}\cdot\mathbf{r}}\,dk_y\,dk_z \tag{10-58}$$

The amplitude function $\mathbf{A}$ is called the *wave number spectrum.* [Compare $\mathbf{A}(\mathbf{k})$ with $P(\cos\alpha)$ from Chapter 6.]

$\mathbf{A}$ can be determined analytically if the tangential field is known at every point on a plane, say, $x = d$:

$$E_y(d, y, z) = \iint_{-\infty}^{\infty} A_y(\mathbf{k})e^{-jk_xd}e^{-j(k_yy + k_zz)}\,dk_y\,dk_z \tag{10-59}$$

$$E_z(d, y, z) = \iint_{-\infty}^{\infty} A_z(\mathbf{k})e^{-jk_xd}e^{-j(k_yy + k_zz)}\,dk_y\,dk_z \tag{10-60}$$

from the **E** of Eq. 10-57 above. Thus, A_y and A_z are two-dimensional FTs of E_y and E_z over $x = d$. This result means that we know, or can find by inversion, A_y and A_z. Therefore, **E** and **H** above can be found anywhere. The integrations are usually evaluated by saddle point or stationary phase methods; the result is

$$\mathbf{E}(\mathbf{r}) = \frac{2\pi j}{r} k_{xo} \mathbf{A}(\mathbf{k}_0) e^{-jkr}, \tag{10-61}$$

where

$$\mathbf{k}_o = k\hat{r} = k(\hat{x} \sin\theta \cos\phi + \hat{y} \sin\theta \sin\phi + \hat{z} \cos\theta).$$

This development takes no account of any interaction that may take place between the scatterer and the probe which measures the near field on $x = d$.

10-13 RADAR RANGES

There are numerous ranges of almost every type in existence today. Knott (1986; also in Currie, 1984) reviews several of them, both indoor and outdoor. Among those with an interesting heyday but no longer in existence

Figure 10-19 Anechoic chamber.

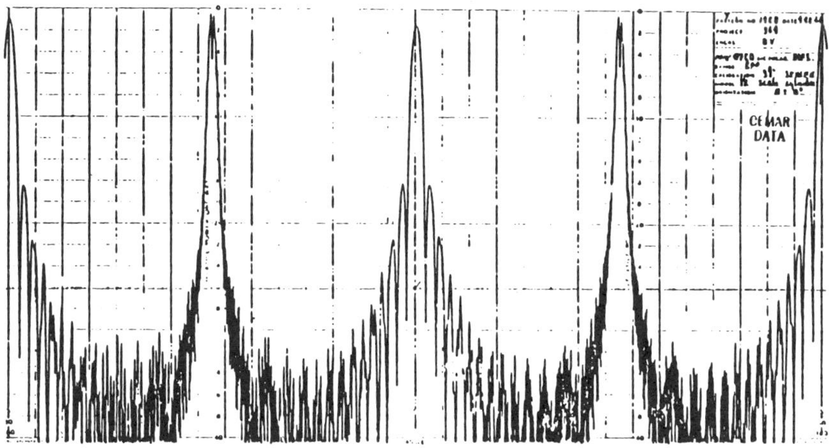

Figure 10-20 A typical measurement pattern.

are the Conductron Corporation ranges, both indoor and outdoor, of the 1960s. Figure 10-19 shows their anechoic chamber (which was the proud possessor of a ferrite RAM floor); together with, in Fig. 10-20, a typical RCS measurement pattern made on their outdoor range.

The pioneering measurement techniques (such as background subtraction) of the Teledyne Micronetics outdoor range are being continued under

Figure 10-21 A low-frequency antenna.

new ownership and the new name of Mission Gorge Range. Their capabilities serve to illustrate desirable features for a modern measurement facility:

> Long pulse RCS measurements throughout the frequency range of 130 MHz [Fig. 10-21] to 18 GHz; stepped frequency coherent measurements in octave bandwidth blocks throughout the same frequency range; Fourier transforms on stepped frequency coherent data to yield high-resolution plots in all bands from VHF through Ku bands; true ground plane geometry; automatic data processing; background subtraction and 80 dB dynamic recording range at all frequencies, together with −70 dBsm sensitivity at band centers above 500 MHz and −55 dBsm at frequency centers below 500 MHz; and bistatic measurement capabilities.

Mission Gorge Range also plans to add, in the immediate future, a rigid support pylon (such as that shown in Fig. 10-17a in addition to the more common type of support (Fig. 10-17b).

One type of indoor range that deserves attention is the Macroscope, developed by *Da*wson and *Re*chlin of DARE Technology, and presently installed at Fort Belvoir. It has a moving target platform that provides the Doppler information necessary to form a RCS image of the target. Since it operates at a high frequency (94 GHz), it is well suited to the study of small models of large targets. In fact, it was so used successfully in the late 1970s to study the RCS of a 1/35th scale model of the hydrofoil ship TUCUMCARI (Paddison *et al.*, 1978).

PART III

PRACTICAL ASPECTS OF RCS

In Chapters 11–15 we emphasize practical methods of estimating the statistical behavior of RCS, concluding with (Chapter 15) a discussion of one important application of these methods.

Beginning in Chapter 11 we examine the RCS of simple shapes, with special consideration for physical optics as a practical method from which fairly straightforward formulas can be obtained for a wide variety of simple shapes. In Chapter 12 we develop a statistical framework in which to estimate RCS behavior for complex targets; this framework forms the heart of our work. The statistical viewpoint provides not only smoothed curves of RCS versus aspect but also estimates of the distribution of RCS in a neighborhood of a given aspect. We find that the β-distribution is the most realistic for describing complex target RCS. The statistical formulas used in finding these β-distributions offer no computational difficulties.†

Chapter 13 contains the proof of our pudding. Here we compute the RCS and distribution of RCS for two well-known complex targets, the 727–100C (at *L*-band) and the Firebee (at *X*-band), and compare these theoretical results with experimentally measured data. The comparisons are gratifyingly good.

It is very often necessary to compare two different representations of RCS; for example, theoretically computed versus experimentally measured data, or statically measured versus dynamically measured data. Chapter 14 suggests a new idea for carrying out such comparisons: a comparison of respective distributions of RCS with the goodness of the comparison being

† A computer code called "RCS" (Copyright 1987, A. L. Maffett and H. W. Klimach) with which to compute and graph all pertinent statistics is described in Chapter 13.

evaluated by a statistical test. The idea is applied to a comparison of theoretical and experimental data for the 727–100C.

Finally, in Chapter 15, we suggest one important application of the distribution of RCS for a complex target, namely, detection theory. Special emphasis is given to the role of the β-distribution of RCS, and its comparison with other postulated distributions of RCS, in detection theory.

11

RCS OF SIMPLE SHAPES

We examine first a set of simple shapes that can be used as a basis for complex target RCS determination. The concept of representing a complex target as a collection of simple scattering centers, or components, is a valid one for all frequency regions (e.g., an aircraft in the Rayleigh region can be modeled as a pair of crossed dipoles). However, it is almost always the individual component size, not the size of the entire target, which determines the frequency region in which the component must be considered.

So we first reexamine the Rayleigh region, which was introduced in Chapter 7, as a basis for a frequency region, from the viewpoint of simple shapes. In particular, we note that elongated bodies and flattened bodies require special modifications of the volume on which their Rayleigh RCS depends. For resonance region components, we are obliged to resort to numerical methods such as the numerical electromagnetic code (Section 9-4).

For the optics region we have chosen to give RCS formulas based largely on physical optics. (Components having edges have already been treated by special diffraction methods in Section 8-2. Special components such as traveling waves, creeping waves, and cavities are discussed separately in the final sections of this chapter.) We have set forth the limitations of PO in Chapter 8. Here we emphasize them further by examining (Section 11-4) in detail the truncated elliptic cone RCS from the PO viewpoint. Then in Section 11-5 we present a carefully delimited set of RCS formulas for a collection of simple shapes, together with graphical illustrations for each. The reader should thus be able to determine exactly when and when not to use PO for the components of a complex target.

11-1 RAYLEIGH THEORY: A DIFFERENT VIEWPOINT

From earlier discussion (Section 7-1) we know that a solution for the vector potential $\mathbf{A}$ (defined by the flux density $\mathbf{B} = \nabla \times \mathbf{A}$) satisfying

$$\nabla^2 \mathbf{A} + \omega^2 \mu \epsilon \mathbf{A} = -\mu \mathbf{J} \tag{11-1}$$

is given by

$$\mathbf{A} = \frac{\mu}{4\pi} \int_V dV' \mathbf{J}(\mathbf{r}') \frac{e^{-jk\|\mathbf{r}-\mathbf{r}'\|}}{\|\mathbf{r}-\mathbf{r}'\|}, \tag{11-2}$$

with the integration extending over all nonzero source points $\mathbf{r}'$. We extend an argument† of Shen and Kong (1983) in the development below to illustrate that an electric dipole suffices as a model from which to obtain the Rayleigh RCS for a dielectric sphere but *not* for a perfectly conducting sphere.

Consider an infinitesimal antenna of length Δz, cross-sectional area ΔA and constant current I. The current density $\mathbf{J}(\mathbf{r}')$ (in amperes/m^2) times ΔA is $\mathbf{I}$. Since $\Delta V' = \Delta A \, \Delta z$,

$$\Delta V' \mathbf{J}(\mathbf{r}') = I \, \Delta z \hat{z}. \tag{11-3}$$

With the antenna at the origin, and setting $\mathbf{r}' = 0$, we get

$$\mathbf{A} = \hat{z} \frac{\mu I \, \Delta z e^{-jkr}}{4\pi r}. \tag{11-4}$$

Using the definitions (Magnus and Oberhettinger, 1954) of divergence, curl, gradient, and Laplacian in spherical coordinates, and noting that

$$\hat{z} = \hat{r} \cos\theta - \hat{\theta} \sin\theta, \tag{11-5}$$

we can evaluate $\mathbf{B}$ and then $\mathbf{E}$ (from Ampère's law):

$$\frac{1}{\mu}\mathbf{B} = \mathbf{H} = \frac{1}{\mu} \nabla \times \mathbf{A} = \hat{\phi} \frac{jkI \, \Delta z e^{-jkr}}{4\pi r} \left(1 + \frac{1}{jkr}\right) \sin\theta \tag{11-6}$$

$$\mathbf{E} = \frac{1}{j\omega\epsilon} \nabla \times \mathbf{H} = \sqrt{\frac{\mu}{\epsilon}} \frac{jkI \, \Delta z e^{-jkr}}{4\pi r} \left[\hat{r}\left(\frac{1}{jkr} + \frac{1}{(jkr)^2}\right) 2\cos\theta + \hat{\theta}\left(1 + \frac{1}{jkr} + \frac{1}{(jkr)^2}\right) \sin\theta\right]. \tag{11-7}$$

Suppose we have two charge reservoirs separated by Δz (Fig. 11-1). At $t = 0$, one reservoir is full of positive charge $+q$, the other of negative charge $-q$. At $t = T/4$, the charges flow, neutralize each other as a result of current flow I. This flow exchanges polarities at $t = T/2$. The process continues at an angular frequency $\omega = 2\pi/T$. This model represents an

† Used with permission of PWS–Kent Publishing Co., Boston.

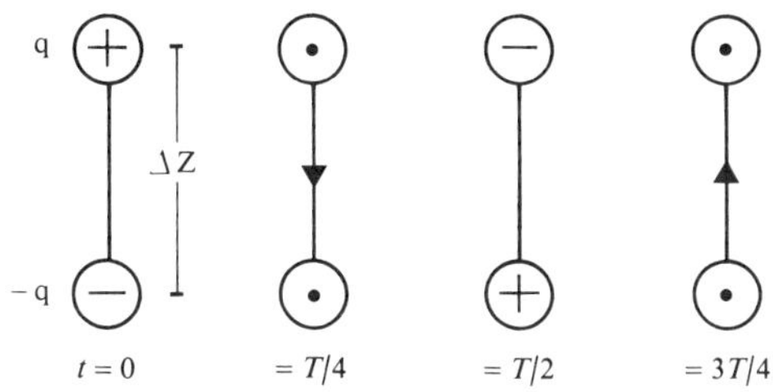

Figure 11-1 Oscillating dipole.

oscillating dipole moment $p = q\,\Delta z$. The time derivative of p is the current $I\,\Delta z$; so we have the Hertzian dipole mentioned earlier ($dp/dt = j\omega p$).

For $kr \gg 1$, the $\hat{r}$-term in the expression for $\mathbf{E}$ (Eq. 11-7) is small enough to be discarded; so we get the expressions previously encountered in Section 4-2.

Suppose we have a dielectric sphere (dielectric constant ϵ) whose radius a is very much smaller than λ. We can take as a set of solutions expressions like those we wrote above for a dipole:

$$H_\phi = \frac{A \sin\theta}{\eta r^3}[jkr + (jkr)^2]e^{-jkr}; \tag{11-8}$$

$$E_r = \frac{2A\cos\theta}{r^3}[1 + jkr]e^{-jkr}; \tag{11-9}$$

$$E_\theta = \frac{A\sin\theta}{r^3}[1 + jkr + (jkr)^2]e^{-jkr}; \tag{11-10}$$

where A is a constant to be determined. Placing the origin at the center of the sphere, we write the incident field as

$$\mathbf{E}^i = E_0(\hat{r}\cos\theta - \hat{\theta}\sin\theta)e^{-j\mathbf{k}\cdot\mathbf{r}} = E_0\hat{z}e^{-j\mathbf{k}\cdot\mathbf{r}}. \tag{11-11}$$

Then near the sphere, where $kr \ll 1$, we get (using the first terms of E_r and E_θ above)

$$\mathbf{E}^s = \frac{A}{r^3}(\hat{r}2\cos\theta + \hat{\theta}\sin\theta). \tag{11-12}$$

Inside the sphere the field is parallel to the incident field: $\mathbf{E}^i = B(\hat{r}\cos\theta - \hat{\theta}\sin\theta)$. Matching boundary conditions at the surface gives two equations:

$$-E_0 + \frac{A}{a^3} = -B$$

$$E_0 + \frac{2A}{a^3} = \frac{\epsilon}{\epsilon_0}B.$$

(Note that on the sphere $\mathbf{k}\cdot\mathbf{r}=ka\ll 1$ so that $e^{-j\mathbf{k}\cdot\mathbf{r}}\approx 1$.) Thus,

$$A=\frac{\epsilon-\epsilon_0}{\epsilon-2\epsilon_0}\,a^3E_0. \tag{11-13}$$

We may think of the incident field as inducing a dipole moment that oscillates at the same frequency as the incident wave.

Far from the sphere, where $kr\gg 1$, we keep only the second term of H_ϕ:

$$H_\phi\approx-\frac{E_0(\epsilon-\epsilon_0)k^2a^3}{\eta(\epsilon+2\epsilon_0)r}\sin\theta e^{-jkr}=\frac{1}{\eta}E_\theta. \tag{11-14}$$

So the scattered field is strongest in the forward and backward directions. The time-averaged Poynting vector for the scattered field is

$$\langle\mathbf{S}^s\rangle=\frac{E_0^2(\epsilon-\epsilon_0)^2k^4a^6}{2\eta(\epsilon+2\epsilon_0)^2r^2}\sin^2\theta\hat{r}. \tag{11-15}$$

Thus, the backscattered RCS is

$$\sigma(\theta)=\frac{4\pi r^2\langle\mathbf{S}^s\rangle\cdot\hat{r}}{E_0^2/2\eta}=4\pi\left[\frac{\epsilon-\epsilon_0}{\epsilon+2\epsilon_0}\right]^2k^4a^6\sin^2\theta; \tag{11-16}$$

in particular,

$$\sigma\left(\frac{\pi}{2}\right)=\frac{64\pi^5a^6}{\lambda^4}\left[\frac{\epsilon-\epsilon_0}{\epsilon+2\epsilon_0}\right]^2, \tag{11-17}$$

or

$$\sigma(\pi/2)/\lambda^2=\pi^5(2a/\lambda)^6[(\epsilon-\epsilon_0)/(\epsilon+2\epsilon_0)]^2. \tag{11-18}$$

This last result agrees with the Rayleigh RCS for a dielectric sphere of radius a and relative dielectric constant ϵ/ϵ_0. If we allow the imaginary part of ϵ/ϵ_0 to become large so that $(\epsilon-\epsilon_0)/(\epsilon+2\epsilon_0)\rightarrow 1$, which is the condition for a perfectly conducting sphere, then the RCS (Eq. 11-18) becomes

$$\sigma=\pi^5(2a)^6\lambda^{-4}=\frac{9k^4V^2}{4\pi}=0.716k^4V^2. \tag{11-19}$$

This result is *not* the correct Rayleigh RCS for a metallic sphere, as we can see by comparing the result of Eq. 7-31 ($\sigma=(81/16\pi)k^4V^2=1.611k^4V^2$ which is larger than the result of Eq. 11-19 by a factor of 9/4). The reason is that 11-19 *must* include, for a metallic sphere, a contribution from a magnetic dipole as well as a contribution from an electric dipole [see also Kerker, 1975, p. 377].

11-2 RAYLEIGH RCS SHAPE FACTOR

Although the Rayleigh RCS depends essentially on the volume of a target, and rather little on its shape, there are nevertheless effects on RCS caused by either an elongation or a flattening of the target. These effects are most easily seen in the axial RCS for a prolate spheroid or an oblate spheroid, respectively, represented in ellipsoidal coordinate systems. (The notation used by most mathematical texts for ellipsoidal systems is ξ, η, ϕ, which we retain here. The context prevents any possible confusion between the coordinate η and the impedance η used elsewhere.)

Consider a *prolate spheroidal* system

$$x = \frac{d}{2} \cos \phi \sqrt{(\xi^2 - 1)(1 - \eta)^2)}, \tag{11-20}$$

$$y = \frac{d}{2} \sin \phi \sqrt{(\xi^2 - 1)(1 - \eta^2)}, \tag{11-21}$$

$$z = \frac{d}{2} \xi \eta, \tag{11-22}$$

where $1 \leq \xi < \infty$, $0 \leq \eta \leq 1$. In the xy-plane the curves $\xi = \text{constant}$ and $\eta = \text{constant}$ form confocal ellipses and hyperbolas, respectively, with common interfocal length d (Fig. 11-2). If these curves are rotated about the z-axis, the result is a set of prolate spheroids and two-sheet hyperboloids, respectively. For a fixed prolate spheroid $\xi = \xi_0$ the Rayleigh RCS is, approximately,

$$\sigma = \frac{4}{\pi} k^4 V^2 F^2(x), \tag{11-23}$$

where x is the ratio minor axis: major axis. V is the spheroid volume:

$$V = \frac{\pi}{6} d^3 \xi_0 (\xi_0^2 - 1); \tag{11-24}$$

and k is the wave number. F is called the shape, actually, in this case,

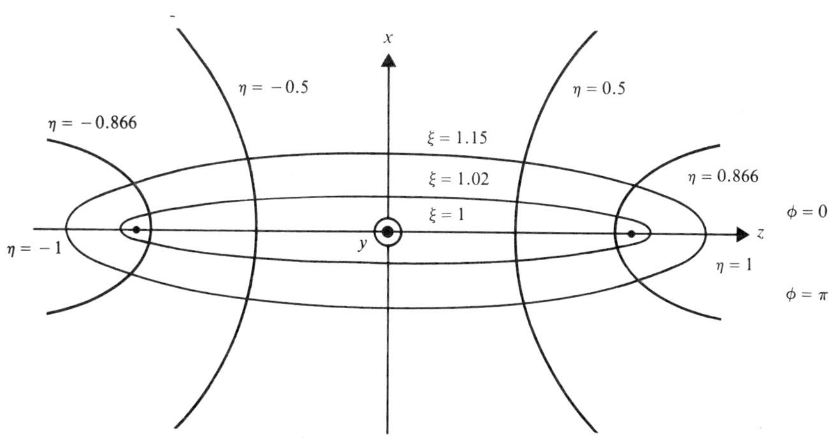

Figure 11-2 The prolate spheroidal system.

elongation, factor for the prolate spheroid:

$$F(x) = 1 + \frac{e^{-x}}{\pi x}. \tag{11-25}$$

For the *oblate spheroid,* consider the system

$$x = \frac{d}{2}\cos\phi\sqrt{(\xi^2+1)(1-\eta^2)}, \tag{11-26}$$

$$y = \frac{d}{2}\sin\phi\sqrt{(\xi^2+1)(1-\eta^2)}, \tag{11-27}$$

$$z = \frac{d}{2}\xi\eta, \tag{11-28}$$

where $0 \le \xi < \infty$, $0 \le \eta \le 1$ (Fig. 11-3). Once again, if the confocal ellipses $\xi = \text{constant}$ and hyperbolas $\eta = \text{constant}$ are rotated about the z-axis, the result is a confocal set of oblate spheroids and one-sheet hyperboloids, respectively, with common interfocal length d (Fig. 11-3). For a fixed oblate spheroid $\xi = \xi_0$ the Rayleigh RCS is, approximately,

$$\sigma = \frac{4}{\pi} k^4 V^2 F^2\left(\frac{1}{x}\right), \tag{11-29}$$

where x and k are as above and the spheroid volume V is

$$V = \frac{\pi}{6} d^3 \xi_0(\xi_0^2 + 1). \tag{11-30}$$

F is the shape—actually, in this case, flattening—factor of the oblate

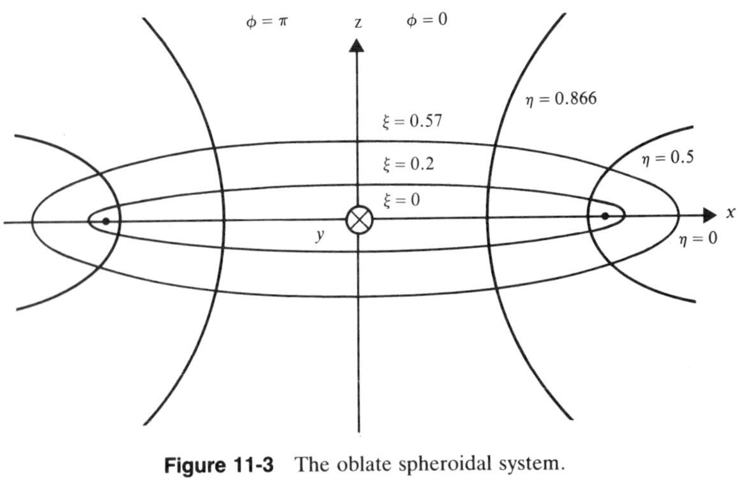

Figure 11-3 The oblate spheroidal system.

spheroid and is defined by

$$F\left(\frac{1}{x}\right) = 1 + \frac{e^{-1/x}}{\pi/x}, \qquad 0 \le x \le 1. \tag{11-31}$$

As $\xi_0 \to 0$, $F \to 1$, and $\sigma \to (1/9\pi)k^4 d^6$, or

$$\frac{\sigma}{\lambda^2} \to \frac{16\pi^3}{9}\left(\frac{2a}{\lambda}\right)^6, \tag{11-32}$$

where $a = d/2$ is the radius of the disk into which the oblate spheroid $\xi = \xi_0$ is flattened. Equation 11-32 is the normal incidence RCS of a disk of radius a.

The shape factors for both the prolate and oblate spheroids are shown in Fig. 11-4. Notice for the former that the shape factor increases without bound as the prolate spheroid becomes increasingly thin. As $x \to 1$, both spheroids approach a sphere, for which Eq. 11-23 and/or Eq. 11-29 yield

$$\sigma = \frac{4}{\pi} k^4 V^2 \left[1 + \frac{1}{e\pi}\right]^2 \approx 1.6 k^4 V^2, \tag{11-33}$$

which compares well with the Rayleigh RCS given by Eq. 7-31 for the sphere.

Siegel (1958, 1963) gives a heuristic argument for extending applicability of the shape factor (11-25) to the Rayleigh RCS of other long, thin shapes.

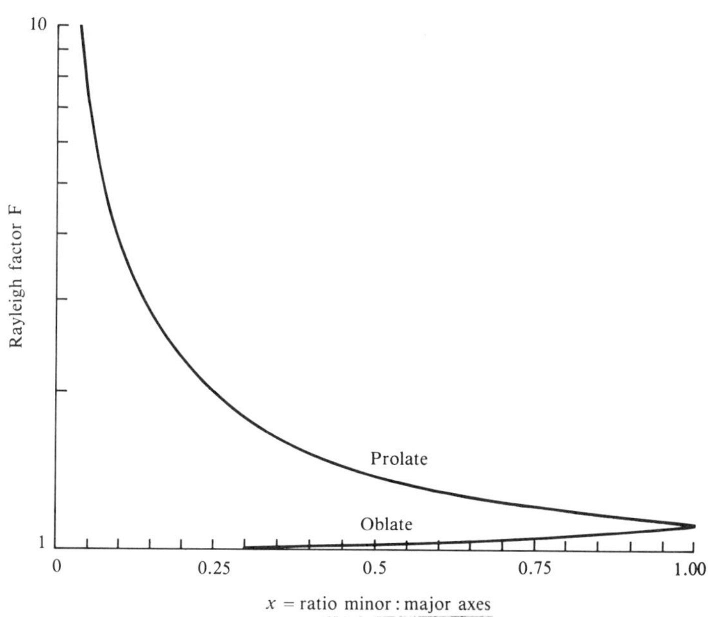

Figure 11-4 Rayleigh shape factors.

For example, the on-axis Rayleigh RCS of a right circular cone is given by Eq. 11-23 with $F(x)$ from Eq. 11-25, x being defined as

$$x = \frac{h}{4a}, \tag{11-34}$$

where h is the cone altitude and a its base radius.

Additional shape factor formulas have been worked out for $\sigma(0)$ by Crispin and Maffett (1965a, p. 836). The shapes are: lens, elliptic ogive, spindle (paraboloidal ogive), finite cylinder, and cone-spheroid.

Figure 11-5 shows how well this theory agrees with measured data as a function of cone base radius a for a variety of cone angles. The agreement extends to larger values of the radius a as the cone becomes "fatter" (i.e., as the cone angle increases). This result is not surprising; for the shape factor simply plays a smaller role as the cone "fattens."

Siegel (1958, 1963) characterized a Rayleigh scatterer as a "discontinuity" seen by the incident field. If this viewpoint is valid, then it should be immaterial whether finite cones are viewed nose-on or base-on. Figure 11-6 shows, for four different cone angles, the maximum size of cone radius a for which this concept holds. Note that this maximum seems rather independent of cone angle.

The on-axis Rayleigh RCS results can be extended to other angles as well. Let θ denote angle from nose-on. Then the results below are available (Crispin and Maffett, 1965a), encompassing both horizontal and vertical polarizations. Horizontal polarization indicates that the E-vector is in the plane of the angle θ; since we deal here only with bodies having axial symmetry, θ can be measured in any plane containing that axis.

The factor F becomes a function of both θ and polarization.

$$F_H = \frac{1}{2}\left[\frac{1}{2-E} + \frac{\cos^2\theta}{E} + \frac{\sin^2\theta}{2-2E}\right] \tag{11-35}$$

$$F_V = \frac{1}{2}\left[\frac{1}{E} + \frac{\cos^2\theta}{2-E} + \frac{\sin^2\theta}{2E}\right]. \tag{11-36}$$

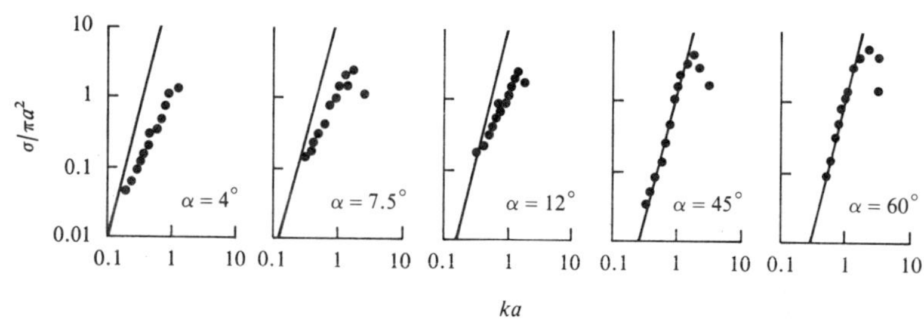

Figure 11-5 Nose-on RCS for finite cones in the Rayleigh region. α = half-cone angle; $2a$ = base diameter. Solid curve: Rayleigh theory; ··: measured.

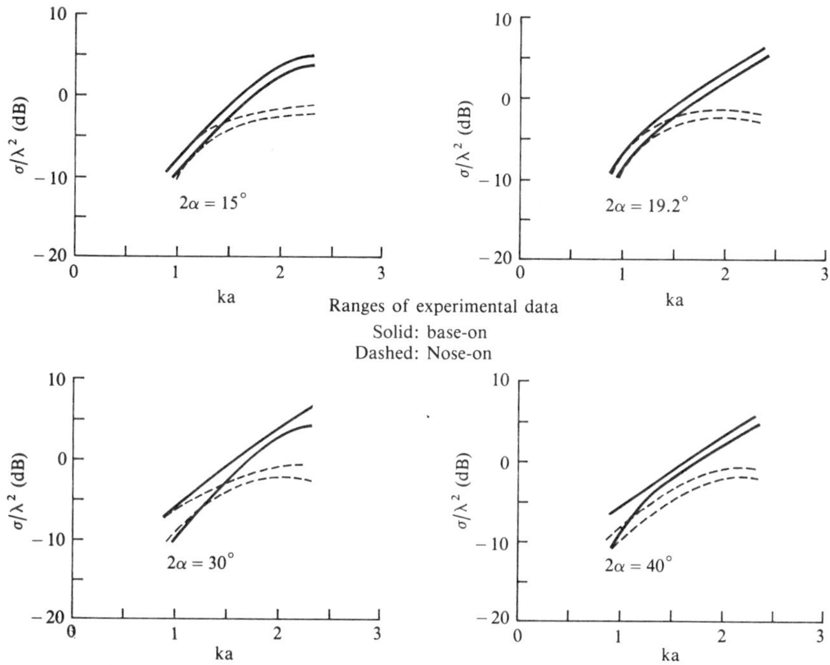

Figure 11-6 Nose-on and base-on RCS for finite cones in the Rayleigh region.

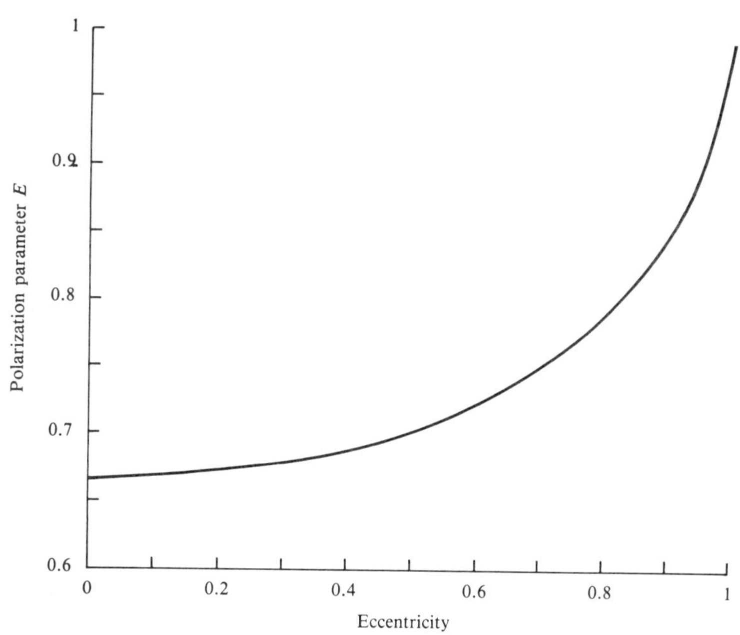

Figure 11-7 Polarization parameter E versus eccentricity for Rayleigh RCS of a prolate spheroid.

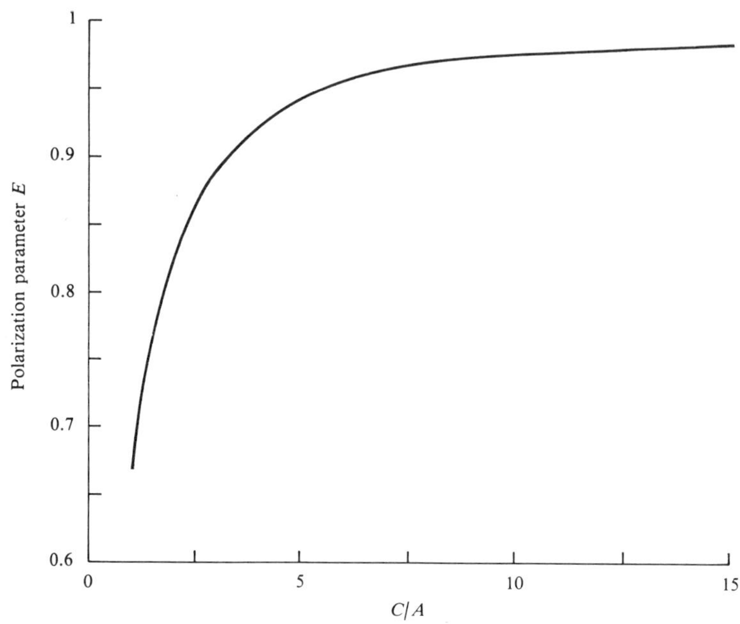

Figure 11-8 Polarization parameter E versus C/A for Rayleigh RCS of a prolate spheroid.

For the spheroids, E takes the following forms:

1. For the prolate spheroid, having axes a, a, $c = a/\sqrt{1-e^2}$, e being eccentricity, we have

$$E = e^{-2} - \frac{1}{2} e^{-3}(1 - e^2) \ln \frac{1+e}{1-e}. \tag{11-37}$$

   Figures 11-7 and 11-8 show E as a function of eccentricity e and as a function of the ratio c/a, respectively.
2. For the sphere, $E = 2/3$.
3. For the oblate spheroid, having axes a, a, $a\sqrt{1-e^2}$, we have

$$E = e^{-3}\sqrt{1-e^2}\,(\arcsin e) - e^{-2}(1 - e^2). \tag{11-38}$$

11-3 ASYMPTOTIC EVALUATION OF SCATTERING INTEGRALS

The method of physical optics (PO) might very well be called the workhorse of high frequency approximation methods. It can be applied easily, and with good results, to many simple shapes. The PO approximation was introduced and defined in Chapter 8 but was not applied to specific examples. We discuss first one of the most common methods of evaluating the PO integral. Then we give some applications, illustrating, in particular, the physical interpretations which can be assigned to the PO integral.

Quite often the PO integral must be approximated with the help of asymptotic methods (steepest descent or stationary phase) whose application we have already encountered in Chapter 6 in connection with scattering by a perfectly absorbing half-plane. Since these asymptotic methods (asymptotic because their appropriateness increases as λ decreases) are so useful in the evaluation of scattering and diffraction integrals, we detour for a brief examination of them.

11-3-1 Evaluation of Integrals by Asymptotic Methods

Consider

$$F(z) = \int e^{zf(w)}\chi(w)\,dw, \tag{11-39}$$

with $f(w)$ and $\chi(w)$ both analytic, along some contour such that the integral converges for $Re(z) > 0$. We wish to investigate F for complex z of large modulus. The path of integration must be deformed so that

1. It goes through a zero w_0 of $f'(w)$; and
2. $Im[f(w)]$ is constant on that path.

11-3-2 Geometric Picture

Let $w = u + jv$, $p = Re[f(w)]$. In the space represented by the variables u, v, p, the height of the surface $p = Re[f(w)]$ above the uv-plane is never a maximum or a minimum, but can be a saddle point (maximum modulus theorem; see Churchill and Brown, 1984; p. 117).

Map this surface by contour lines into the uv-plane. The curves of this map on which $q = Im[f(w)] = \text{constant}$ are orthogonal to the contour lines and as such are therefore projections of paths of *steepest descent* (or ascent). Modify the integration path so as to go along a steepest descent path, starting at the saddle point w_0. On this path $f(w) = f(w_0) - \tau$, where τ is real, positive, and increases indefinitely as the path is traversed. Thus

$$F(z) = e^{zf(w_0)} \int_0^{\infty} e^{-z\tau}\chi(w)\frac{dw}{d\tau}\,d\tau, \tag{11-40}$$

when z is of large modulus. To obtain an asymptotic evaluation of the integral we may apply a lemma of Watson (Whittaker and Watson, 1958, p. 151), which permits the integrand to be expanded in a series of which only the first term need be retained. On this basis we can write a saddle point formula (with one saddle point z_0 between A and B) as:

$$F(z) = \frac{\chi(w_0)\exp[zf(w_0)]\sqrt{2\pi}\exp(j\alpha)}{\sqrt{|zf''(w_0)|}} \tag{11-41}$$

where α is chosen so that ρ is positive between w_0 and B in $w - w_0 = \rho e^{j\alpha}$.

11-3-3 Stationary Phase

We may now specialize the Steepest Descent method by taking the path of integration along the real axis and by taking $Re[f(w)] = 0$;

$$F(t) = \int_A^B \chi(x) e^{jt\psi(x)}\, dx = \int_{x=A}^{B} \frac{\chi(x)}{\psi'(x)} e^{jt\theta}\, d\theta, \tag{11-42}$$

where we have set $\psi(x) = \theta$. The contributions to the integral of Eq. 11-42 occur where $\psi'(x_0) = 0$ and consequently

$$F(t) \approx \sqrt{\frac{2\pi}{t\,|\psi''(x_0)|}}\, \chi(x_0) \exp[jt\psi(x_0) + \frac{j\pi}{4} \operatorname{sgn} \psi''(x_0)] + 0\left(\frac{1}{t}\right), \tag{11-43}$$

where the appearance of the sign function is a result of the α choice above. The idea can be extended to two dimensions, in which case we have

$$\begin{aligned} F(K) &= \int_c^d \int_a^b f(x, y) e^{jKh(x,y)}\, dx\, dy \\ &\approx \frac{\pm 2\pi j \exp[jKh(x_0, y_0]f(x_0, y_0)}{K\sqrt{r_0 t_0 - s_0^2}} + O(K^{-3/2}) \end{aligned} \tag{11-44}$$

as $K \to \infty$. We assume $f(x, y)$ to be separable and analytic and that there exists only one stationary phase point in the integration region, say (x_0, y_0), such that

$$p_0 = q_0 = 0, \qquad r_0 t_0 - s_0 \neq 0. \tag{11-45}$$

In Eq. 11-45 we use

$$p = h_x, \qquad q = h_y, \qquad r = h_{xx}, \qquad s = h_{xy}, \qquad t = h_{yy}, \tag{11-46}$$

where the subscript denotes differentiation with respect to that variable. If $h(x, y) = X(x)Y(y)$, then

$$\frac{X(x)}{g_x(x, y)} \quad \text{and} \quad \frac{Y(y)}{g_y(x, y)}$$

are of bounded variation in the integration region (Siegel et al., 1955); that is to say, they are well enough behaved to carry out the asymptotic approximation to the integral.

11-3-4 Application: Recovery of GO for a Smooth Convex Surface

The PO monostatic RCS of a smooth surface is (Chapter 8)

$$\sigma = \frac{k^2}{\pi} \|(\mathbf{r} \cdot \mathbf{f})\hat{a}\|^2, \tag{11-47}$$

where

$$\mathbf{f} = \int \exp[jk\mathbf{r}' \cdot (2\hat{k})]\hat{n}' \, dS, \tag{11-48}$$

the integration being over the illuminated portion of the surface S. Choose the coordinate system so that the vector area element is projected on the incidence plane of constant phase. Assume that, in this system, S can be represented as

$$S: z = h(x, y), \tag{11-49}$$

where h has well-behaved second derivatives. Then $\mathbf{r} \cdot \mathbf{f}$ can be written

$$I = \int e^{2jkh(x,y)} \, dx \, dy. \tag{11-50}$$

Now use the stationary phase method on this integral to get

$$I \approx \frac{2\pi j \exp(2jkh(x_0, y_0))}{2k\sqrt{r_0 t_0 - s_0^2}} \tag{11-51}$$

whence we have

$$\sigma = \frac{k^2}{\pi} \frac{\pi^2}{k^2} \left| \frac{\exp(2jkh(x_0, y_0))}{\sqrt{r_0 t_0 - s_0^2}} \right|^2 = \pi R_1 R_2, \tag{11-52}$$

where the principal radii of curvature R_1 and R_2 at the stationary phase point (x_0, y_0) are such that

$$R_1 R_2 = \frac{(1 + p_0^2 + q_0^2)^2}{r_0 t_0 - s_0^2}. \tag{11-53}$$

At the stationary phase point (x_0, y_0) we have, of course,

$$p_0 = q_0 = 0$$

and so

$$R_1 R_2 = \frac{1}{r_0 t_0 - s_0^2}. \tag{11-54}$$

11-4 INTERPRETATION OF PO INTEGRAL—TRUNCATED ELLIPTIC CONE

We now apply a stationary phase evaluation to the PO integral for a truncated elliptic cone. We recall that the PO RCS is given by (Eq. 8-10)

$$\sigma = \frac{4\pi}{\lambda^2} \|(\hat{r} \cdot \hat{h})\mathbf{f} - (\hat{r} \cdot \mathbf{f})\hat{h}\|^2,$$

where

$$\mathbf{f} = \int \exp[jk\mathbf{r}' \cdot (\hat{r} - \hat{k})]\hat{n}'\, dS,$$

the integration being over the illuminated portion of the target.

For the monostatic case $\hat{r} \cdot \hat{h} = 0$; so if the vector area element is projected on the incidence plane of constant phase, the resulting scalar form for $\mathbf{f}$ can be written $f = \int e^{-2jk\rho}\, dA$. We develop the integral expression f for the truncated elliptic cone, using the following facts:

Cone Equation

$$x = z \tan\alpha \cos\phi_1,$$

$$y = \frac{z}{\eta} \tan\alpha \sin\phi_1,$$

$$z = z.$$

Here η is the ratio of major axis to minor axis of the elliptic section of the cone and α is the cone half-angle.

Surface Area Element

$$\begin{aligned}\mathbf{dS} &= [(-\hat{x}\sin\phi_1 + \hat{y}\eta^{-1}\cos\phi_1)z\tan\alpha\, d\phi_1] \\ &\quad \times [(\hat{x}\tan\alpha\cos\phi_1 + \hat{y}\eta^{-1}\tan\alpha\sin\phi_1 + \hat{z})\, dz], \\ &= \frac{z\tan\alpha}{\eta}(\hat{x}\cos\phi_1 + \hat{y}\eta\sin\phi_1 - \hat{z}\tan\alpha)\, d\phi_1\, dz.\end{aligned}$$

Here the operator "×" denotes cross-product.

Projection dA of **dS** Along the Incidence Direction

$$\begin{aligned}dA &= \frac{z\tan\alpha}{\eta} d\phi_1\, dz(\hat{x}\cos\phi_1 + \hat{y}\eta\sin\phi_1 - \hat{z}\tan\alpha) \\ &\quad \cdot (\hat{x}\sin\theta\cos\phi + \hat{y}\sin\theta\sin\phi + \hat{z}\cos\theta), \\ &= \frac{z\tan\alpha}{\eta} d\phi_1\, dz(\sin\theta\cos\phi\cos\phi_1 \\ &\quad + \sin\theta\sin\phi\sin\phi_1 - \cos\theta\tan\alpha).\end{aligned}$$

Phase Factor On the Cone Surface

$$\begin{aligned}e^{-2jk\rho} &= \exp\Big[2jk(\hat{x}\sin\theta\cos\phi + \hat{y}\sin\theta\sin\phi + \hat{z}\cos\theta) \cdot (x\hat{x} + y\hat{y} + z\hat{z})\Big] \\ &= \exp\Big[2jkz(\sin\theta\cos\phi\tan\alpha\cos\phi_1 + \frac{1}{\eta}\sin\theta\sin\phi\tan\alpha\sin\phi_1 + \cos\theta)\Big].\end{aligned}$$

The PO integral f has a stationary phase point defined by (the second stationary phase point is on the shadowed side of the cone)

$$\frac{\partial}{\partial\phi_1}\left(\sin\theta\tan\alpha\cos\phi_1+\frac{1}{\eta}\sin\theta\sin\phi\tan\alpha\sin\phi_1+\cos\theta\right)$$
$$=\sin\theta\tan\alpha\left(-\cos\phi\sin\phi_1+\frac{1}{\eta}\sin\phi\cos\phi_1\right)=0,$$

whence

$$\sin\phi_1=\frac{\sin\phi}{\sqrt{\sin^2\phi+\eta^2\cos^2\phi}};\qquad \cos\phi_1=\frac{\eta\cos\phi}{\sqrt{\sin^2\phi+\eta^2\cos^2\phi}}.$$

Evaluation of the ϕ_1-integration by the stationary phase method gives (setting $G=\sqrt{\sin^2\phi+\eta^2\cos^2\phi}$):

$$f=\int\sqrt{\frac{\pi\eta z\tan\alpha}{kG\sin\theta}}\left(\frac{\sin\theta}{G}-\frac{1}{\eta}\cos\theta\tan\alpha\right)$$
$$\times\exp\left[2jkz\left(\frac{G}{\eta}\sin\theta\tan\alpha+\cos\theta\right)+\frac{\pi j}{4}\right]dz.$$

For $(G/\eta)\sin\theta\tan\alpha+\cos\theta\neq 0$ (normal incidence), the z-integration may be carried out by parts, neglecting the new integral with regard to the constant terms; in this way two contributions come from the curves $z=L_1$

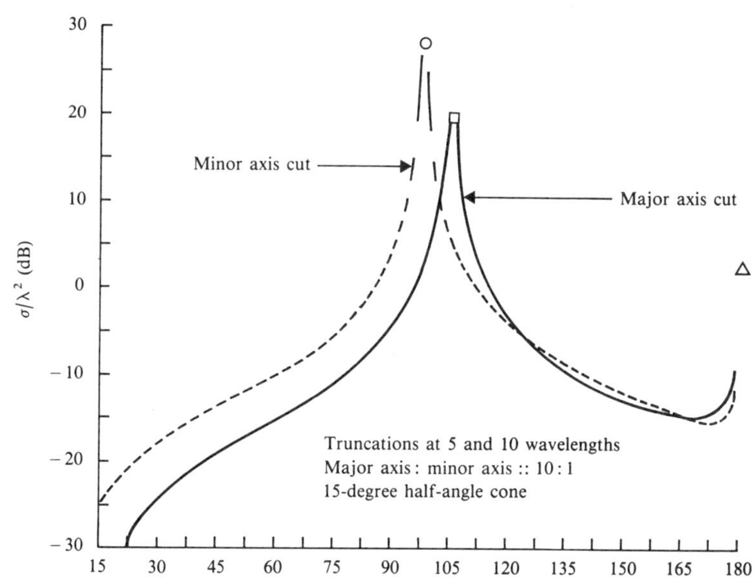

Figure 11-9 Truncated elliptic cone RCS.

and $z = L_2$:

$$f = \sqrt{\frac{\pi\eta z \tan\alpha}{kG\sin\theta}}\left(\frac{\sin\theta}{G} - \frac{1}{\eta}\cos\theta\tan\alpha\right)\frac{\exp[2jkzQ + (\pi j/4)]}{2jkQ}\Bigg|_{L_1}^{L_2},$$

where $Q = (G/\eta)\sin\theta\tan\alpha + \cos\theta$; $2jkQ$ is actually equal to $\hat{k}\cdot\mathbf{r}$, where $\mathbf{r}$ is the vector along a generator to the integration point. So when the incidence direction $\hat{k}$ is perpendicular to $\mathbf{r}$, Q is 0. ($\phi = 0$ implies $\theta = 90 + \alpha$; $\phi = 90$ implies $\tan\theta = -\eta\tan\alpha$.)

Figure 11-9 shows two plots of RCS, normalized to λ^2, versus aspect, in a plane containing the cone axis. One aspect cut contains the major axis, the other cut contains the minor axis. The aspect range begins at the half-cone angle and continues to 180°, it being assumed the cone has a solid flat back.

11-5 MISCELLANEOUS SIMPLE SHAPES

In this section we give formulas and graphs to illustrate the approximate RCS behavior of an assortment of simple shapes:

cavity;
ellipsoid (encompassing prolate and oblate spheroids and sphere);
elliptic cylinder;
flat plates (rectangular and circular);
dihedral (90°);
corners (triangular, square, and circular);
torus;
and paraboloid.

All RCS curves, excepting that for the cavity, are based on PO theory. The cavity RCS curve shown is based on an empirical study; but a more careful approach to that not-so-simple shape is suggested in the following section. The wedge, and edges in general, are treated in Section 8-2.

For the simple shapes chosen we give *approximate RCS formulas based on physical optics*. These formulas are approximate in the sense that the PO integrals on which they are based are evaluated by asymptotic methods, usually stationary phase (cf. Section 11-4). These formulas, together with the ones appearing elsewhere for the truncated cone, wedge, traveling and creeping waves, and so on, are used later (Chapter 13) in the formulation of a computer code for calculating the RCS statistics of a complex target. Therefore, the format we use to present the PO RCS formulas is one that can be easily adapted to usage in that code. The phrase "RCS formula"

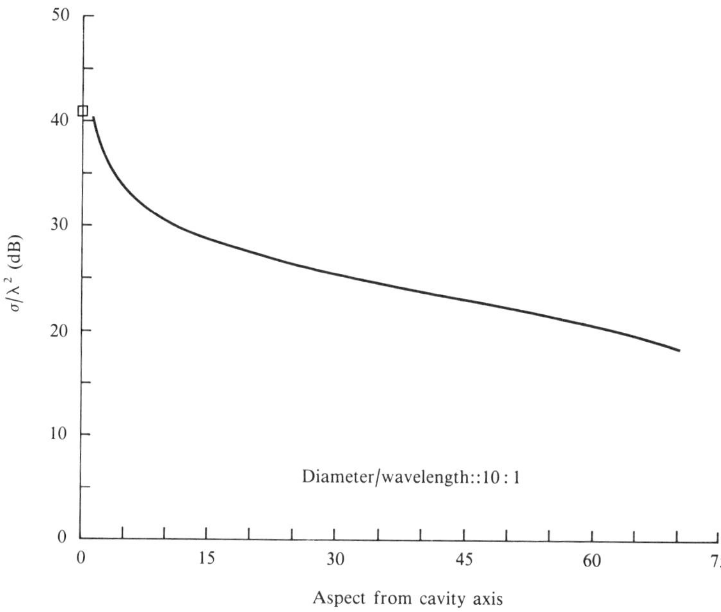

Figure 11-10 Circular cavity RCS (empirical).

always refers to RCS for the aspect θ', ϕ', where the prime refers to the generic axes for that particular scatterer. (We reserve unprimed notation for the coordinate system in which to describe a complex target which we consider in Chapter 12.)

Cavity. The RCS formulas here are empirical and are applicable to a cavity with (a) circular aperture of diameter d or (b) an arbitrary aperture of area A.

Inputs: d or A.

RCS Formulas: (Fig. 11-10)

(a)
$$\begin{cases} 0.05(kd)^3\lambda^2, & \theta' = 0; \\ 0.05(2\pi d)^2 \cot\theta', & 0 < \theta' < \approx 70°. \end{cases}$$

(b)
$$\begin{cases} 0.4(4\pi A/\lambda^2)^{3/2}\lambda^2, & \theta' = 0, \quad A \geq 2\lambda^2, \\ 4\pi A^2/\lambda^2, & \theta' = 0, \quad A < 2\lambda^2; \\ 0.8\pi A \cot\theta', & 0 < \theta' < \approx 70°. \end{cases}$$

Remarks: The off-axis RCS, in either case, must never exceed the on-axis RCS.

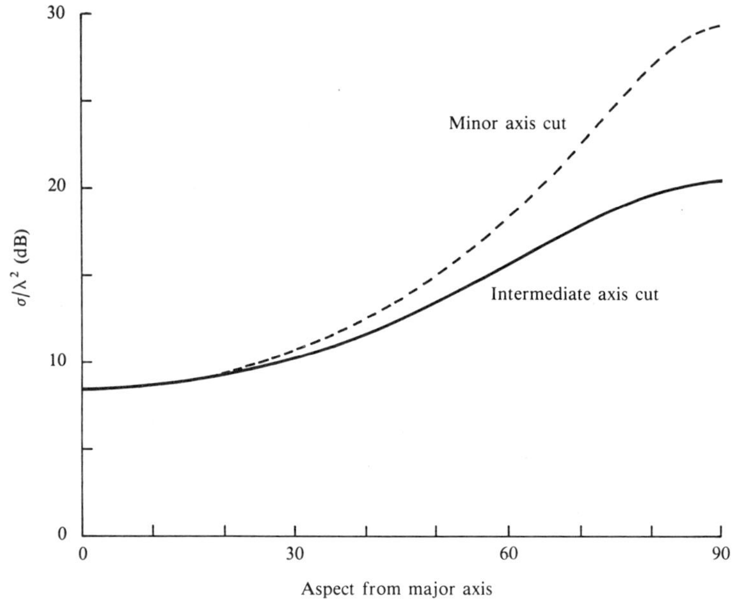

Figure 11-11 Ellipsoid RCS. Major axis: intermediate axis: minor axis:: 10: 5: 3λ.

Ellipsoid

Inputs: a, b, c (semi-axes along x', y', z');
RCS formula:

$$\frac{\pi(abc)^2}{(a^2 \sin^2 \theta' \cos^2 \phi' + b^2 \sin^2 \theta' \sin^2 \phi' + c^2 \cos^2 \theta')^2},$$

all θ, ϕ. (Figure 11-11).
Prolate spheroid: $a = b < c$;
Oblate spheroid: $a = b > c$;
Sphere: $a = b = c$.

Elliptic Cylinder

Inputs: a, b (semi-major and minor axes), L, length;
RCS formulas:

$$\frac{2\pi(Lab)^2}{\lambda(a^2 \cos^2 \phi' + b^2 \sin^2 \phi')^{3/2}}, \qquad \theta = \pi/2;$$

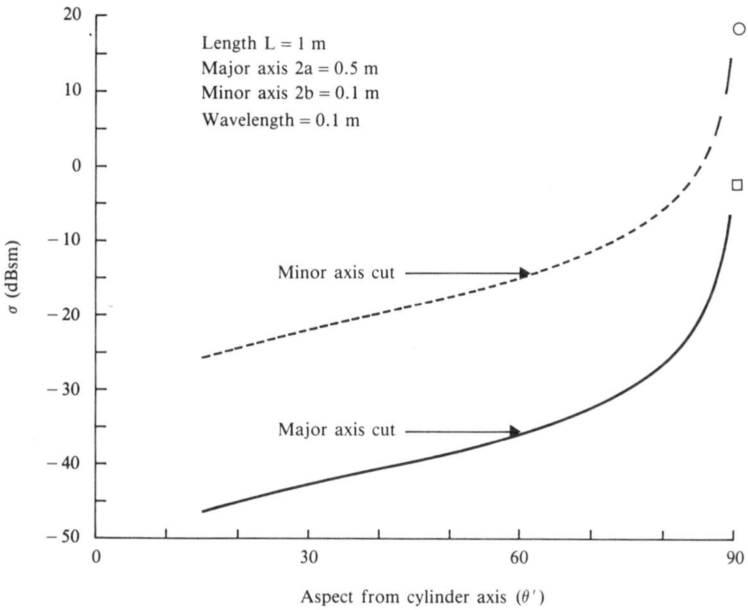

Figure 11-12 Elliptic cylinder RCS.

for $\theta \neq \pi/2$, there are two components, each given by

$$\frac{\lambda(ab)^2 \sin\theta'}{8\pi \cos^2\theta'[a^2\cos^2\phi' + b^2\sin^2\phi']^{3/2}}.$$

(Figure 11-12.)

Circular cylinder: $b = a$.

Remarks: Off-normal RCS must *never* exceed the normal incidence RCS. To achieve a smooth curve, logarithmic fairing may be used.

Ogive. This three-dimensional body is defined as the minor arc of a circle of radius R_1 rotated about its chord (resulting in a football-shaped solid with pointed ends). The maximum distance a from chord to arc is the ogive radius. Refer to Section 11-7 for remarks concerning the combination of physical optics and traveling wave theories for the ogive; see also Fig. 11-26.

Inputs: R_1, a. Determine additional geometrical parameters α and L; L is ogive length, α is ogive half-angle and subtends half the chord which forms the ogive axis:

$$\cos\alpha = 1 - (a/R_1), \qquad 0° < \alpha < 90°;$$
$$L = 2[R_1^2 - (R_1 - a)^2]^{1/2}.$$

Note that

$$R_1 = \frac{a}{1 - \cos\alpha}.$$

In a wz-system with $\hat{z}$ along the ogive axis, the ogive has equation

$$(w+h)^2 + z^2 - R_1^2 = 0,$$

$$|z| \le \sqrt{R_1^2 - h^2} = L/2,\ h \le w + h \le R_1.$$

RCS Formulas:

$$\sigma(\theta') = \frac{\lambda^2 \tan^4\alpha}{16\pi \cos^6\theta'(1 - \tan^2\alpha \tan^2\theta')^3}, \quad 0° \le \theta' < 90° - \alpha;$$

$$\sigma(90° - \alpha) = (R_1^2 \sin^2\alpha)/4\pi, \qquad \theta' = 90° - \alpha;$$

$$\sigma(\theta') = \pi R_1^2[1 - (R_1 - a)/R_1 \sin\theta'], \qquad 90° - \alpha < \theta' \le 90°.$$

Comment: Results for the first equation must be bounded by those for the second and third equations.

Flat Plates. We consider two types:

Rectangular: $2a \times 2b$ along the x' and y' axes, respectively, with normal along the z' axis.

Inputs: a, b.

RCS Formulas: $64\pi a^2b^2/\lambda^2$, $\theta' = 0$; (Fig. 11-13).

In $x'z'$-plane ($\phi' = 0$ or π), there are two returns, each $b^2\cos^2\theta'/4\pi\sin^2\theta'$, $0 < \theta' < \pi/2$;

In $y'z'$-plane ($\phi' = \pm\pi/2$), there are two returns, each $a^2\cos^2\theta'/4\pi\sin^2\theta'$, $0 < \theta' < \pi/2$;

For $\phi' \ne 0, \pm\pi/2, \pi$, there are four returns, each $\lambda^2\cos^2\theta'/64\pi^3\sin^4\theta'\sin^2\phi'\cos^2\phi'$, $0 < \theta < \pi/2$.

Remarks: None of the last three formulas may exceed the *first.* To achieve smoothness we may adopt fairing as suggested in the preceding. If a more careful evaluation of the PO integral is desired, the following formula is appropriate:

$$\frac{64\pi a^2b^2}{\lambda^2}\left[\frac{\sin(kb\sin\theta'\sin\phi')}{kb\sin\theta'\sin\phi'}\right]^2\left[\frac{\sin(ka\sin\theta'\cos\phi')}{ka\sin\theta'\cos\phi'}\right]^2$$

Circular: With radius a and normal along the z'-axis.

Inputs: a

RCS Formulas: $\dfrac{4\pi}{\lambda^2}(\pi a^2)^2$, $\theta' = 0$ (Fig. 11-14);

Off-normal, there are two returns, each

$$\frac{a\lambda}{8\pi\sin\theta'\tan^2\theta'}, \qquad 0 < \theta' < \pi/2.$$

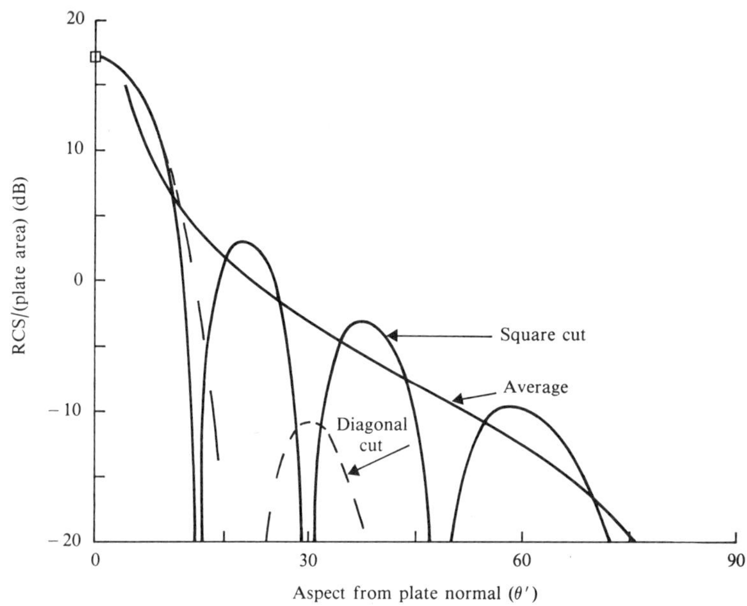

Figure 11-13 Square flat-plate RCS (edge = 2 wavelengths).

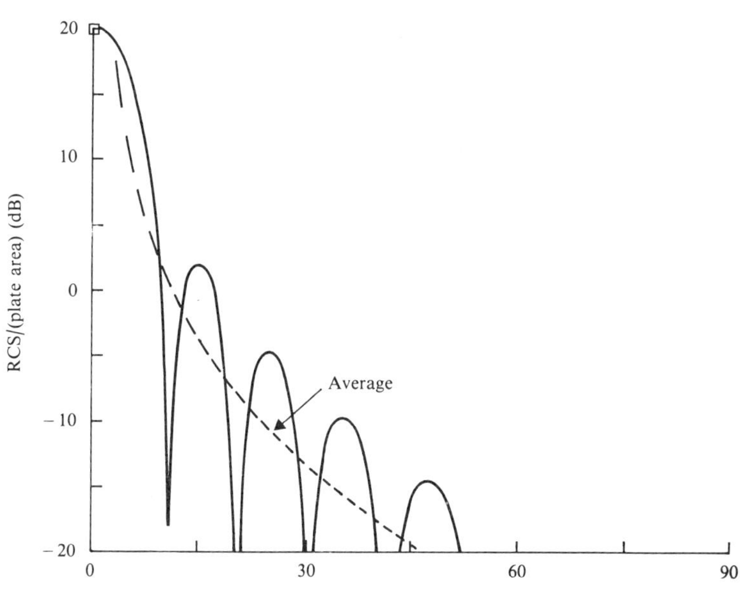

Figure 11-14 Circular flat-plate RCS (disk circumference/wavelength:: 10: 1).

Remarks: Off-normal RCS must never exceed normal RCS; fair if necessary.

90-degree Dihedral. Edges a and c oriented along x'- and y'-axes; edge b along z'-axis.

Inputs: a, b, c.

Compute: Auxiliary parameters γ and δu, where:

$$\gamma = \arctan(c/a); \qquad \delta u = \frac{\pi}{2} - \theta' \text{ (in radians, <1).}$$

RCS Formulas: (Fig. 11-15).

$$\text{(a) } 4\pi\lambda^{-2}(ab)^2(2\sin\phi')^2, \qquad \theta' = \pi/2, \qquad 0 < \phi' \le \gamma,$$

$$\text{(b) } 4\pi\lambda^{-2}(cb)^2(2\cos\phi')^2, \qquad \theta' = \pi/2, \qquad \gamma \le \phi' < \pi/2.$$

$$\frac{\lambda^2 \cdot \text{(a) or (b)}}{8\pi^2 b^2(\delta u)^2}, \qquad \theta' \ne \pi/2, \qquad \delta u < 1,$$

$$0, \ \theta' \ne \pi/2, \qquad \delta u \ge 1.$$

Corners. Of three kinds: triangular, rectangular, and elliptical. For the aspect θ', ϕ', $0 < \theta' < \pi/2$, $0 < \phi' < \pi/2$, compute the direction cosines

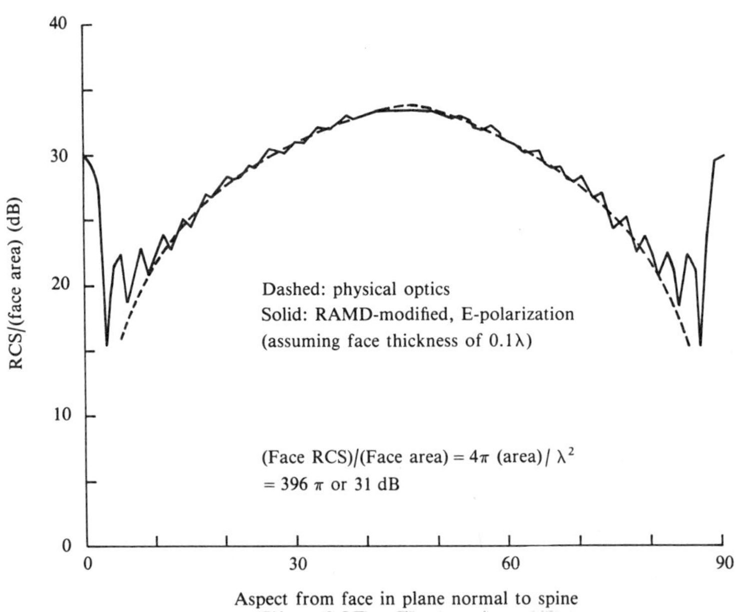

Figure 11-15 RCS for 90° dihedral with square faces $10\lambda \times 10\lambda$.

that the incidence direction $\hat{k}$ makes with the x', y', z' axes:

$$x'\text{: } \sin\theta' \cos\phi,$$
$$y'\text{: } \sin\theta' \sin\phi',$$
$$z'\text{: } \cos\theta';$$

rearrange these in order of *magnitude* from the smallest to the largest. Name the smallest magnitude h, the next largest magnitude m, and the largest magnitude n; that is,

$$h \le m \le n.$$

For the edges of the corner choose designations a, b, and c, in such a way that $a \ge b \ge c$. Set $p = h/a$, $q = m/b$, $r = n/c$. We then have $p \le q \le r$. (*This requirement is a very important one and must be treated with care.*) In all cases the (PO) formula for RCS is $4\pi A^2/\lambda^2$, where aperture area A is given for each type of corner in terms of h, m, n, and the dimensional parameters a, b, and c of the particular corner.

General Triangular. a, b, c are edge lengths (Fig. 11-16).

Inputs: p, q, r.

$$A = \begin{cases} \dfrac{4hmc}{p+q+r}, & p+q \le r \\ abc\left[p+q+r-2\dfrac{p^2+q^2+r^2}{p+q+r}\right], & p+q \ge r. \end{cases}$$

Standard Triangular: $a = b = c$ (Fig. 11-16).

Example 1.† An example is instructive. Suppose we desire to shift the pattern peak of a triangular corner up or down with respect to the plane formed by the edges a and b. We can accomplish this goal by increasing or decreasing, respectively, the length of c. In so doing we also increase or decrease, respectively, the RCS relative to that of a triangular corner with equal edges. Figure 11-16b shows these results; note that we also express edge c as a multiple of edge a (= edge b).

General Rectangular. a, b, c are edge lengths.

Inputs: p, q, r.

$$A = \begin{cases} hb[4c - (n/q)], & q \ge r/2, \\ 4hmc^2/n, & q \le r/2. \end{cases}$$

Standard Rectangular: $a = b = c$ (Fig. 11-17).

† Elmer L. Johansen very kindly suggested this example and furnished the data for Fig. 11-16b.

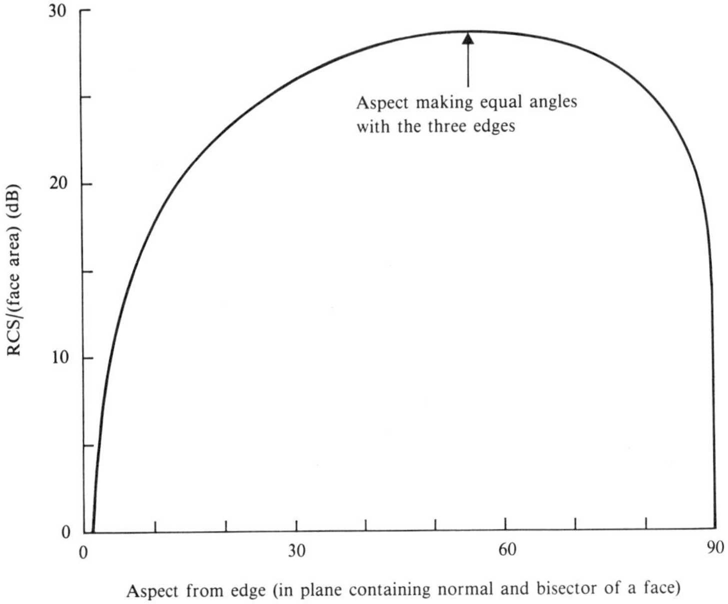

Figure 11-16a RCS of a triangular corner with equal (10λ) edgelengths.

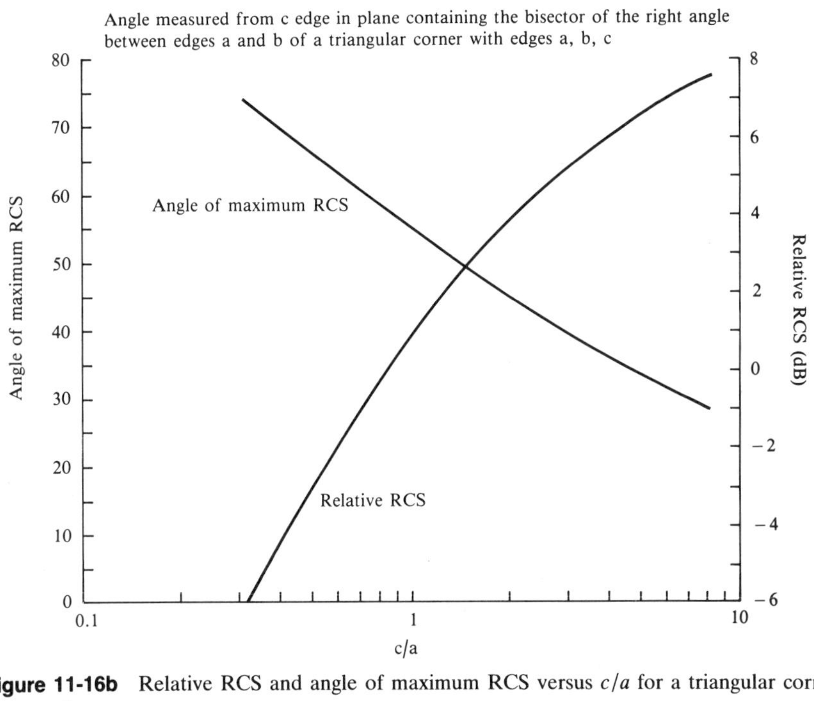

Figure 11-16b Relative RCS and angle of maximum RCS versus c/a for a triangular corner with $a = b$.

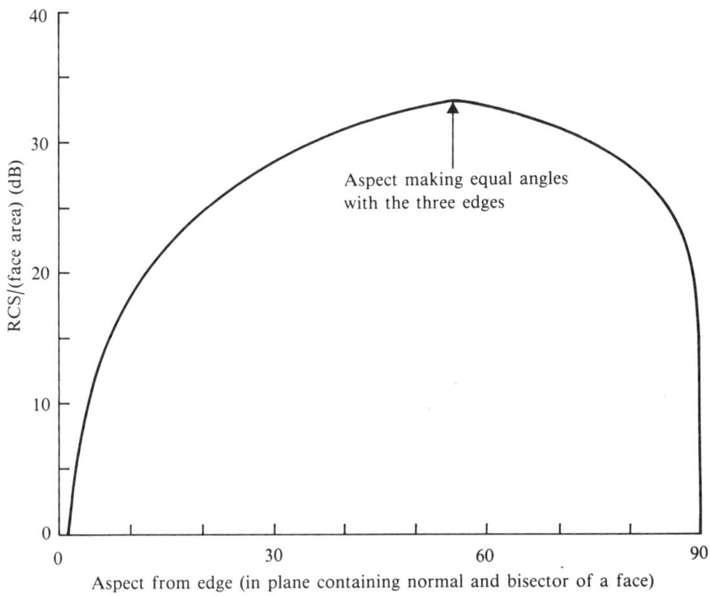

Figure 11-17 RCS of a square corner with equal (10λ) edgelengths.

Elliptical. a, b, c are semi-axis lengths.

Inputs: p, q, r.

$$\text{Set } H = p^2, \qquad M = q^2, \qquad N = r^2; \qquad H \le M \le N.$$

$$A = \begin{cases} abc\left[\sqrt{N}\tan^{-1}\dfrac{(H+M)^2 - N^2}{4N\sqrt{HM}} + \sqrt{M}\tan^{-1}\dfrac{(H+N)^2 - M^2}{4M\sqrt{HN}} \right. \\ \left. + \sqrt{H}\tan^{-1}\dfrac{(M+N)^2 - H^2}{4H\sqrt{MN}}\right], \qquad N \le H + M; \\ abc\left[\sqrt{M}\tan^{-1}\dfrac{2\sqrt{HN}}{M+N-H} + \sqrt{H}\tan^{-1}\dfrac{2\sqrt{MN}}{H-M+N}\right], \\ N \ge H + M. \end{cases}$$

Circular. $a = b = c$ (Fig. 11-18).

Torus. The z'-axis is normal to the plane of the torus. $a + b$ is the larger radius of the torus; $a - b$, the smaller.

Inputs: a, b.
RCS Formulas: $8\pi^3 ba^2/\lambda$, $\theta' = 0$ (Fig. 11-19).

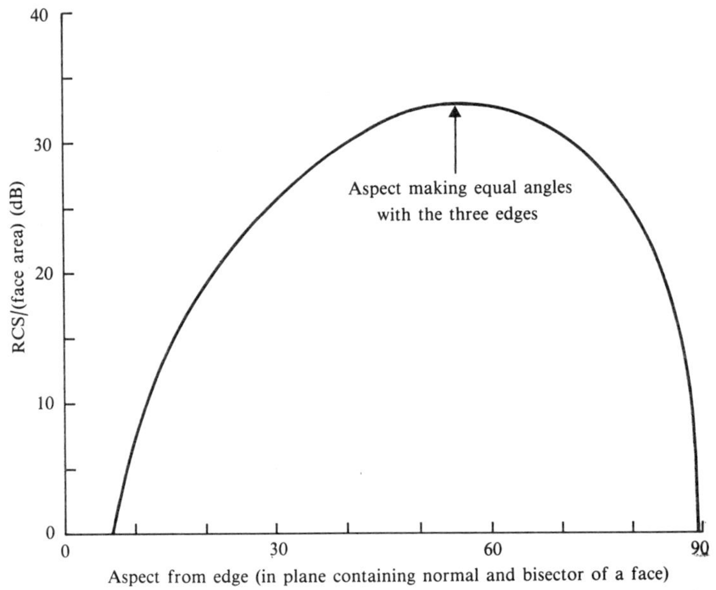

Figure 11-18 RCS of a circular corner with equal (10λ) edgelengths.

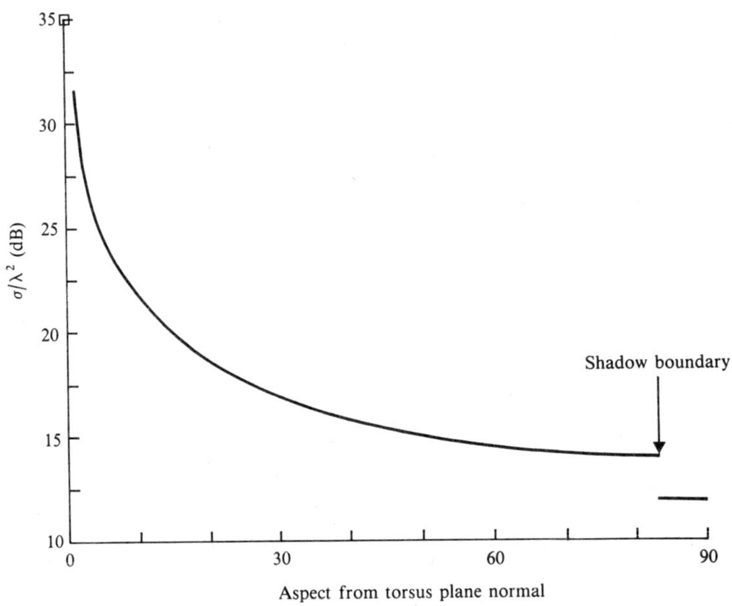

Figure 11-19 Torus RCS. 5λ outer radius, 3λ inner radius.

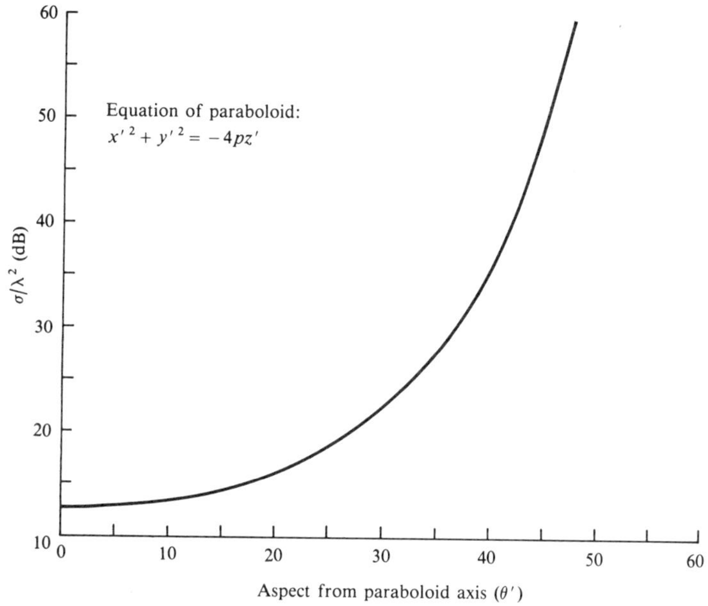

Figure 11-20 RCS of the exterior surface of a paraboloid of revolution, convex side only, $p = \lambda$.

Otherwise ($\theta' > 0$), there are two contributions as follows:

$$\text{1st: } \pi\left(\frac{ba}{\sin\theta'} + b^2\right);$$

$$\text{2nd: } \begin{cases} \pi\left(\dfrac{ba}{\sin\theta'} - b^2\right), \\ 0, \qquad \text{for } 0 \le \cos\theta' \le b/2a. \end{cases}$$

Paraboloid. Vertex at origin, $+z'$ along symmetry axis on convex side. Latus rectum $= 4p$.

Inputs: $p(>0)$.
RCS Formulas: $4\pi p^2 \sec^4\theta'$, $\theta' < \pi/2$ (Fig. 11-20).

11-6 RCS OF A CAVITY

Although the estimate of RCS for a cavity given by Fig. 11-10 is surprisingly good and therefore quite useful (to estimate, for example, the RCS of engine inlets and exhausts, as well as the RCS of cavity volumes covered by radomes and cockpits), it has no mathematical basis other than the accummulation of a large amount of measured data. But a more rigorous development could offer improved results reflecting, say, the effect of the cavity interior. The following sections suggest a possible approach.

11-6-1 Statement of Problem

The problem we wish to solve is as follows: Compute the far fields scattered by a cavity illuminated by a plane wave incident at an arbitrary angle and with arbitrary polarization. The cavity may be empty or it may contain a variety of objects. The sizes of cavity aperture and depth, as well as the sizes of any of its contents, are large in respect to the illuminating wavelength λ. The scattered fields, once found, are easily converted into RCS.

11-6-2 Method of Solution

The method of solution is based on the assumption that a viable ray-tracing program is in hand. Such a program must also keep track of pertinent quantities such as distance traveled along a ray, location of the ray intersection with a given surface, as well as curvature properties of that surface at the intersection point. In addition, the ray-tracing program must be modified so that a ray can carry certain attributes with it; in the present case, these attributes can be taken to be the electromagnetic fields (i.e., polarization) associated with the rays.

At the mouth of the cavity place an aperture or integration plane S and divide this plane into increments ΔS so that the ith incident ray is associated with the ith increment ΔS_i. The phase of the field is measured from S as the ray is bounced into the cavity and returns to S (but not necessarily to the same ΔS_i). All fields present in each ΔS_i are summed for each returning ray that has passed through the increment; the scattered field due to these summed fields (which include the incident fields) is found from the Stratton–Chu far scattered field integral. From the far scattered field the RCS is easy to obtain. These main results are given in Eqs. 11-67 and 11-68.

The fields suffer changes in amplitudes as they are bounced inside the cavity. Two such changes are taken into account below:

1. The constitutive parameters of the cavity walls and contents affect the field amplitudes by means of the Fresnel reflection coefficients (Section 1-6-2); and
2. The curvatures of the cavity walls and contents produce their effects on the amplitudes through the Fock factors (Section 11-6-4). The Fresnel and Fock factors impose their effects cumulatively on the field amplitudes as the ray is reflected to and fro from cavity walls and contents.

11-6-2-1 Incident Field

Let $\hat{a}$, with components (a_x, a_y, a_z), be a unit vector in the incident ray direction. The field of the incident plane wave is

$$\mathbf{E}^0 e^{-j\phi}, \qquad \mathbf{H}^0 e^{-j\phi} \tag{11-47}$$

where $\mathbf{E}^0$ and $\mathbf{H}^0$ are constant vectors and

$$\phi = k\psi = k(xa_x + ya_y + za_z) \tag{11-48}$$

is the phase of the wave at the point (x, y, z) in space; k is the wave number $(= 2\pi/\lambda$, λ being wavelength; $j = \sqrt{-1})$. For points $\mathbf{x}(x, y, z)$ along the ray direction, the scalar product $\mathbf{x} \cdot \hat{a}$ measures distance s along the ray. The direction of $\mathbf{E}^0$ defines the direction of incident polarization (to be specified) and, of course, $\mathbf{E}^0$, $\mathbf{H}^0$, and $\hat{a}$ are mutually perpendicular and form a right-handed system.

11-6-2-2 Fresnel Reflection

Let $\phi_0 = k(x_0 a_x + y_0 a_y + z_0 a_z)$ denote the phase of the incident field at the surface of the body, that is, the cavity wall or its contents, on which it impinges. The reflected wave at the body surface is

$$\mathbf{E}^* e^{-jk\psi_0}, \qquad \mathbf{H}^* e^{-jk\psi_0} \tag{11-49}$$

where $\mathbf{E}^*$ and $\mathbf{H}^*$ are related to $\mathbf{E}^0$ and $\mathbf{H}^0$ by the Fresnel reflection coefficients. Let $\hat{n}$ be a unit normal to the surface, positive toward the side from which the wave impinges. The plane determined by $\hat{n}$ and $\hat{a}$ is called the incidence plane. Resolve the incident and reflected fields at the body surface into two components: $\mathbf{E}^{0,*}$, parallel to the incidence plane, and $\mathbf{E}^{0,*}$, perpendicular to the incidence plane.

The parallel and perpendicular components of the incident and reflected fields are then related by the parallel and perpendicular Fresnel reflection coefficients†:

$$R_{\parallel} = \frac{\epsilon \cos\theta - \sqrt{\epsilon\mu - \sin^2\theta}}{\epsilon \cos\theta + \sqrt{\epsilon\mu - \sin^2\theta}}, \tag{11-50}$$

$$R_{\perp} = \frac{\mu \cos\theta - \sqrt{\epsilon\mu - \sin^2\theta}}{\mu \cos\theta + \sqrt{\epsilon\mu - \sin^2\theta}}, \tag{11-51}$$

where θ is the incidence angle measured in the incidence plane from the surface normal $\hat{n}$ to the incidence direction $\hat{a}$. The constitutive parameters ϵ and μ are, respectively, the permittivity and permeability of the reflecting surface. For a perfect conductor (infinite conductivity) μ is unity and ϵ becomes infinite. Thus,

$$R_{\parallel} = 1 \qquad \text{and} \qquad R_{\perp} = -1$$

in this circumstance.

† Compare Section 3-2. For example, to obtain the perpendicular reflection coefficient above, use Eq. 3-28 setting $\mu_1 = \epsilon_1 = 1 = N_1$.

Therefore, the incident and reflected field components at the reflecting surface are related as follows:

$$\mathbf{E}_{\parallel}^{*} = R_{\parallel}\mathbf{E}^{0}, \qquad \mathbf{E}_{\perp}^{*} = R_{\perp}\mathbf{E}^{0}, \tag{11-52}$$

$$\mathbf{H}_{\parallel}^{*} = R_{\parallel}\mathbf{H}^{0}, \qquad \mathbf{H}_{\perp}^{*} = R_{\perp}\mathbf{H}. \tag{11-53}$$

11-6-2-3 Pseudo-focus

A pencil of rays that strikes a convex surface is reflected as a cone of rays. The broadening of the pencil into a cone is describable in terms of the principal radii of curvature of the surface at the point of reflection. Fock (1957) gives expressions for pencil broadening that can also be used to describe the corresponding reduction in reflected field amplitudes.

When the reflecting surface is concave, the reflected rays converge, almost to a focus, and thereafter diverge as they would have diverged had they struck the convex side of the same surface at the same angle of incidence. As the incident pencil approaches zero radius, the convergence point of the reflected cone, call it the pseudo-focus, approaches a limit at a finite distance from the surface.

Let R_0 be the radius of curvature of the surface in the incidence plane and let (x_0, y_0) be the intersection of the pencil axis with the surface in that plane. Then, the intersection point (x_0, y_0), the radius of curvature R_0, and the incidence angle θ are related by

$$x_0 = R_0 \cos\theta, \qquad y_0 = R_0 \sin\theta. \tag{11-54}$$

Let one of the pencil generators that lies in the incidence plane strike the reflecting surface at (x_1, y_1) with

$$x_1 = R_0 \cos(\theta + \Delta\theta), \qquad y_1 = R_0 \sin(\theta + \Delta\theta). \tag{11-55}$$

Then, the intersection of the rays reflected from (x_0, y_0) and (x_1, y_1) is the pseudo-focus (x_I, y_I). It can be shown that

$$\lim_{\Delta\theta\to 0} \frac{x_I}{R_0} = \frac{1}{2}(\sin\theta \sin 2\theta + \cos\theta) \tag{11-56}$$

and

$$\lim_{\Delta\theta\to 0} \frac{y_I}{R_0} = \sin^3\theta, \qquad 0° \le \theta \le 90°, \tag{11-57}$$

and from these values that the distance s_0 from (x_0, y_0) to (x_I, y_I) is

$$\frac{s_0}{R_0} = \frac{1}{2}\cos\theta. \tag{11-58}$$

11-6-2-4 Fock Amplitudes

In a plane perpendicular to a ray reflected from a convex surface, the expression for beam broadening derived by Fock (1957) is $D(s)/D(0)$,

where

$$D(s) = \cos\theta + 2s[(R_1^{-1} + R_2^{-1})\cos^2\theta + R_0^{-1}\sin^2\theta] + 4s^2R_1^{-1}R_2^{-1}\cos\theta \quad (11\text{-}59)$$

R_0, as used above, is the radius of curvature of a normal cross-section of the reflecting surface by the plane of the incident ray. R_1 and R_2 are the principal radii of curvature of the normal cross-section of the surface. Then, according to Fock's development for the case of the convex surface, the electric and magnetic fields must be modified as follows:

$$\mathbf{E} = \mathbf{E}^* \sqrt{\frac{D(0)}{D(s)}}\, e^{-jk(\psi_0+s)}, \quad (11\text{-}60)$$

$$\mathbf{H} = \mathbf{H}^* \sqrt{\frac{D(0)}{D(s)}}\, e^{-jk(\psi_0+s)}, \quad (11\text{-}61)$$

where $\mathbf{E}^*$ and $\mathbf{H}^*$ are the reflected fields at the body surface obtained from the Fresnel formulas (Section 11-6-2-2).

For a concave surface the results above must be modified. The reflected pencil first converges until the pseudo-focal point (x_I, y_I), at distance s_0, is reached; thereafter, the reflected rays diverge into the same size of cone they would have diverged into had they impinged onto the convex side of the same surface. The initial convergence of the reflected rays is accounted for by increasing the electric field by the factor $D(s)/D(0)$, as long as $0 \le s \le s_0$, where $D(s)$ is given by Eq. 11-59, and s_0 by Eq. 11-58. Then the subsequent divergence of the reflected rays is accounted for by decreasing the reflected fields by the additional factor $D(0)/D(s-s_0)$ when $s \ge s_0$ (i.e., the same effect applied in Eqs. 11-60 and 11-61 for a divergent cone of rays beginning at $s = s_0$). Thus, the total factor applied to the reflected fields beyond $s = s_0$ is $[D(s_0)/D(0)][D(0)/D(s-s_0)]$, or $D(s_0)/D(s-s_0)$. Thus, the fields of Eqs. 11-60 and 11-61 become

$$\mathbf{E}' = \begin{cases} \mathbf{E}^*\sqrt{D(s)/D(0)}\, e^{-jk(\psi_0+s)}, & 0 \le s \le s_0, \\ \mathbf{E}^*\sqrt{D(s_0)/D(s-s_0)}\, e^{-jk(\psi_0+s)}, & s \ge s_0; \end{cases} \quad (11\text{-}62)$$

$$\mathbf{H}' = \begin{cases} \mathbf{H}^*\sqrt{D(s)/D(0)}\, e^{-jk(\psi_0+s)}, & 0 \le s \le s_0, \\ \mathbf{H}^*\sqrt{D(s_0)/D(s-s_0)}\, e^{-jk(\psi_0+s)}, & s \ge s_0. \end{cases} \quad (11\text{-}63)$$

In going from the first surface struck by a ray to the second one, it is assumed that Eqs. 11-62 and 11-63 can be reapplied at the second surface just as though that surface were being illuminated by a plane wave. Such an approximation is obviously not true globally but may be usable locally.†

† I. J. LaHaie has suggested to the writer that in order to remove the implicit assumption that the cavity is cylindrical the divergence factors must be modified. Such divergence factors have been derived by T. B. A. Senior in a University of Michigan Radiation Laboratory memorandum dated February 17, 1972 and entitled "Divergence Factors"; they are not introduced here.

11-6-2-5 Aperture Field Distribution

At the aperture of the scattering cavity, establish a reference plane containing an origin of coordinates. The actual aperture region S in the reference plane may be slightly in front of, and larger than, the cavity mouth so that other scattering effects can be taken into account subsequently (e.g., diffraction from edges of the cavity lip, or scattering from the immediate neighborhood of the cavity mouth).

Divide S into incremental pieces so that there is one piece ΔS_i for each ray piercing S. If the ΔS_i are roughly square, they should not be larger than $\lambda/5$ on a side (and preferably smaller, but not smaller than $\lambda/10$).

Suppose that a steady state has been established for returning rays reflected from the cavity interior. A ray is traced from the time it pierces S as an incident ray until it repierces S as a returning ray† (it of course need not return to the same ΔS_i through which it entered). The fields entering and leaving each ΔS_i are then summed, creating a field distribution on the aperture region S. This field distribution is then used in the Stratton–Chu formula to produce an estimate for the far scattered field $\mathbf{H}^s$.

11-6-3 Assembly of Subdivisions of the Problem

Since the modification of the field from its passage through the reference surface as an incident field to its passage through the reference surface as an emitted field is completely described in Section 11-6-2, it is now possible to assemble the results to produce the far scattered field.

11-6-3-1 Integration: Scattered Field

Let S_0 be any region of the reference surface S from which the scattered energy is to be found. The far scattered field $\mathbf{H}^s$ due to S_0 is obtained from the Stratton–Chu formulation (Stratton, 1941) as modified by Jones (1964) for applicability to an open surface:

$$\mathbf{H}^s = \frac{1}{4\pi}\,\mathbf{curl}\int_{S_0} (\hat{n}' \times \mathbf{H}')g(R')\,dS' + \frac{j}{4\pi\omega\mu}(\mathbf{grad}\,\mathrm{div} + k^2)\int_{S_0} (\hat{n}' \times \mathbf{E}')g(R')\,dS', \tag{11-64}$$

where $g(R')$ is (once again) the free-space Green's function

$$\frac{e^{-jkR'}}{R'}$$

and $R' = \|\mathbf{r} - \mathbf{r}'\|$ is the distance between observation point $\mathbf{r}$ and integration point $\mathbf{r}'$; $\hat{n}'$ is the unit normal (positive toward the side of S on which

† It may be decided, after an examination of the sizes of successive divergence factors and/or reflection coefficients, that a ray can be abandoned after some minimal number of reflections.

energy is incident) to the surface at the integration point. The integration is over the region S_0 of the reference surface S; and $\mathbf{H}'$ and $\mathbf{E}'$ are the total fields at the integration point. If R is the distance from the origin in S to the observer, then one can make the approximation

$$R' \approx R - \hat{a} \cdot \mathbf{r}' \tag{11-65}$$

($\hat{r}$ being direction of observation) to get the far scattered magnetic field:

$$\begin{aligned} \mathbf{H}^s_{S_0} = & \frac{-jke^{-jkR}}{4\pi R} \hat{r} \times \int_{S_0} (\hat{n}' \times \mathbf{H}')e^{-jk\hat{r}\cdot\mathbf{r}'} \, dS' \\ & - \frac{j\omega\epsilon}{4\pi R} e^{-jkR} \left[\hat{r} \times \hat{r} \times \int_{S_0} (\hat{n}' \times \mathbf{E}')e^{-jk\hat{r}\cdot\mathbf{r}'} \, \mathrm{dS}' \right]. \end{aligned} \tag{11-66}$$

Finally, the integrals may be approximated by sums taken over all ΔS_i belonging to S_0, the region of interest. The primes mean that those quantities must be evaluated at some convenient point within each ΔS_i. For example, $\mathbf{H}'$ is the vector sum of all incident and reflected fields associated with rays that pass through that ΔS_i within the interval of interest. If S_0 were metallic (an unlikely circumstance, since S_0 is a region of a reference surface), then $\hat{n}' \times \mathbf{E}' = 0$ and so the second integral of Eq. 11-66 vanishes.

11-6-3-2 RCS

The far field RCS, which is a measure of energy observable from the region S_0, is really the quantity of interest here. We recall from 5-3 that it is defined as

$$\sigma_{S_0} = \lim_{R\to\infty} 4\pi R^2 \, \|\mathbf{H}^s_{S_0}\|^2 / \|\mathbf{H}^0\|^2 \tag{11-67}$$

and is a function of incidence direction and observation direction, wavelength, and constitutive parameters of the scattering walls and contents of the cavity. Since the amplitude $\|\mathbf{H}^0\|$ of the incident field will appear, undisturbed, as a factor in the amplitude $\|\mathbf{H}^s_{S_0}\|$ of the scattered field, it can be taken to be unity. The limit in R causes no difficulty since R exactly cancels from the expression for σ_{S_0}. Thus RCS becomes

$$\begin{aligned} \sigma_{S_0} = \frac{k^2}{4\pi} \Bigg\| & \hat{r} \times \int_{S_0} (\hat{n}' \times \mathbf{H}') \exp(-jk\hat{r} \cdot \mathbf{r}') \, \mathrm{dS}' \\ & - \sqrt{\frac{\epsilon}{\mu}} \left[\hat{r} \times \hat{r} \times \int_{S_0} (\hat{n}' \times \mathbf{E}') \exp(-jk\hat{r} \cdot \mathbf{r}') \, dS' \right] \Bigg\|^2. \end{aligned} \tag{11-68}$$

11-7 TRAVELING WAVES

Peters (1958) has observed that the RCS of a long thin body may be written in terms of the gain $G(\theta, \phi)$ it would have when viewed as a long, thin antenna:

$$\sigma = \gamma^2 \frac{G^2(\theta, \phi)}{4\pi} \lambda^2. \tag{11-69}$$

This observation follows from the definition

$$\sigma = \lim_{R \to \infty} 4\pi R^2 \frac{S_r}{S_i} \tag{11-70}$$

and the fact that the power intensity S_r of the wave radiated in the direction of the transmitter is

$$S_r = P_i G(\theta, \phi)/4\pi R^2 = \left(\frac{\gamma G(\theta, \phi)\lambda}{4\pi R}\right)^2 S_i, \tag{11-71}$$

where γ is the current reflection coefficient at the end of the antenna. (The preceding equation can also be written to serve the bistatic case.)

The derivation of $G(\theta, \phi)$ can be found in an appendix given by Peters (1958). The gist of his result is that

$$\sigma = \frac{(\gamma\lambda)^2}{\pi Q^2} \left(\frac{\sin\theta}{1 - p\cos\theta} \sin\left[\frac{kL}{2p}(1 - p\cos\theta)\right]\right)^4 \tag{11-72}$$

where p is the relative phase velocity of surface current and L is the length of the long, thin body. The parameter Q is a rather complicated function of p and L involving the sine integral (Abramowitz and Stegun, 1964, p. 231)

$$\mathrm{Si}(x) = \int_0^x \frac{\sin t}{t}\, dt \tag{11-73}$$

and the modified cosine integral

$$\mathrm{Cin}(x) = \int_0^x \frac{1 - \cos t}{t}\, dt. \tag{11-74}$$

$$Q = -\frac{2}{p^2} + \frac{\mathrm{Cin}(x) - \mathrm{Cin}(y)}{p^3} + \frac{1}{2p^3}[(p-1)\cos(x)] + (p+1)\cos(y)$$
$$+ (p^2 - 1)\frac{kL}{p}[\mathrm{Si}(x) - \mathrm{Si}(y)], \tag{11-75}$$

where

$$x = \frac{kL}{p}(1+p), \qquad y = \frac{kL}{p}(1-p).$$

In Eq. 11-75, if p is near unity and $kL > 4$, a simpler approximate expression can be written:

$$Q \approx 2.35 \log(2kL) - 0.55. \tag{11-76}$$

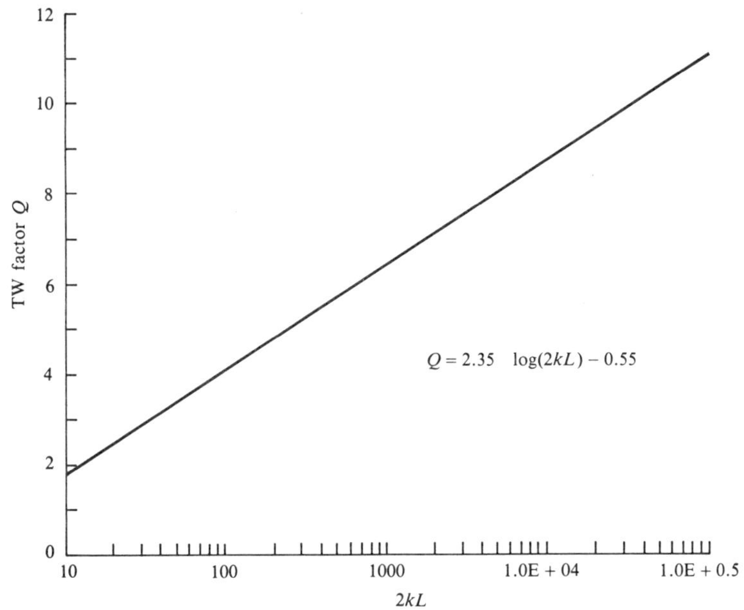

Figure 11-21 Traveling wave amplitude factor Q.

This latter, plotted in Fig. 11-21, is quite adequate for most applications. In the formula for σ above, the factor $\sin\theta/(1-p\cos\theta)$ is a slowly varying function of θ, while

$$\sin\left[\frac{kL}{2p}(1-p\cos\theta)\right]$$

is a rapidly varying one. So the former can be considered as an envelope of the latter and a very good idea of traveling wave RCS can be gotten by examining this envelope. A sampling of slowly varying envelopes is given in Fig. 11-22. And the relation between the slowly varying envelope and the rapidly varying function of Eq. 11-72 (the latter factor) is shown in Fig. 11-23.

Polarization. It is essential to observe that a traveling wave is generated on a long, thin body when the E-vector of the incident wave lies in the plane formed by the incidence direction and the long axis of the body. Conditions are usually such that this polarization is referred to as *horizontal.* Figures 11-24 and 11-25 show traveling wave effects, and lack thereof, for a 10:1 prolate spheroid for the horizontal and vertical polarizations, respectively.

A traveling wave computation can be combined with the PO result given for the ogive in previous section to obtain a more accurate estimate of RCS for the H-polarization case. Figure 11-26 shows a simple power addition

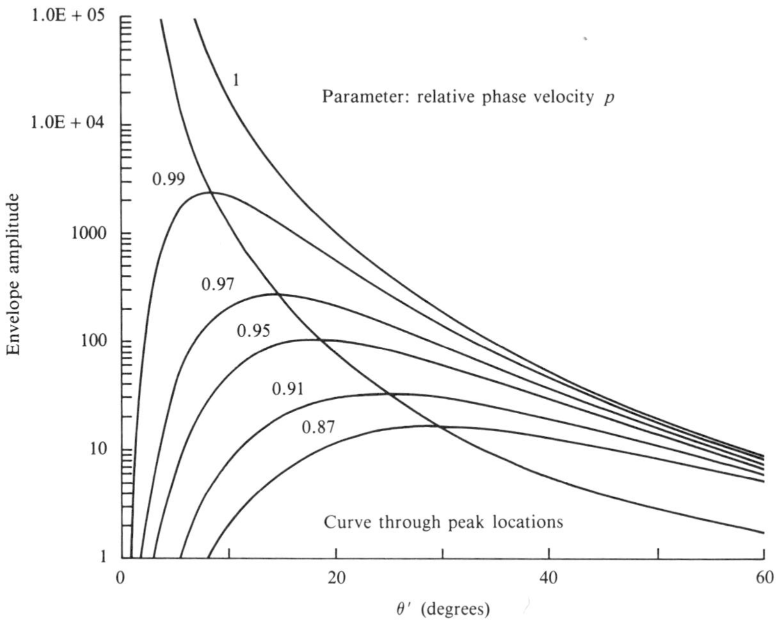

Figure 11-22 Envelopes of traveling wave RCS.

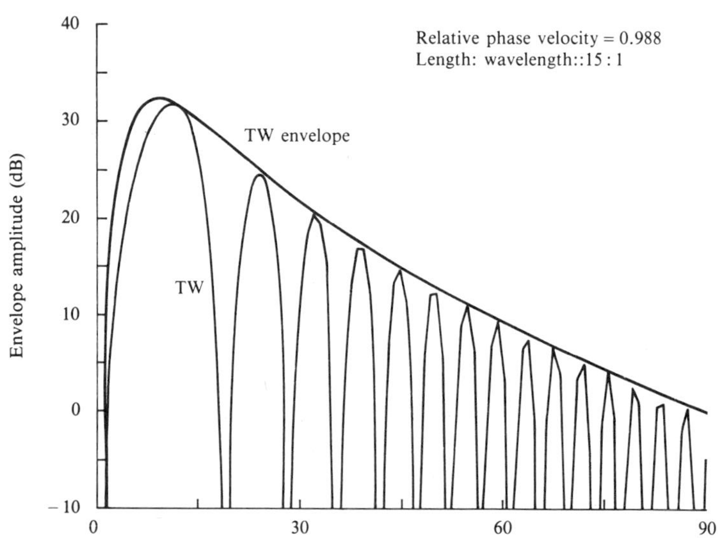

Figure 11-23 Envelope of traveling wave RCS.

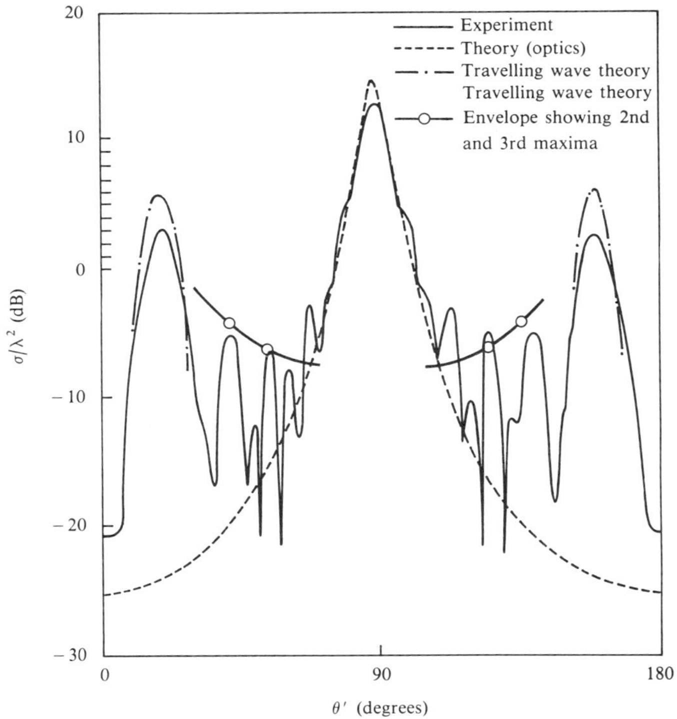

Figure 11-24 RCS of a 10:1 prolate spheroid of length 5.55λ for polarization in the plane formed by the incidence direction and the spheroid axis.

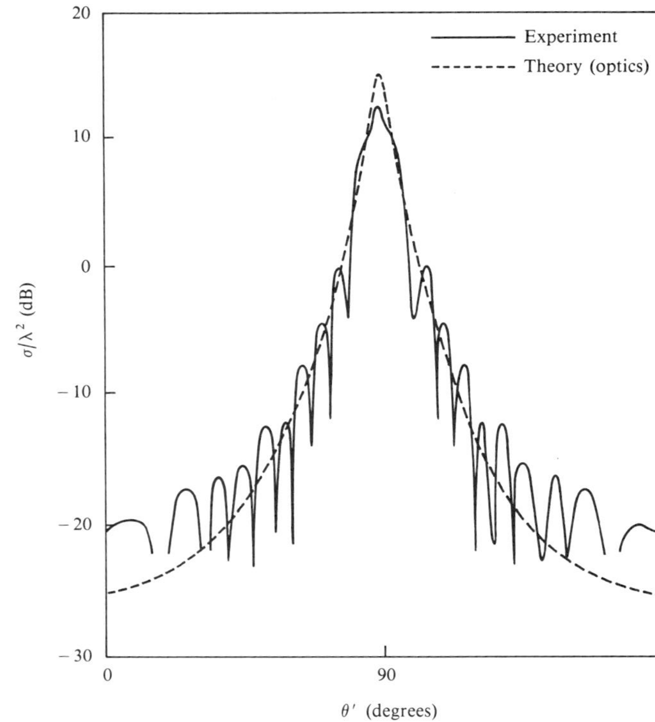

Figure 11-25 RCS of a 10:1 prolate spheroid of length 5.55λ for polarization normal to the plane formed by the incidence direction and the spheroid axis.

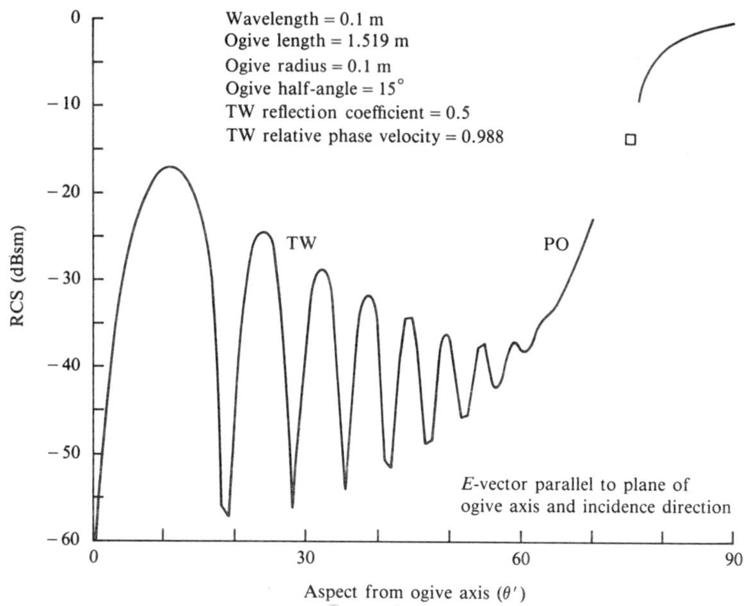

Figure 11-26 Ogive RCS—estimate based on physical optics and traveling wave theories.

of these two results to give a composite estimate of H-polarization RCS for the ogive.

11-8 CREEPING WAVES

The traveling wave phenomenon can also be observed in the return from flat-backed cones. If the flat base is replaced by a spherical cap, the reflection coefficient of the rim singularity is substantially reduced, and with a cone-sphere the waves pass rather freely over the join to produce currents supported by the spherical cap. These are attenuated as they progress through the shadow and are referred to as *creeping waves.* (It is of interest to note that, for rear-on incidence, the effect of a traveling wave reflection at the tip is observable even when the entire cone surface is in shadow.)

These effects can also be observed in the numerical results for cylinders obtained from the two-dimensional code discussed in Section 11-9 below. Indeed, the remarks made above for the behavior of currents in the shadow regions are explicitly exhibited for cylinders having wedge shapes terminated by flat planar backs and by circular backs.

It is sometimes possible to detect the creeping wave return with pulsed illumination, although it is usually quite difficult to create a pulse sufficiently short to separate the creeping wave return from other effects at lower frequencies where the former is most prominent. However, some success in this regard has been achieved on the Ohio State University ElectroScience Laboratory compact range (Young and Walton, 1984). Theoretically, the creeping wave return can be separated from other effects by Fourier methods, providing it is possible to find or measure RCS (both amplitude and phase) over a sufficiently wide frequency interval. Rheinstein (1965) has given such a result for a sphere and for a cone sphere.

We review briefly a theoretical formulation of creeping wave expressions for a sphere. Suppose, following Senior (1965), we decompose the far field amplitude S of a sphere return into optics S^0 and creeping wave S^c components:

$$S = S^0 + S^c, \tag{11-77}$$

where

$$S^0 = -\frac{ka}{2} e^{-2ika}\left[1 - \frac{i}{2ka} + O((ka)^{-4})\right] \tag{11-78}$$

and

$$S^c = \frac{\tau^4}{\beta[Ai(-\beta)]^2}\left[1 + e^{i\pi/3}\frac{8\beta}{15\tau^2}\left(1 + \frac{9}{32\beta^3}\right) + O(\tau^{-4})\right]$$
$$\times \exp\left[i\pi ka - \frac{2\pi i}{3} - e^{-i\pi/6}\beta\tau\pi - e^{i\pi/6}\frac{\beta^2\pi}{60\tau}\left(1 - \frac{9}{\beta^3}\right) + O(\tau^{-3})\right] \tag{11-79}$$

with

$$\tau = \left(\frac{ka}{2}\right)^{1/3}, \qquad \beta = 1.01879\ldots, \quad \text{and} \quad Ai(-\beta) = 0.535657\ldots.$$

$Ai(\cdot)$ is an Airy integral involving Bessel functions of order $\frac{1}{3}$; see Magnus and Oberhettinger (1954). If we now write

$$W = \frac{2}{ka} e^{2ika} S, \tag{11-80}$$

with W^0 and W^c defined similarly, then

$$\frac{\sigma}{\pi a^2} = |W|^2. \tag{11-81}$$

For $ka = 1$ the estimate of W (comparing with the exact series result of Mie) is 3% in error in amplitude and 2.7° in phase. For $ka = 2$, these respective numbers are 3% and 0.2°. Even for $ka = 0.7$ the amplitude error is only 11%; this is actually superior to the Rayleigh formula here, for RCS purposes.

Using W^0 as computed from S^0 above, to subtract from the Mie result, Bechtel (1962) gets an "actual" W^c. This result is compared with W^c as obtained from S^c above in Fig. 11-27.

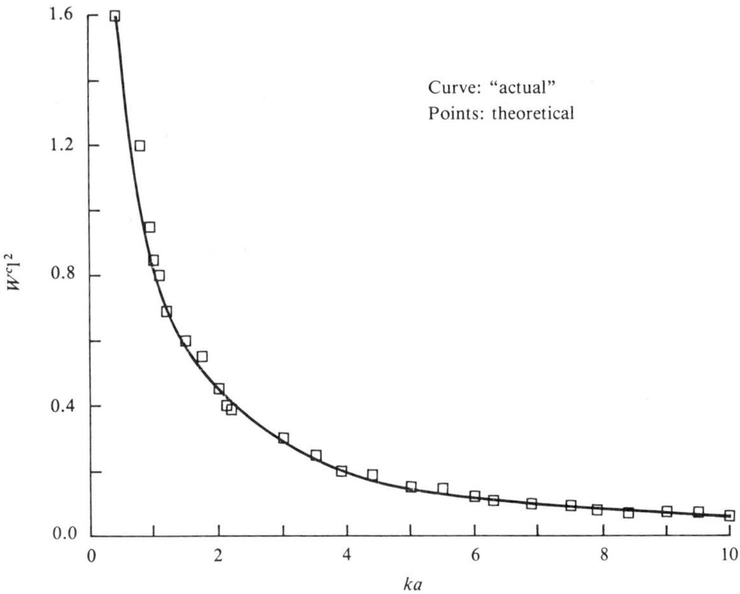

Figure 11-27 Creeping wave RCS contributions for a sphere of radius a. (Data from Bechtel, 1962; Senior, 1965.)

We note that the MoM (Chapter 9), since it is based on a surface current determination, encompasses both traveling wave and creeping wave effects.

11-9 TWO-DIMENSIONAL SCATTERING EXAMPLES

In Section 9-6 we introduced and reviewed the subject of two-dimensional scattering, finishing with an example of a wedge-circle-cylinder. We made the observation there that certain types of scattering seemed to be observable in the RCS results. Now that we have considered such secondary scatterers above as edge, traveling wave, and creeping wave (a primary scatterer being specular), we can attempt to identify the genesis of these phenomena in the two-dimensional results.

Consider edge and traveling wave scattering first. In Section 9-6 we remarked that traveling wave peaks seemed clearly identifiable for the wedge-circle-cylinder at 28 and 139° from the edge-on axial direction. A check with the Peters theory shows that these aspects are indeed appropriate locations and that the amplitudes there check out as well. But the two-dimensional theory allows additional light to be cast on the matter by examining the surface current distributions for these aspects.

Figure 11-28 shows the surface current amplitude versus segment location at the 28° aspect for both E- and H-polarizations. For E-polarization the

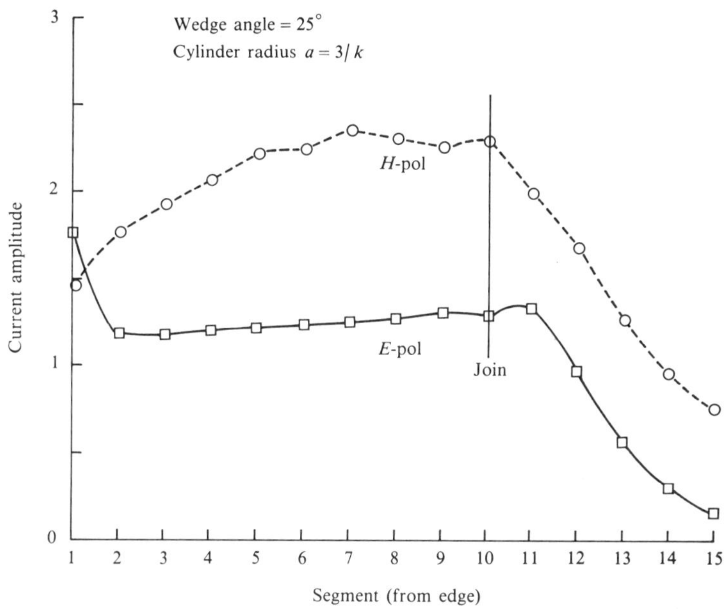

Figure 11-28 Wedge cylinder surface currents for E- and H-polarizations; incidence, from edge-on: 28°.

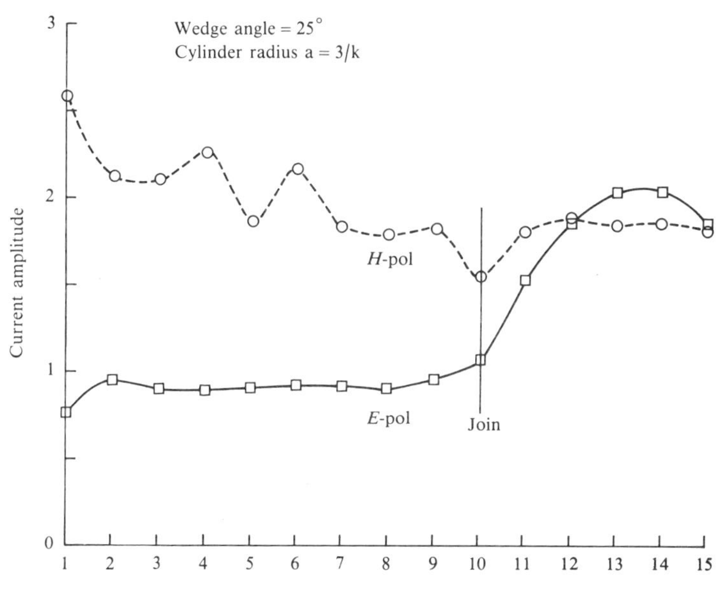

Figure 11-29 Wedge cylinder surface currents for E- and H-polarizations; incidence, from edge-on: 139°.

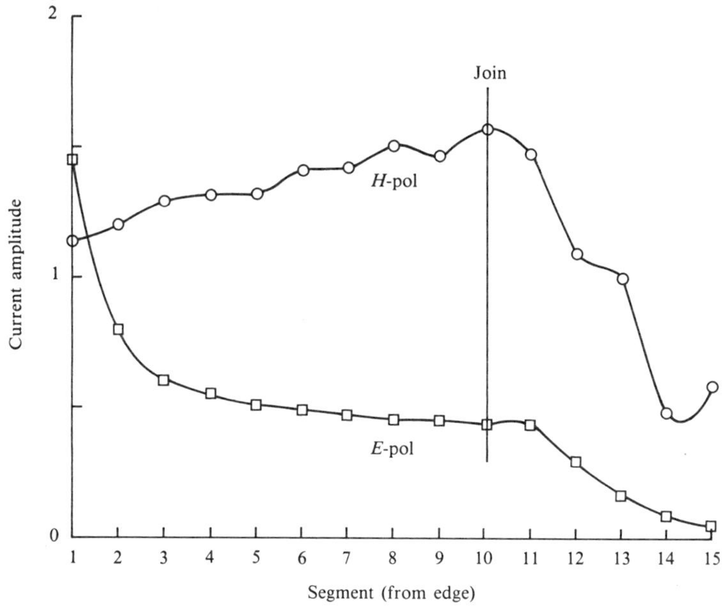

Figure 11-30 Wedge cylinder surface currents for E- and H-polarizations and edge-on incidence.

current amplitude is highest at the wedge edge and falls off rapidly thereafter, especially in the shadowed region aft of the join. For H-polarization the current amplitude builds gradually as the segment number approaches the join. This current distribution does not persist for nearby aspects. Clearly, then, this accumulation of current near the join (which marks a discontinuity in surface contour curvature) is the cause of the traveling wave RCS peak. We can also note that the higher current amplitude near the wedge edge for E-polarization is the cause of enhanced edge scattering (and larger RCS) for that polarization orientation.

Figure 11-29 tells a similar story with the leading edge now serving as the discontinuity from which the traveling wave return arises for H-polarization. Indeed, for E-polarization, the current amplitude on the wedge surface is very small in comparison with that for H-polarization. However, the E-polarization current amplitude on the circular portion of the cylinder, in the vicinity of segments 13 and 14, which is responsible for the specular return, is clearly evident.

Figure 11-30, for edge-on incidence, shows the high current amplitude present on the shadowed circular cylinder surface for H-polarization. To this current we may ascribe the generation of the creeping wave. This high current amplitude persists for aspects neighboring edge-on and the resulting creeping wave return combines with edge returns and traveling wave side lobes to cause the H-polarization pattern oscillation near edge-on (Fig. 9-13).

12

RCS OF COMPLEX SHAPES

It is often desirable to obtain the probability distribution of RCS for use in detection and defense system design studies. The statistics used are averaged over, and apply within, a solid angular region in which returns from individual scatterers are nearly constant in amplitude. The region must contain several target lobes for this approach to be useful. Any regularly spaced components must be treated as a compound component with its own statistics. These must then be folded into the statistics of other parts of the target. Note that some anomalies may arise along certain planar or conical "cuts" since phases here may be more restricted than throughout the solid angular region. For example, top and bottom vertical tail fins may have to be treated as one component in a horizontal cut but as independent scatterers for a vertical or a rolled or a pitched cut.

The linearity of operations with fields allows us to view a complex target as a collection of simple scatterers (often called components, or scattering centers). Suppose the field of the ith simple scattering component of a complex target is proportional to

$$a_i \exp(j\psi_i), \tag{12-1}$$

with both a_i and ψ_i real for all $i = 1, \ldots, N$, N being the number of component scatterers. Then the field of the complex target is proportional to

$$A = \sum_{i=1}^{N} a_i \exp(j\psi_i). \tag{12-2}$$

The quantity a_i is proportional to $\sqrt{\sigma_i}$, where σ_i is the RCS of the ith component. Each σ_i is of course a function of aspect (of both transmitter and receiver, which may be coincident or separated in monostatic or bistatic situations, respectively), frequency, polarization, and the constitutive parameters of the complex scatterer. Therefore,

$$\sigma = AA^*, \quad \text{or} \quad |A|^2. \tag{12-3}$$

Equation 12-3 is often referred to as the relative phase RCS of the complex target.

Suppose that the ψ_i are not known or are known only very approximately. Then on the basis stated in the opening remarks of this section, we may still proceed toward an estimate of RCS by taking the phases to be random variables each having a uniform distribution over the interval 0, 2π. If we further assume the collection of simple scatterers to be mutually independent, then we may average over each ψ_i individually.† Thus, for example, the mean RCS (i.e., the mean value of Eq. 12-3) is

$$m = \langle AA^* \rangle, \tag{12-4}$$

where $*$ means complex conjugate and $\langle \cdot \rangle$ indicates the multiple averaging process just defined.

As an example, consider $N = 2$. Then

$$\begin{aligned}
\langle AA^* \rangle &= \left(\frac{1}{2\pi}\right)^2 \int_0^{2\pi} \int_0^{2\pi} (a_1 e^{j\psi_1} + a_2 e^{j\psi_2})(a_1 e^{-j\psi_1} + a_2 e^{-j\psi_2})\, d\psi_1\, d\psi_2, \\
&= \left(\frac{1}{2\pi}\right)^2 \int_0^{2\pi} \int_0^{2\pi} (a_1^2 + a_2^2 + 2a_1 a_2 \cos(\psi_1 - \psi_2))\, d\psi_1\, d\psi_2, \\
&= a_1^2 + a_2^2 + a_1 a_2 \left(\frac{1}{2\pi}\right)^2 \int_9^{2\pi} [\sin(2\pi - \psi_2) + \sin \psi_2]\, d\psi_2, \\
&= a_1^2 + a_2^2 + a_1 a_2 \left(\frac{1}{2\pi}\right)^2 \int_0^{2\pi} 0\, d\psi_2 = a_1^2 + a_2^2 = \sigma_1 + \sigma_2.
\end{aligned}$$

12-1 FIRST ORDER STATISTICS OF RCS

So in general, we have

$$m = \langle AA^* \rangle = \left\langle \sum_{i,j} a_i a_j \exp[j(\psi_i - \psi_j)] \right\rangle = \sum_i \sigma_i \equiv S_1 \tag{12-5}$$

† In defining the independent random variables ψ_i, the implication is that all physical parameters on which they depend [such as aspect, frequency, polarization, and target constitutive parameters], in particular, aspect, are held fixed. But, in fact, we may assume that the region of interest is defined by a small solid angle neighborhood about a particular aspect direction; it is in such a neighborhood that the ψ_i may take on random values. This viewpoint then makes it convenient to deal with measured data even though the experimental (static or dynamic) aspect region may be confined to just a plane traverse of a solid angle neighborhood about a direction of interest. In fact, this is just the condition we encounter later in dealing with static measured data.

The deviation from the mean is

$$F = AA^* - m = 2\sum_{i<j} a_i a_j \cos(\psi_i - \psi_j); \tag{12-6}$$

so the variance is $m_2 = \langle F^2 \rangle$, or

$$m_2 = 4\left\langle \sum_{i<j} \sum_{k<q} a_i a_j a_k a_q \cos(\psi_i - \psi_j)\cos(\psi_k - \psi_q) \right\rangle$$

$$= 2\sum_{i<j} \sigma_i \sigma_j = \left(\sum_i \sigma_i\right)^2 - \sum_i \sigma_i^2 \equiv 2S_2. \tag{12-7}$$

Using these first order statistics of RCS we are now afforded an estimate of central tendency m of RCS, together with a deviation about m in the form of $\sqrt{m_2}$. Thus,

$$m \pm \sqrt{m_2} \tag{12-8}$$

provides a range of RCS about its central level. The precise nature of this range we shall see later when we have examined the distribution of RCS. It should be remembered that both m and m_2 are functions of aspect, frequency, polarization, and target constitutive parameters.

But we are interested in more than just a central level of RCS and some deviation about that level. We want to know the distribution of RCS and for this we must turn to higher order statistics of RCS.

12-2 HIGHER ORDER STATISTICS OF RCS

We use a method, similar to one described by Pearson (1902), which requires the special statistics, skewness and kurtosis, of RCS. And to make these definitions we must have the third and fourth moments of RCS.

The third moment of RCS is $m_3 = \langle F^3 \rangle$, or

$$m_3 = 8\left\langle \sum_{i<j} \sum_{k<q} \sum_{m<n} a_i a_j a_k a_q a_m a_n \cos(\psi_i - \psi_j) \times \cos(\psi_k - \psi_q)\cos(\psi_m - \psi_n) \right\rangle$$

$$= 12 \sum_{i<j<k} \sigma_i \sigma_j \sigma_k \equiv 12 S_3. \tag{12-9}$$

The fourth moment of RCS is $m_4 = \langle F^4 \rangle$, or

$$m_4 = 16\left\langle \sum_{i<j} \sum_{k<q} \sum_{m<n} \sum_{g<h} a_i a_j a_k a_q a_m a_n a_g a_h \times \cos(\psi_i - \psi_j)\cos(\psi_k - \psi_q)\cos(\psi_m - \psi_n)\cos(\psi_g - \psi_h) \right\rangle$$

$$= 6\sum_{i<j} \sigma_i^2 \sigma_j^2 + 24 \sum_{i \neq j < k} \sigma_i^2 \sigma_j \sigma_k + 216 \sum_{i<j<k<q} \sigma_i \sigma_j \sigma_k \sigma_q$$

$$= 6S_2^2 + 12 S_1 S_3 + 132 S_4. \tag{12-10}$$

12-3 SPECIAL STATISTICS OF RCS

Statisticians like to look at dimensionless combinations of the statistical moments about the mean to determine the appropriate probability distributions for representing data. Here the mean RCS, m, is a significant parameter. So we choose the following three statistics of RCS to work with:

$$\textit{Width: } w = \frac{m_2}{m^2}; \tag{12-11}$$

$$\textit{Modified skewness: } g = m_3/mm_2; \tag{12-12}$$

$$\textit{Kurtosis: } K = \frac{m_4}{m_2} - 3. \tag{12-13}$$

The locations of measured or theoretical data points in the w, g and w, K planes enable us to describe distribution of RCS.†

The definition of RCS distribution is governed by the nature of the right half of the w,g-plane (Fig. 12-1). The region below the line

$$g = w - 1 \tag{12-14}$$

is nonphysical, that is, there is no nonnegative probability distribution of positive quantities that can place a point (w, g) in this region. The line

$$g = 2w \tag{12-15}$$

is the locus of the χ^2- (or γ-) distribution (with Swerling cases marked). The Rayleigh point is

$$(w, g) = (1, 2). \tag{12-16}$$

The locus of the lognormal is shown in the upper half-plane of w,g-space.

The simplest distribution which maps into the physical half-plane below $g = 2w$ is the β:

$$\beta(x; a, b) = \frac{\Gamma(a+b)}{\Gamma(a)\Gamma(b)} x^{a-1}(1-x)^{b-1}, \qquad a, b > 0, \qquad 0 \le x \le 1, \tag{12-17}$$

which provides a *1–1 map between points* (a, b) *and points* (w, g). The β-distribution also provides a unique 1–1 map between the points (a, b) and the points (w, K). (Verification of these two statements is given in Section 12-4.) The variate x in Eq. 12-17 is related to RCS σ by

$$x = \frac{\sigma - \sigma_{\min}}{\sigma_{\max} - \sigma_{\min}}$$

or, to a sufficient approximation, by $x = \sigma/\sigma_{\max}$.

† We might also consider a modified kurtosis; but we find the concept not as satisfying as the concept of modified skewness. In modifying either the first term or the entire expression for K by, say, a multiplier fraction m_2/m^2, we should have to decide against or for the retention of a zero kurtosis for the Gaussian distribution.

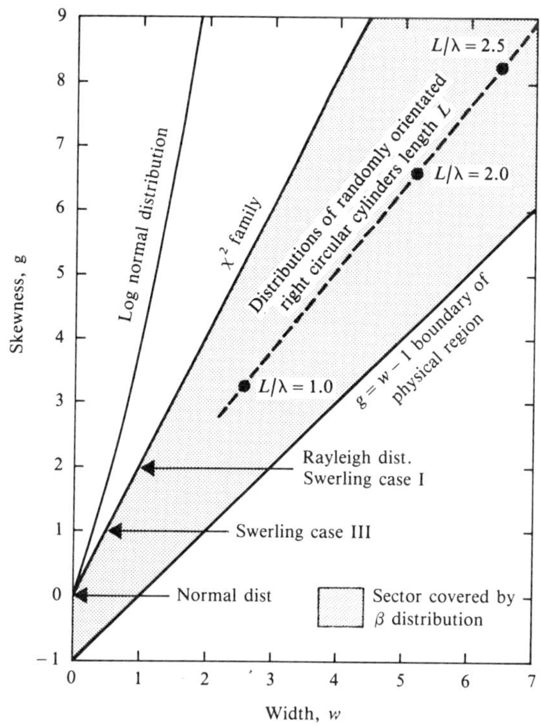

Figure 12-1 Width-skewness plane representations for some distributions used to describe RCS scintillations.

The upper part of the w,g-plane is also covered by the modified β-distribution, except for a small piece near the g-axis:

$$\beta(x; a, b) = \frac{\Gamma(a+b)}{\Gamma(a)\Gamma(b)} x^{a-1}/(1+x)^{a+b}. \qquad a, b > 0 \qquad 0 \leq x \leq 1. \quad (12\text{-}18)$$

For these distributions we have

$$m = \begin{cases} a/(a+b) & \text{[regular beta]}, \\ a/(b-1) & \text{[modified beta]}; \end{cases} \quad (12\text{-}19)$$

and if

$$n = \begin{cases} a+b+1 & \text{[regular beta]}, \\ 2-b & \text{[modified beta]}, \end{cases} \quad (12\text{-}20)$$

then

$$w + 1 = (n-1)(a+1)/na, \quad (12\text{-}21)$$

$$g + 2 = 2n(w+1)/(n+1), \quad (12\text{-}22)$$

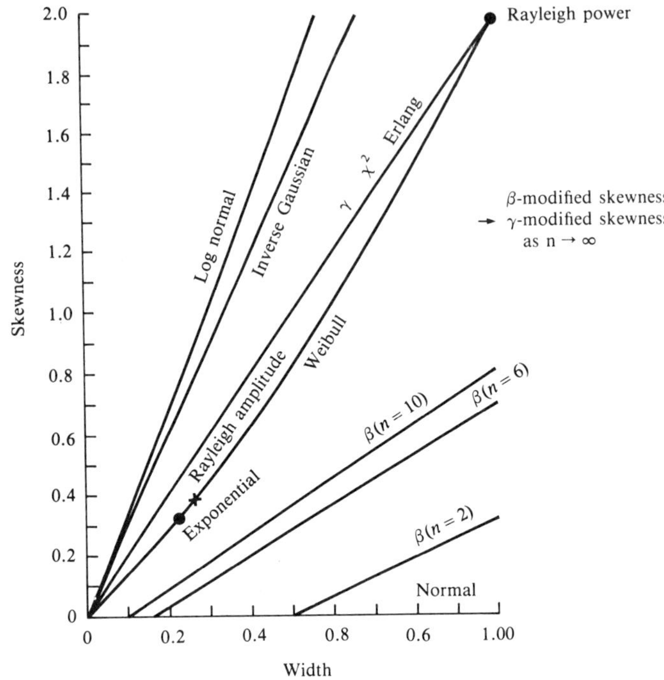

Figure 12-2 Width-skewness curves for some standard distributions.

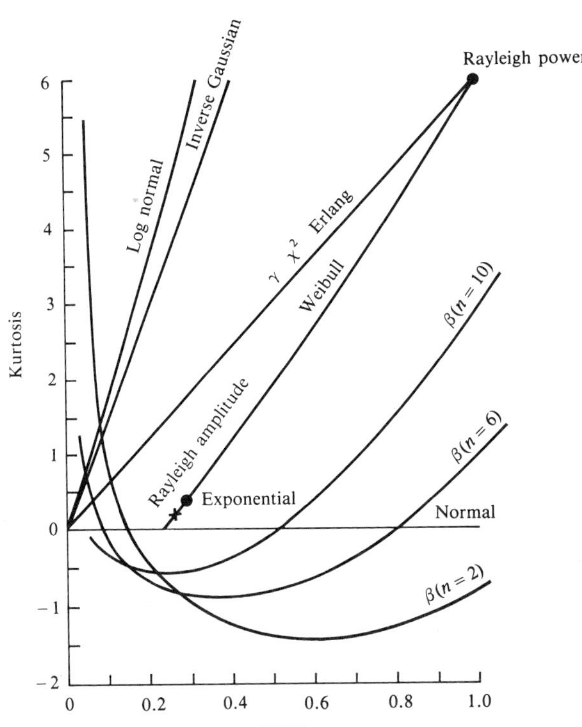

Figure 12-3 Width-kurtosis curves for some standard distributions.

and

$$K + 6 = \frac{6(w+1)}{(n+1)(n+2)}\left(n^2 + \frac{1}{w}\right). \tag{12-23}$$

Figure 12-2 shows the w, g curves for several standard distributions as well as for the β-distribution for a range of n-values. Figure 12-3 shows corresponding and corroborating w, K information.

12-4 β-PARAMETERS

To find the β-parameters a and b, given, say, the special statistics of RCS w and g (which may be computed from either experimental data or a theoretical model), we first find

$$n = \frac{g+2}{2w-g}, \tag{12-24}$$

and

$$p = \frac{1}{nw}. \tag{12-25}$$

Then we solve the pair of equations

$$n = a + b + 1 \tag{12-26}$$

and

$$p = \frac{a}{b}, \tag{12-27}$$

for a and b, obtaining

$$a = \frac{p(n-1)}{p+1}, \tag{12-28}$$

$$b = \frac{n-1}{p+1}; \tag{12-29}$$

or, directly in terms of width w and skewness g:

$$a = \frac{2(g-w+1)}{gw+4w-g},$$

$$b = \frac{2w(g-w+1)(g+2)}{(2w-g)(gw+4w-g)}.$$

These equations verify our claim concerning the 1–1 map between points (a, b) and (w, g). So for every statistical pair w, g we have a unique β-distribution of RCS with parameters a, b.

Note from the above that we can find n in terms of g, whence a

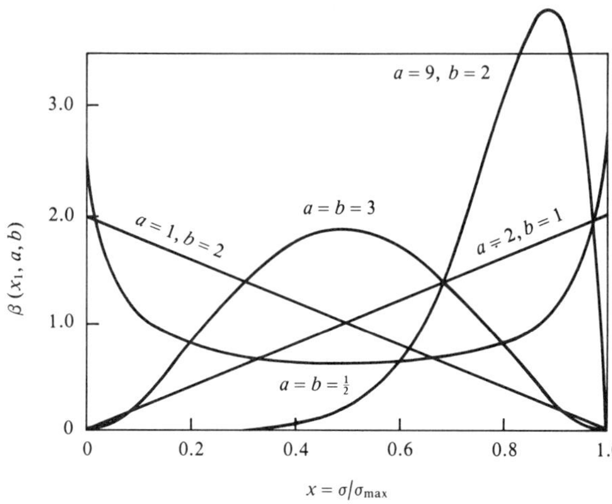

$$\beta(x;\ a, b) = \frac{\Gamma(a+b)}{\Gamma(a)\Gamma(b)}\, x^{a-1}(1-x)^{b-1}$$

$$\text{mean } m = \frac{a}{a+b}$$

$$\text{width } w = \frac{b}{a(a+b+1)}$$

$$\text{modified skewness } g = \frac{2(b-a)}{a(a+b+2)}$$

$$\text{Kurtosis } K = \frac{6(a+b+1)}{(a+b+2)(a+b+3)}\left(\frac{a}{b} + \frac{b}{a} - \frac{1}{a+b+1} - 3\right)$$

Figure 12-4 Examples of the two-parameter β-distribution.

corresponding value of K can be calculated. Thus, we may also verify the 1–1 map between points (a, b) and (w, K). This relation is used later to compare measured third and fourth moments.

Figures 12-4 and 12-5 suggest the manner in which the two parameters a and b affect the behavior of the β-density. Note again that if we wish to have the abscissa portray RCS, we must use the maximum RCS as a normalizing factor. There is then some comfort in noting that the β-distribution does not require infinite RCS.

12-5 RANDOM AMPLITUDES

Up to this point the component amplitudes a_i have been taken to be constant, or at least slowly varying, near the aspect of interest. Now suppose we assume each to be random in such a way that, for the kth component, the mean return is

$$\langle a_k^2 \rangle = \sigma_k, \tag{12-30}$$

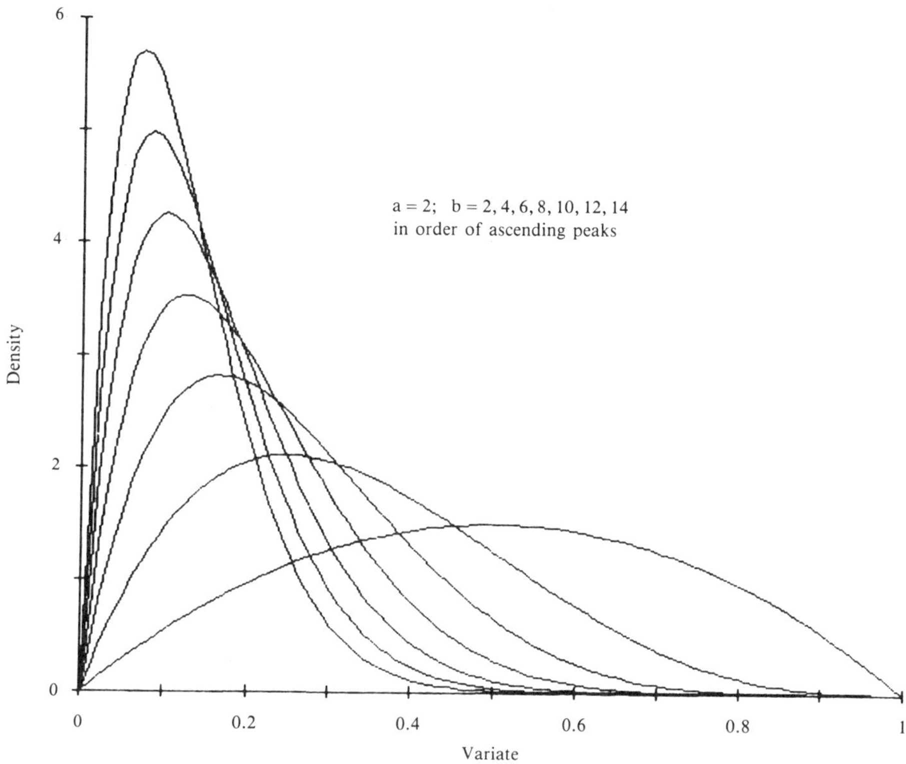

Figure 12-5 β-densities for a variety of parameter values.

and the moments about the mean are

$$\langle (a_k^2 - \sigma_k)^2 \rangle = \mu_k, \tag{12-31}$$

$$\langle (a_k^2 - \sigma_k)^3 \rangle = \gamma_k, \tag{12-32}$$

and

$$\langle (a_k^2 - \sigma_k)^4 \rangle = \beta_k. \tag{12-33}$$

Then, with the sums S_i, $i = 1, 2, 3, 4$, as given above, we find, for the complex target,

$$\text{Mean RCS: } m = S_1; \tag{12-34}$$

$$\text{Variance of RCS: } m_2 = 2S_2 + \sum_k \mu_k; \tag{12-35}$$

$$\text{Third moment of RCS: } m_3 = 12S_3 + 6\sum_{i \neq k} \sigma_i \mu_k + \sum_k \gamma_k; \tag{12-36}$$

and

$$\text{Fourth moment of RCS: } m_4 = 132S_4 + 12S_1S_3 + 6S_2^2 + \sum_k \beta_k + 12 \sum_{i \neq k} \gamma_i \sigma_k + 6 \sum_k \mu_k(S_1^2 - \sigma_k^2) + 36 \sum_{i<k} \mu_i \mu_k + 72 \sum_{k \neq i<j} \mu_k \sigma_i \sigma_j. \quad (12\text{-}37)$$

Notice that these formulas are somewhat more complicated than those given above for constant or slowly varying amplitudes.

12-6 ALLOWABLE REGIONS

By manipulating the moments shown above (which depend on the individual component scattering characteristics), it can be shown that, for a complex target comprised of these components, the points:

$$(w, g) \qquad \text{and} \qquad (w, K)$$

are restricted to lie within bounded regions (for arbitrary, but fixed, aspect,

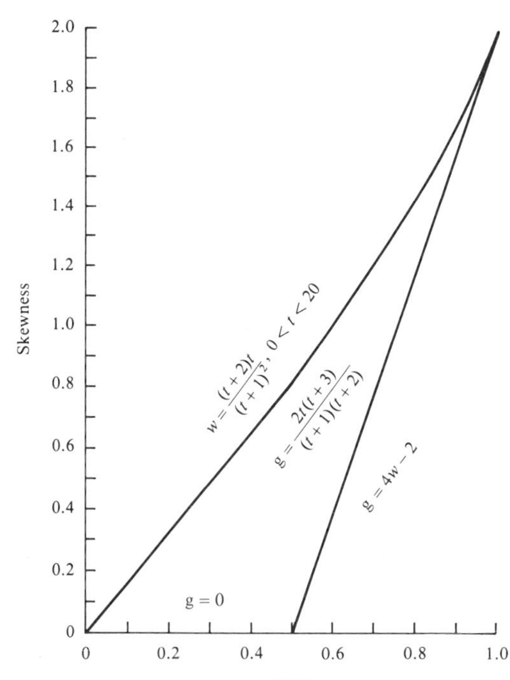

Figure 12-6 Allowable region of width-skewness space for a complex target.

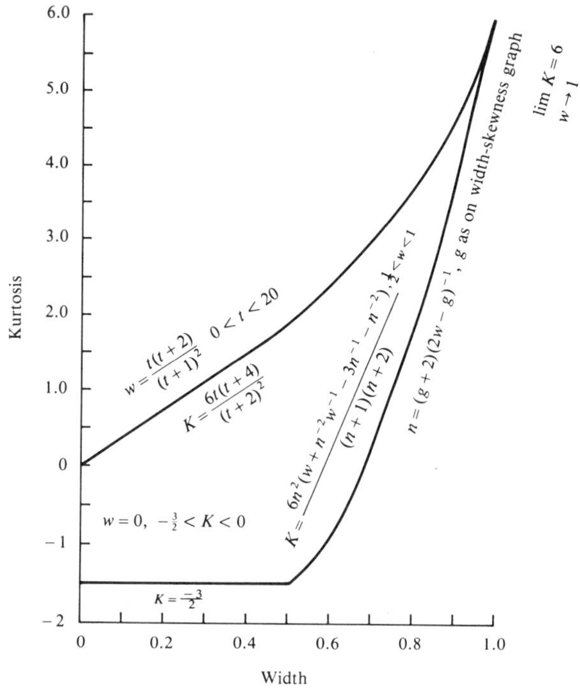

Figure 12-7 Allowable region of width-kurtosis space for a complex target.

frequency, and polarization). These regions are called *allowable regions* (see Fig. 12-6 and Fig. 12-7 as well as the appendix to this section).

In width-skewness space the allowable region is bounded by the following curves (Fig. 12-6):

Right boundary: $g = 4w - 2, \qquad 1/2 \leq w \leq 1;$ (12-38)

Lower boundary: $g = 0, \qquad 0 \leq w \leq 1/2;$ (12-39)

Left boundary: $w = t(t+2)/(t+1)^2,$ (12-40)

$g = 2t(t+3)/(t+1)(t+2), \qquad 0 \leq t \leq 20.$

The allowable region is roughly triangular.

In width-kurtosis space the allowable region is bounded by (Fig. 12-7):

Right boundary: $K = (9/w) - 36 + 33w, \qquad 1/2 \leq w \leq 1;$ (12-41)

Lower boundary: $K = -3/2, \qquad 0 \leq w \leq 1/2;$ (12-42)

Left boundary: $w = 0, \qquad -3/2 \leq K \leq 0;$ (12-43)

Upper boundary: $w = t(t+2)/(t+1)^2,$ (12-44)

$K = 6t(t+4)/(t+2)^2, \qquad 0 \leq t \leq 20.$

It can be shown that, if the component moments μ_k, γ_k, and β_k correspond to points in these allowable regions, then the full complex target scintillations will also lie in these regions.

Figures 12-2 and 12-6 should be compared to see that only w, g curves for the β-distribution (and the Rayleigh point) fall inside the w, g allowable region. Figures 12-3 and 12-7 can be compared for corroborating w, K evidence.

12-6-1 Appendix: Follin Sums of RCS Products

Follin (unpublished Johns Hopkins University Applied Physics Laboratory memorandum, modified with permission) has derived a number of alternative expansions and representations for the sums S_i. Since the allowable regions are based on some on these formulas, we include an abbreviated set of them below:

Summation	Number of Terms in Summation	
$S_1 = T_1 = \sum_{i}^{N} \sigma_i$	N	(1)
$S_2 = \sum_{i<j}^{N} \sigma_i \sigma_j$	$N(N-1)/2$	(2)
$S_3 = \sum_{i<j<k}^{N} \sigma_i \sigma_j \sigma_k$	$N(N-1)(N-2)/6$	(3)
$S_4 = \sum_{i<j<k<h}^{N} \sigma_i \sigma_j \sigma_k \sigma_h$	$N(N-1)(N-2)(N-3)/24$	(4)
$T_k = \sum_{i}^{N} \sigma_i^k, \quad k = 1, 2, 3, 4$	N	(5)
$S_1^2 = 2S_2 + T_2$		(6)
$S_1^3 = 6S_3 + T_3 + 3\sum_{i \neq j}^{N} \sigma_i^2 \sigma_j$	$N(N-1)$	(7)
$S_1^4 = 24S_4 + T_4$		
$+ 6\sum_{i \neq j \neq k} \sigma_i^2 \sigma_j \sigma_k$	$N(N-1)(N-2)$	
$+ 6\sum_{i<j} \sigma_i^2 \sigma_j^2$	$N(N-1)/2$	
$+ 4\sum_{i \neq j} \sigma_i^3 \sigma_j$	$N(N-1)$	(8)
$S_1 S_2 = 3S_3 + \sum_{i \neq j} \sigma_i^2 \sigma_j$	$N(N-1)$	(9)

Summation	Number of Terms in Summation	
$S_1S_3 = 4S_4 + \sum_{i\neq j<k} \sigma_i^2\sigma_j\sigma_k$	$N(N-1)(N-2)/2$	(10)
$S_2^2 = 6S_4 + \sum_{i<j} \sigma_i^2\sigma_j^2 + 2\sum_{i\neq j<k} \sigma_i^2\sigma_j\sigma_k$	$2N(N-1)(N-2)/2$	(11)

$$S_2T_2 = \sum_{i\neq j} \sigma_i^3\sigma_j + \sum_{i\neq j<k} \sigma_i^2\sigma_j\sigma_k \tag{12}$$

$$\left.\begin{aligned} T_1 &= S_1 \\ T_2 &= S_1^2 - 2S_2 \\ T_3 &= S_1^3 - 3S_1S_2 + 3S_3 \\ T_4 &= S_1^4 - 4S_1^2S_2 + 2S_2^2 + 4S_1S_3 - 4S_4 \end{aligned}\right\} \tag{13}$$

$$\left.\begin{aligned} 2S_2 &= T_1^2 - T_2 \\ 6S_3 &= T_1^3 - 3T_1T_2 + 2T_3 \\ 24S_4 &= T_1^4 - 6T_1^2T_2 + 3T_2^2 + 8T_1T_3 - 6T_4 \end{aligned}\right\} \tag{14}$$

$$T_2S_4 = \sum_{i\neq j<k<h<m} \sigma_i^2\sigma_j\sigma_k\sigma_h\sigma_m + \sum_{i\neq j<k<h} \sigma_i^3\sigma_j\sigma_k\sigma_h \tag{15}$$

$$\begin{aligned} S_2^3 = 6S_2S_4 &+ 4\sum_{i<j\neq k<h} \sigma_i^2\sigma_j\sigma_k\sigma_h + \sum_{i<j} \sigma_i^3\sigma_j^3 + 3\sum_{i\neq j\neq k} \sigma_i^3\sigma_j^2\sigma_k \\ &+ 12\sum_{i\neq j<k<h<m} \sigma_i^2\sigma_j\sigma_k\sigma_h\sigma_m + 6\sum_{i\neq j<k<h} \sigma_i^3\sigma_j\sigma_k\sigma_h \end{aligned} \tag{16}$$

If one additional scatterer is added, then:

$$\begin{aligned} S_1(n+1) &= S_1(n) + \sigma_{n+1} \\ S_2(n+1) &= S_2(n) + S_1(n)\sigma_{n+1} \\ S_3(n+1) &= S_3(n) + S_2(n)\sigma_{n+1} \\ S_4(n+1) &= S_4(n) + S_3(n)\sigma_{n+1}. \end{aligned} \tag{17}$$

12-6-2 Appendix: Some Inequalities for the Allowable Regions

Many of the formulas for the basic sums S_i and T_i summarized in the preceding section are required to establish the inequalities for the allowable regions. Some of the developments are straightforward.

$$0 < w = m_2/m^2 = \frac{2S_2}{2S_2 + T_2} < 1,$$

$$0 \le g = m_3/mm_2 = \frac{12S_3}{6S_3 + 2\sum_{i \ne j} \sigma_i^2 \sigma_j} < 2 \qquad \text{(from Eq. (9) above).}$$

From (10) and (11) above:

$$m_3 m/m_2^2 = \frac{12S_3S_1}{4S_2^2} < 2, \quad \text{or} \quad m_3/mm_2 < 2w.$$

In addition,

$$m_4/m_2^2 = \frac{3}{2} + \frac{3S_1S_3}{S_2^2} + \frac{33S_4}{S_2^2} \ge \frac{3}{2}$$

and

$$m_4 - 12S_2^2 = 144S_4 - 6\sum_{i<j} \sigma_i^2 \sigma_j^2.$$

Then, using (15) and (16)

$$\begin{aligned} S_1^2(m_4 - 12S_2^2) < 144S_1^2S_4 &= 288S_2S_4 + 144T_2S_4 \\ &= 288S_2S_4 + 144 \sum_{i \ne j<k<h<m} \sigma_i^2 \sigma_j \sigma_k \sigma_h \sigma_m \\ &\quad + 144 \sum_{i \ne j<k<h} \sigma_i^3 \sigma_j \sigma_k \sigma_h < 48S_2. \end{aligned}$$

$$m_4/m_2^2 = K + 3 < 3 + 6w < 9, \quad \text{or} \quad K < 6.$$

From (10), (11), and (12):

$$\begin{aligned} 3S_1S_3 - 4S_2^2 + S_1^2S_2 &= \sum_{i \ne j} \sigma_i^3 \sigma_j - 2\sum_{i<j} \sigma_i^2 \sigma_j^2 \\ &= \sum \sigma_i \sigma_j (\sigma_i - \sigma_j)^2 \ge 0; \end{aligned}$$

and so $2w > g > 4w - 2$.

12-6-3 Appendix: Skewness and Kurtosis for Some Standard Densities

$$\beta: \frac{x^{a-1}(1-x)^{b-1}}{B(a, b)}; \qquad x, a, b \ge 0, \qquad B(a, b) = \frac{\Gamma(a)\Gamma(b)}{\Gamma(a+b)};$$

$$g = \frac{2(b-a)(a+b+1)^{1/2}}{(a+b+2)(ab)^{1/2}};$$

$$K = \frac{3(a+b)(a+b+1)(a+1)(2b-a)}{ab(a+b+2)(a+b+3)} + \frac{a(a-b)}{a+b};$$

$$w = \frac{b}{a(a+b+1)}.$$

Normal:

$$\frac{1}{\sqrt{2\pi m_2}} \exp\left[-\frac{(x-m)^2}{2m_2}\right]; \qquad g = 0; \qquad K = 3; \qquad w = m_2/m^2.$$

χ^2*:*

$$2^{-m/2} x^{(m-2)/2} e^{-x/2} / \Gamma(m/2), \qquad x \geq 0; \qquad g = 2^{1/2} m^{-1/2};$$

$$K = 3 + \frac{12}{m}; \qquad w = \frac{2}{m}.$$

γ*:*

$$(x/b)^{c-1} \frac{\exp(-x/b)}{b\Gamma(c)}, \qquad x \geq 0; \qquad g = 2c^{-1/2};$$

$$K = 3 + \frac{6}{c}; \qquad w = \frac{1}{c}.$$

Lognormal:

$$\frac{1}{x\sigma\sqrt{2\pi}} e^{-[\log(x/\mu)]^2/2\sigma^2}, \qquad x > 0,$$

$$\text{median } \mu > 0, \qquad u = e^{\sigma^2}.$$

$$g = (u+2)(u-1)^{1/2}; \qquad K = u^4 + 2u^3 + 3u^2 - 3;$$

$$w = u - 1.$$

Rayleigh: is a special case of χ^2 with $m = 2$ or of γ with $c = 1$.

12-7 STATISTICS OF RCS SCINTILLATION; LOBE STRUCTURE

When a complex target is illuminated by a plane wave, energy is scattered in all directions with numerous lobes in the backward direction and a very broad lobe in the forward direction. Our statistical description of RCS, which is based on the assumption of random phases among component scatterers of the complex target, appears, up to this point at least, to have lost all information concerning the nature of the lobing structure of the pattern. Our goal in this section then is to regain some, quite likely statistical, information about the RCS pattern lobes for a complex target.

Let some salient target feature, such as its center of gravity, be placed at the origin of a ξ-, η-, ζ-coordinate system (see Fig. 12-8) whose $\xi\zeta$-plane contains the illuminator/observer. The relative phase expression ψ_i in Eq. 12-2 is then given by

$$\psi_i = k[\zeta_i(1 + \cos\beta) + \xi_i \sin\beta], \tag{12-45}$$

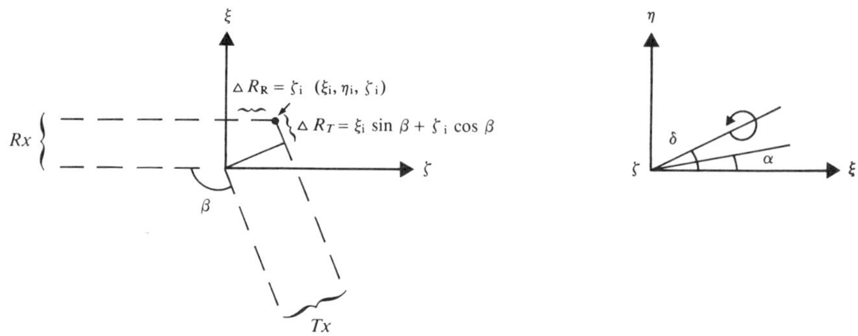

Figure 12-8 Transmitter, receiver, and target in a bistatic situation.

where k is the wave number, i refers to the ith component scatterer,† and β is the bistatic angle between illuminator and observer with apex at the origin.‡ Many of the a_i vary slowly because of the small dimensions of the scatterer or because of asymptotic methods used to estimate their RCS (Chapter 11).

The lobe structure of the RCS pattern relates to the curvature of the autocorrelation function:

$$Q = \langle F(x)F(x+\Delta)\rangle, \tag{12-46}$$

where F is deviation from the mean as given in Eq. 12-6 and x denotes symbolically the dependence of F on the N random phases.

Using the zero-crossing theorem of Rice (1945), we may obtain the lobing rate, or reciprocal of lobe width,

$$N_\Delta = \sqrt{\frac{1}{m_2}\frac{\partial^2 Q}{\partial \Delta^2}}\Bigg|_{\Delta=0}. \tag{12-47}$$

The results obtained are the number N_f of lobes of RCS per Hertz of frequency:

$$N_f = \frac{1}{c\sqrt{m_2}}\sqrt{\sum_{i\neq k} \sigma_i\sigma_k[(\zeta_i - \zeta_k)(1+\cos\beta) + (\xi_i - \xi_k)\sin\beta]^2}, \tag{12-48}$$

where c is the speed of light.

Since the lobe pattern rotates with the target about the illumination direction (neglecting polarization complications), the lobe statistics must be derived in a manner independent of choice of the coordinate system.

† That is, (ξ_i, η_i, ζ_i) is the scattering center location of the ith component.

‡ For the monostatic case, $\beta = 0$, the orientation of $\hat{\xi}$ and $\hat{\eta}$ axes is ambiguous. So we define $\hat{\xi} = d\hat{\zeta}/d\theta$ and $\hat{\eta} = \hat{\zeta} \times \hat{\xi}$.

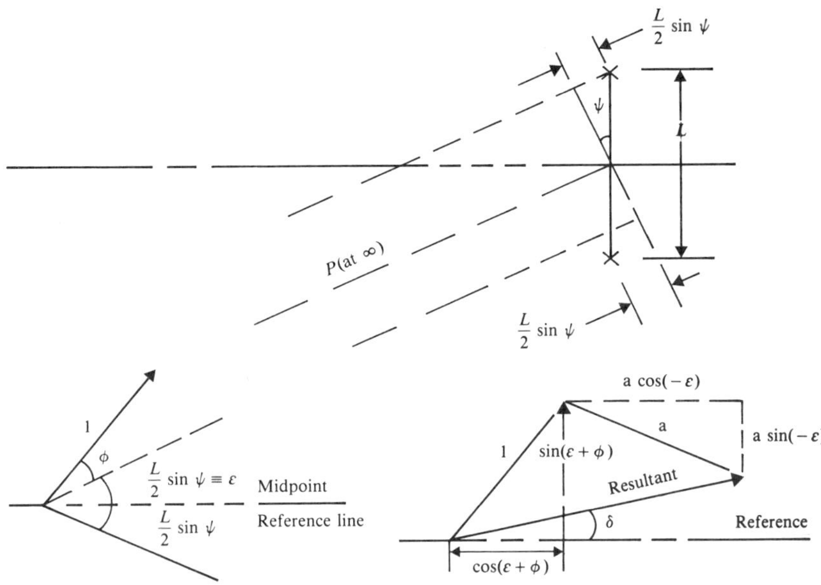

Figure 12-9 Phase behavior for two scatterers separated by a distance L.

Define the two-dimensional tensor

$$A_{rs} = \sum \sigma_i \sigma_k [(\xi_i - \xi_k)(1 + \cos\beta) - (\zeta_i - \zeta_k)\sin\beta]^{4-r-s}$$

$$\times [(\eta_i - \eta_k)(1 + \cos\beta)]^{r+s-2} \qquad r,\ s = 1, 2. \quad (12\text{-}49)$$

Then the number of lobes of RCS per radian in traverse, relative to a rotation axis at an angle δ with the ξ-axis in the $\xi\eta$-plane *is* (Fig. 12-8)

$$N_\delta = \frac{1}{\lambda\sqrt{m_2}} \sqrt{\sum_{r,s} A_{rs}\, d_r\, d_s}\,, \qquad (12\text{-}50)$$

where $d_1 = \sin\delta$ and $d_2 = -\cos\delta$. In the monostatic case N_δ gives a target rotation about a vertical axis.

12-8 GLINT

Glint is a scattering phenomenon that may induce an error in measured target position. It usually refers (Wright, 1984) to an error, as measured by a radar tracker, in apparent target position, caused by constructive and destructive interference of backscattered energy from more than one scattering center. It may occur in both range and angle (cross-range); we confine our attention to the latter in this work.

Glint is generally taken to be a random phenomenon; however, it can be a deterministic function of target aspect angles for a rigid target. In this case it can be measured with repeatability on a radar range. Glint can also be computed theoretically, *provided* that amplitudes and relative phases of *all* target scattering centers are known. But this knowledge is precluded by our fundamental assumption of randomness of relative phases. So we shall evolve *glint statistics* (Section 12-8-3), based (1) on our assumption of random phases among scattering centers and (2) on an interpretation of Howard (1959) that glint can be represented as a tilt, or distortion, of the scattered wavefront.

12-8-1 Two-Component Case

Consider a target composed of only two components separated by a distance L having a relative phase ϕ and an amplitude ratio a (Fig. 12-9). We wish to compute the phase angle δ of the resultant signal of this two-component target. Indeed, we really wish to compute the derivative δ' of this phase angle with respect to aspect ψ, which will represent the tilt of the phase front of the signal from the two components (or the apparent displacement of the phase center of the two components).

We find, letting ψ represent the aspect angle from the normal to the segment connecting the two components (see figure preceding), ϕ an

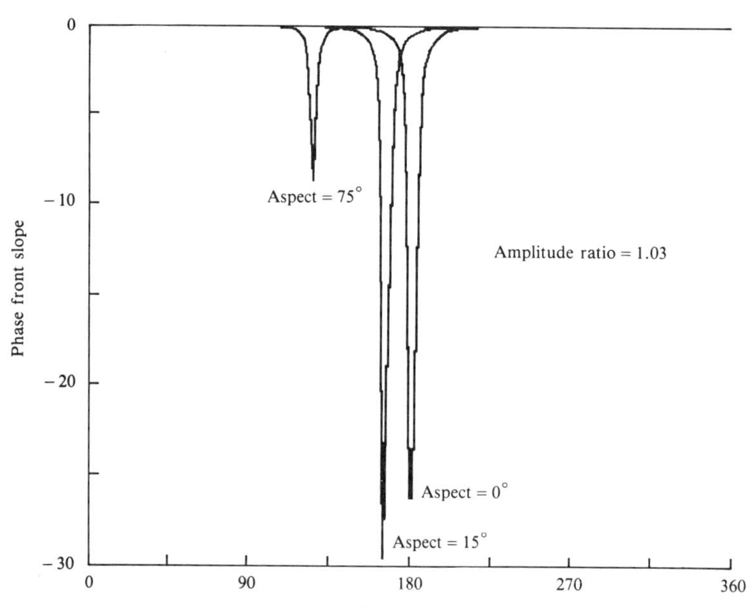

Figure 12-10 Glint: phase front tilt.

arbitrary constant phase and a the amplitude ratio of the two components:

$$\delta = \tan^{-1}\left[\frac{\sin\left(\phi + \dfrac{kL}{2}\sin\psi\right) + a\sin\left(\dfrac{kL}{2}\sin\psi\right)}{\cos\left(\phi + \dfrac{kL}{2}\sin\psi\right) + a\cos\left(\dfrac{kL}{2}\sin\psi\right)}\right]. \qquad (12\text{-}51)$$

Howard (1959) notes that the phase angle δ as a function of ψ can be thought of as the phase that would result if a single component were located at the center of the segment separating the two components. The slope δ' of the phase front is found, as was mentioned above, by differentiating δ with respect to ψ. The result is

$$\delta' = \frac{kL\cos\psi}{2}\,\frac{1-a^2}{1+a^2+2a\cos(\phi + kL\sin\psi)}. \qquad (12\text{-}52)$$

Howard further points out that the expression of δ' for $\psi = 0$ agrees with an earlier result obtained by Meade (1955). Meade referred to his result as the apparent location of the dual component target with respect to the midpoint of the separation L between the components.

These results are illustrated in Figs. 12-10 and 12-11 which show δ' versus phase ϕ for several parameter values. The first shows the apparent displacement of the phase center for a fixed component amplitude ratio of

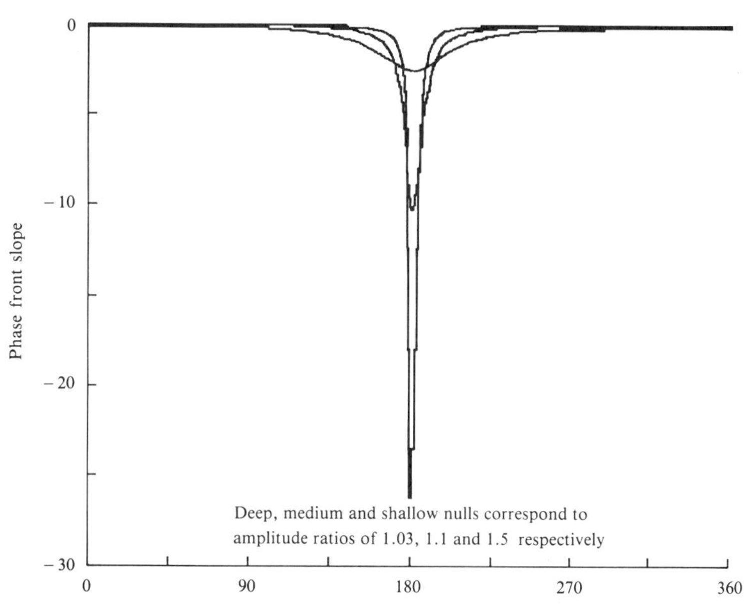

Figure 12-11 Glint: phase front tilt.

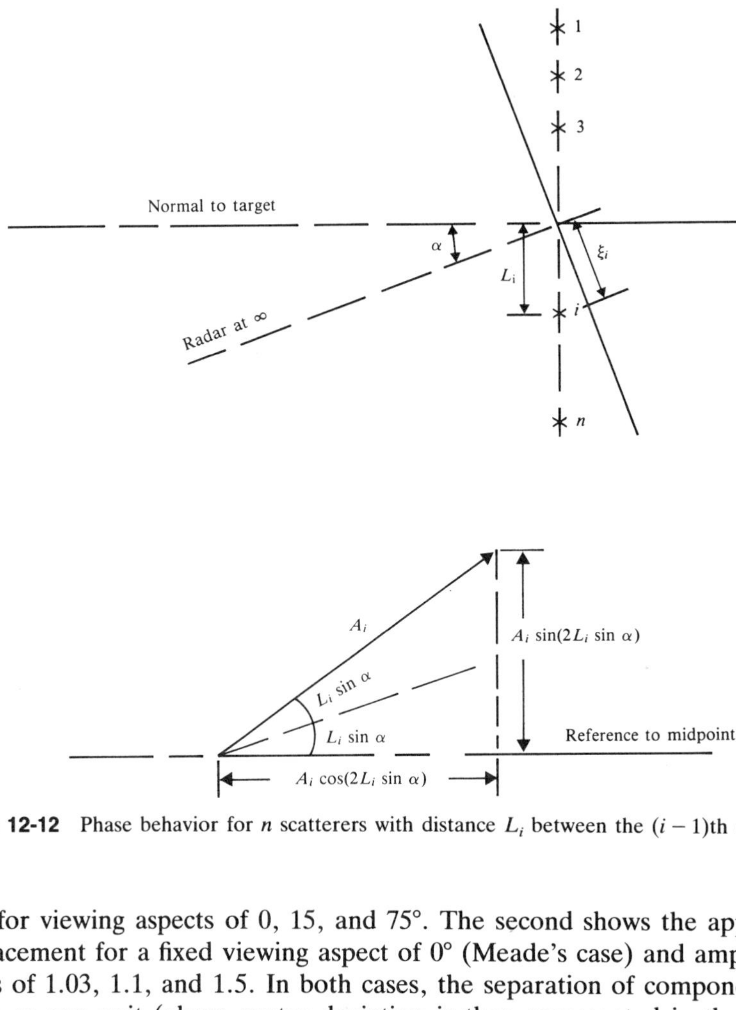

Figure 12-12 Phase behavior for n scatterers with distance L_i between the $(i-1)$th and the ith.

1.03 for viewing aspects of 0, 15, and 75°. The second shows the apparent displacement for a fixed viewing aspect of 0° (Meade's case) and amplitude ratios of 1.03, 1.1, and 1.5. In both cases, the separation of components is taken as one unit (phase center deviation is then represented in the same units).

12-8-2 *n*-Component Case

Let n components having amplitudes a_i be arranged along a line at distances L_i from the center. Let α be the aspect angle measured from the normal to the line (see Fig. 12-12). Then the phase of the resultant signal is

$$\delta = \tan^{-1}\left[\frac{\sum_{i=1}^{n} A_i \sin(2kL_i \sin \alpha)}{\sum_{i=1}^{n} A_i \cos(2kL_i \sin \alpha)}\right]. \tag{12-53}$$

Differentiating with respect to α yields

$$\delta' = \frac{\sum_{i=1}^{n} \sum_{j=1}^{n} A_i A_j \xi_i \cos(\omega_i - \omega_j)}{\sum_{i=1}^{n} \sum_{j=1}^{n} A_i A_j \cos(\omega_i - \omega_j)} \tag{12-54}$$

where $L_i \cos \alpha$ has been set equal to ξ_i and $2kL_i \sin \alpha$ has been put equal to ω_i.

12-8-3 Glint Statistics

The extension of the Poynting vector of the scattered field (as measured by the receiver) to $\zeta = 0$ gives the instantaneous RCS centroid vector

$$\xi, \eta = \frac{\sum_{i,j} (\xi_j, \eta_j) a_i a_j \cos(\psi_i - \psi_j)}{\sum_{i,j} a_i a_j \cos(\psi_i - \psi_j)} \tag{12-55}$$

which is a linear approximation near the center of a tracking beam. The denominator is instantaneous mean RCS (note the sum is over all i, j); its possible singularity may be due to the impossibility of defining the arrival direction of a nonexistent signal. A tracking radar would tend to deflect the vector on the order of a beam width but the actual direction depends on tracking circuit configuration, signal-to-noise ratio, and duration of fade and circuit time constants.

The above equation is intractable for calculation of average centroid.† But if the denominator is averaged separately (slow automatic gain control, or AGC), then the integrals over relative phase can be carried out:

$$\bar{\xi}, \bar{\eta} \approx \frac{1}{m} \sum_j (\xi_j, \eta_j) \sigma_j. \tag{12-56}$$

Suppose an observation axis is located in the $\xi\eta$-plane (Fig. 12-8) at an angle α with the ξ-axis. Setting

$$a_1 = \cos \alpha, \qquad a_2 = \sin \alpha, \tag{12-57}$$

(The angle α is shown in Fig. 12-8), the *standard deviation of glint in the α-direction* can be shown to be

$$S_\alpha = \frac{1}{2m} \sqrt{\sum_{m,n} B_{mn} a_m a_n}, \tag{12-58}$$

† If one component dominates RCS, then that component is tracked, with small variance, just as an FM receiver tracks the strongest station. For three components the mean and variance are known, but there are no results for most targets.

in the random phase approximation; the summand here represents just a rotation of coordinates of the expression

$$B_{mn} = \sum_{i \neq j} \sigma_i \sigma_j (\xi_i + \xi_j - 2\bar{\xi})^{4-m-n} (\eta_i + \eta_j - 2\bar{\eta})^{m+n-2}, \qquad m, n = 1, 2 \tag{12-59}$$

to the α-direction.

Since $\langle \sum_{i,j} a_i a_j \cos(\psi_i - \psi_j) \rangle$ (the denominator of Eq. (12-56) is always positive and no doubt enhances the variance when small, it cannot change sign; thus the lobing frequencies which follow also apply for fast AGC.

The number of *lobes of glint/Hertz* relative to the α-axis of observation is given by

$$N_{\alpha,f} = \frac{1}{2cS_\alpha} \sqrt{\sum_{m,n} C_{m,n} a_m a_n}, \tag{12-60}$$

where

$$C_{mn} = \sum_{i \neq j} \sigma_i \sigma_j (\xi_i + \xi_j - 2\bar{\xi})^{4-m-n} (\eta_i + \eta_j - 2\bar{\eta})^{m+n-2} \times [(\zeta_i - \zeta_j)(1 + \cos\beta) + (\xi_i - \xi_j)\sin\beta]^2, \qquad m = 1, 2. \tag{12-61}$$

For the glint lobing rate caused by target rotation, both rotation axis orientation δ and observation axis orientation α enter the act as follows:

$$N_{\alpha,\delta} = \frac{1}{2\lambda S_\alpha} \sqrt{\sum_{m,n,r,s} D_{mnrs} a_m a_n \, d_r \, d_s} \tag{12-62}$$

where the fourth rank tensor D_{mnrs} is given by

$$D_{mnrs} = \sum_{i \neq j} \sigma_i \sigma_j (\xi_i + \xi_j - 2\bar{\xi})^{4-m-n} (\eta_i + \eta_j - 2\bar{\eta})^{m+n-2} \times [(\xi_i - \xi_j)(1 + \cos\beta) - (\zeta_i - \zeta_j)\sin\beta]^{4-r-s} \times [(\eta_i - \eta_j)(1 + \cos\beta)]^{r+s-2}, \qquad m, n, r, s = 1, 2. \tag{12-63}$$

In all of the above formulas for lobing rates, the summation inside the square root may be written as

$$Q'' = 2 \sum_{i<j} \sigma_i \sigma_j V_{ij} \tag{12-64}$$

by summing over the tensor indices first. The variance of this expression may be defined in the usual way as expectation of the square minus the square of the expectation. The result is then

$$\mathrm{VAR}\, Q'' = 4 \sum_{i \neq j < k} \sigma_i^2 \sigma_j \sigma_k (V_{ij}^2 + 4V_{ij}V_{ik} + V_{ik}^2) + 4 \sum_{i<j<k<h} \sigma_i \sigma_j \sigma_k \sigma_h [4(V_{ij}V_{kh} + V_{ik}V_{hj} + V_{ih}V_{jk}) \times (V_{ij} + V_{ik} + V_{ih} + V_{jk} + V_{jh} + V_{kh})^2]. \tag{12-65}$$

13

EXAMPLES: 727 AT 0.94 GHz; FIREBEE AT 9.0 GHz

We now examine our theory of the β-distribution of complex target RCS at both L- and X-bands, since measured data at these frequencies are available to us. At L-band we examine the Boeing 727–100C (Section 13-1); at X-band, the Teledyne Ryan Firebee (Section 13-2). Since there are no great differences between data for HH and VV polarizations, we choose to discuss, almost exclusively, the HH cases.

We use a computer code, RCS, developed by the writer (in cooperation with H. W. Klimach of Environmental Research Institute of Michigan and F.C. Paddison and J.W. Follin, Jr. of the Johns Hopkins University Applied Physics Laboratory) to compute and graph all first and higher order statistics of RCS. Our examples are discussed in terms of the input quantities required by the code and the output statistics of RCS yielded by the code.

13-1 THE 727–100C

The first of our targets, the 727–100C, was measured by Thomas Dalby of the Boeing Airplane Company as a 1/100 scale model at a modeling frequency of 94 GHz. Figures 13-1a and 13-1b show a picture† and a drawing of the target, reespectively.

13-1-1 Coordinate Transformations and Target Definition

The 727 target is decomposed into a set of 50 component scatterers. For each of these it is necessary to describe a (linear) transformation from the

† Courtesy Northwest Orient Airlines.

Figure 13-1a Photo of a Boeing 727–100C aircraft.

generic coordinate system of the component to the coordinate system of the complex target. The transformation is called TRANS; it consists of nine direction angles whose direction cosines are given in the form:

$$\begin{matrix} \hat{x}\cdot\hat{x}' & \hat{x}\cdot\hat{y}' & \hat{x}\cdot\hat{z}' \\ \hat{y}\cdot\hat{x}' & \hat{y}\cdot\hat{y}' & \hat{y}\cdot\hat{z}' \\ \hat{z}\cdot\hat{x}' & \hat{z}\cdot\hat{y}' & \hat{z}\cdot\hat{z}', \end{matrix}$$

where the caret denotes a unit vector and the primed and unprimed variables refer, respectively, to component and complex target systems.

A visible, or nonshadowed, region must be prescribed for each component. This instruction is accomplished with the command VALID, followed by limitations on the polar angle θ and limitations on the azimuth angle ϕ. Two such sequences are possible; for example, θ_1, θ_2, ϕ_1, ϕ_2, θ_3, θ_4, ϕ_3, ϕ_4, where subscript order indicates relative angle size. If order is reversed (e.g., θ_2, θ_1), then that region is *excluded.* These instructions are phrased in the target coordinate system. In addition, an instruction, called LEGAL, is included. It allows the user to define a usable region for each component in its generic coordinate system. The procedure is identical with the procedure for VALID.

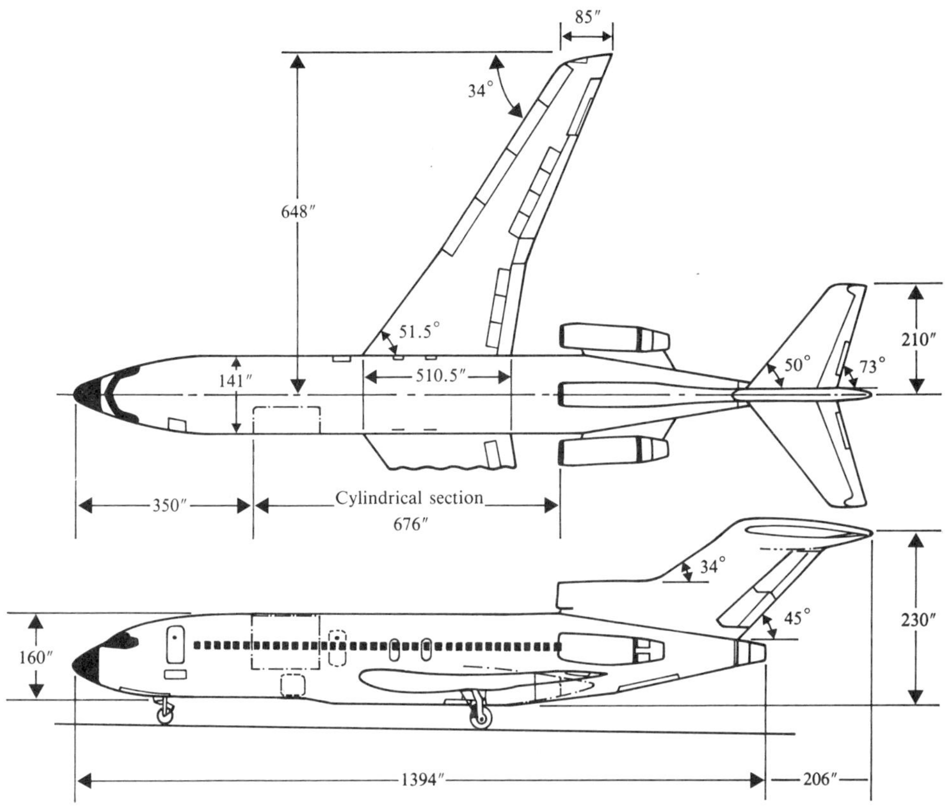

Figure 13-1b Drawing of RCS model of the Boeing 727–100C, with full-scale dimensions.

Since we wish to compute lobe and glint statistics, we must give a location of the scattering center of each component. This requirement is accomplished with the command ORGN, after which is given a three-number set giving the scattering center location in the complex target coordinate system.

Finally, the characteristics of the individual scattering component are specified. Most of the components used rely on physical optics results, which also include dihedral and trihedral (or corner) scatterers. However, the treatment of wedges and, in fact, of edges (both straight and curved) in general is based on Ufimtsev's physical theory of diffraction. It would also be useful to incorporate numerical results (such as NEC) for the treatment of small regions that cannot be well modeled by simpler methods. At present the code offers about 25 simple scatterer shapes to choose from for component models.

TABLE 13-1 RCS File with Sample of 727–100C Components

```
COMM 15
************************
************************
1/17/84, 8/25/85, 1/29/86
727-100C
RCS STATISTICS
HORIZONTAL POLARIZATION
0.94 GHz
************************
In PRINT3 mode STATS 3,n, n an integer, will yield 1st order stats
at every THETA value and HO stats at every nth value of THETA. 1st
order glint stats are also recorded in GLINT1 mode. If a minus sign
preceeds the n3 parameter of STATS, cumulative rather than density
curves are computed.
************************
************************
DATE 1
POLAR 2
WLEN .319149
LOBE 1
DELTA -150,150,60
GLINT 0
STAT 3,30,5,30,60,90,120,150
RPRCS 0
RADAR 3,0,2,0,1,0
PHIR -90
THETA -180,180,1
COMM 1
NOSE
TRANS 0,90,90,90,0,90,90,90,0
VALID 45,90,-90,90
ORGN .73,0,-7.32
SOLID 1.83,1.9,5.49
COMM 1
******
PORT WING
TRANS 90,0,90,114,90,24,24,90,66
VALID 0,90,-180,0
ORGN -.55,-15,-24.9
CONE 15,7.29,4,8.97,15.92
TRANS 88,2,90,113.984,89.187,24,24.078,91.827,66
VALID 113,180,-180,0,150,180,45,180
LEGAL 1,179,1,-5
ORGN -.55,-16.47,-27.45
WEDGE 356,10.98,2570
******
IV. INLETS
STBD (5)
TRANS 89.817,3.5,86.505,177.006,90,87.006,87.012,93.5,4.605
VALID 0,80,-45,105
ORGN 1.1,6.77,-24,52
CAVD 1
******
GRAPH 3,-180,180,7,3,-10,50,7,2
BOEING 727-100C
```

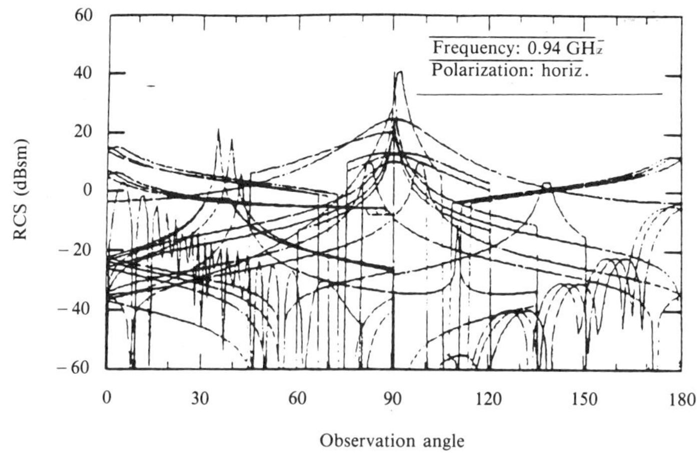

Figure 13-2 RCS of individual components representing the Boeing 727–100C.

A file illustrating some of the commands TRANS, VALID, ORGN, LEGAL, and COMPONENT for several components of the 727 is recorded in Table 13-1 (the final two lines are graphing commands). An RCS pattern is computed for each of the 727 components, and the results overlaid on one graph in Fig. 13-2. These patterns form the ingredients from which we obtain the statistics of RCS.

13-1-2 First Order Statistics

These statistics have been around for some 30 years. They include the mean RCS and standard deviation of RCS (m and $\sqrt{m_2}$ above). We should expect (from the β-distribution) that some 60% of measured RCS values would fall in the interval from $m - \sqrt{m_2}$ to $m + \sqrt{m_2}$. For example, taking $a = 2$, $b = 3$ gives $m = 2/5$, $\sqrt{m_2} = 1/5$. Then

$$\frac{\Gamma(5)}{\Gamma(2)\Gamma(3)} \int_{1/5}^{3/5} x(1-x)^2\, dx = 0.64.$$

This hypothesis is borne out by the experiences of 30 years. Figures 13-3, 13-4, and 13-5 show that the measured data and the theoretical statistical RCS estimates for the 727–100C compare very well.

13-1-3 Statistics for Lobe Structure

Included on the theoretical statistical graph (Fig. 13-3) is a set of computed statistics describing the lobe structure (refer to Section 12-7) of the 727–100C at 30° intervals. The lobe characteristics are shown pictorially by

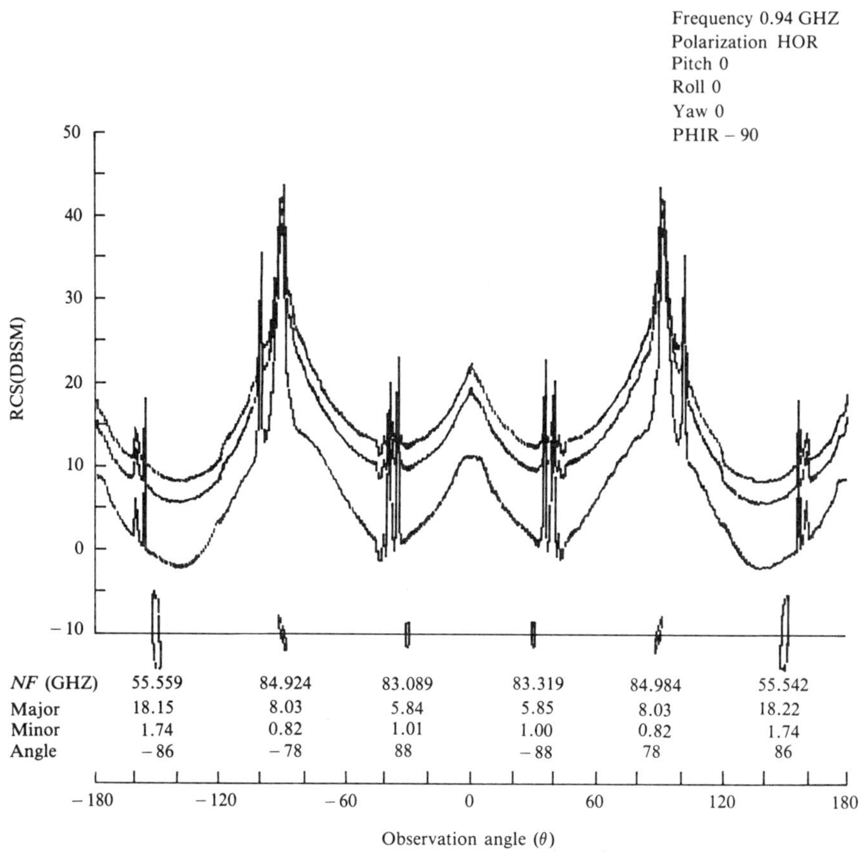

Figure 13-3 Boeing 727–100C monostatic full-scale RCS (computed) for horizontal polarization at 0.94 GHz.

ovals that are always closed (even though some are truncated so as not to interfere with the RCS curves).

The lobe characteristics are also described numerically by the sets of numbers under the horizontal axis. The length of the observation axis intercepted by an oval gives the *average* lobe width at that aspect in the observation plane. The maximum lobe width and its appropriate orientation plane of traverse are indicated by the maximum dimension of the oval figure and its tilt with respect to the observation axis. The first line of numbers NF (GHz) gives the number of lobes per GHz of frequency change.

13-1-4 Higher Order Statistics

These statistics are actually the special statistics of width, skewness, and kurtosis of RCS which are all based on the second, third, and fourth

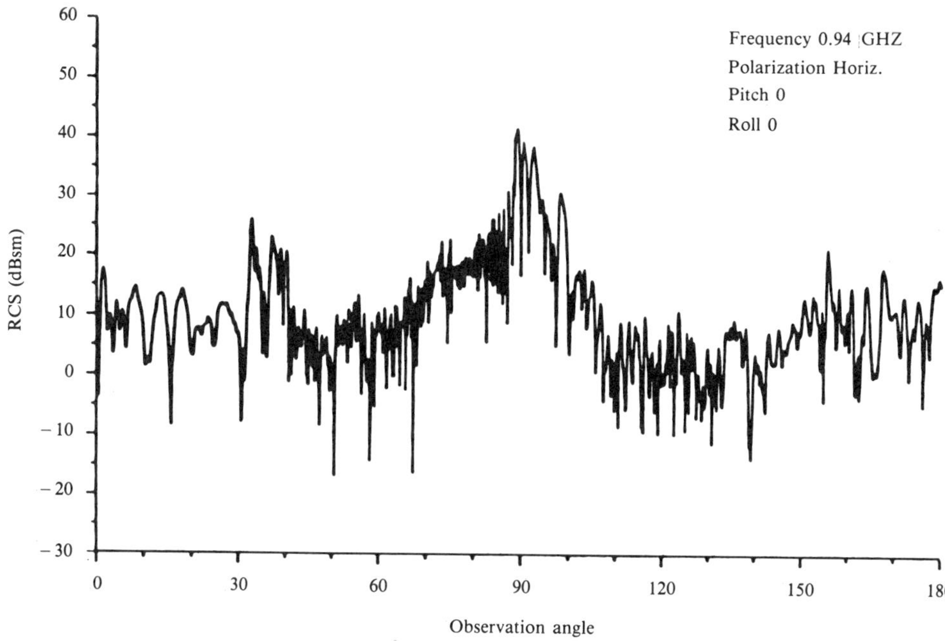

Figure 13-4 Boeing 727–100C monostatic full-scale RCS (measured) for horizontal polarization at 0.94 GHz.

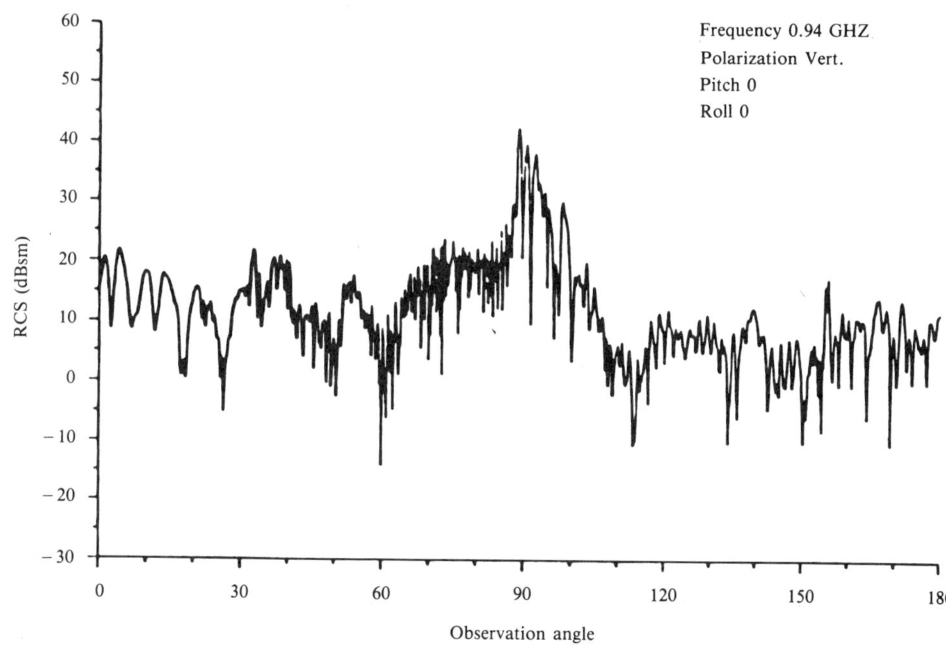

Figure 13-5 Boeing 727–100C monostatic full-scale RCS (measured) for vertical polarization at 0.94 GHz.

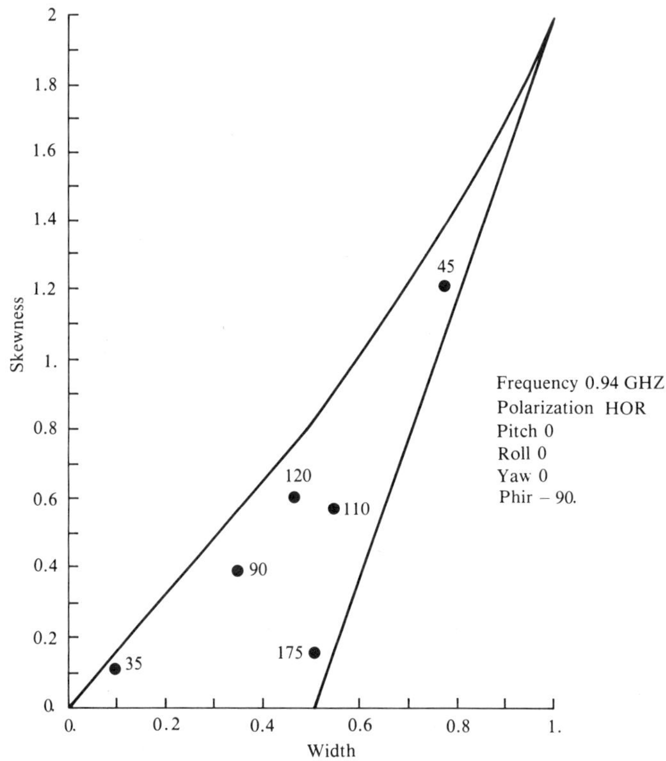

Figure 13-6. Selected width-skewness points for the Boeing 727–100C.

moments of RCS. Figures 13-6 and 13-7 show a selection of (w, g) and (w, K) points that delineate their range for the complex 727–100C target. In fact, for this target, the bulk of width-skewness and width-kurtosis points is concentrated between the 45 and 110° points near the right border of each allowable region. Thus, similar, and very Rayleigh-like, RCS distributions obtain for the 727 at almost all aspects in the horizontal plane. (Recall that a β-density resembles a Rayleigh density at those aspects of a complex target where the component ensemble consists of many members with very similar amplitudes. But recall also the very important point that the β-density extends only to the maximum RCS of the target while the Rayleigh density extends to infinite RCS.)

Densities that look Rayleigh (e.g., Fig. 13-10) correspond to points throughout the region along the right border of the allowable regions, beginning at the Rayleigh point (the topmost point of the allowable regions)

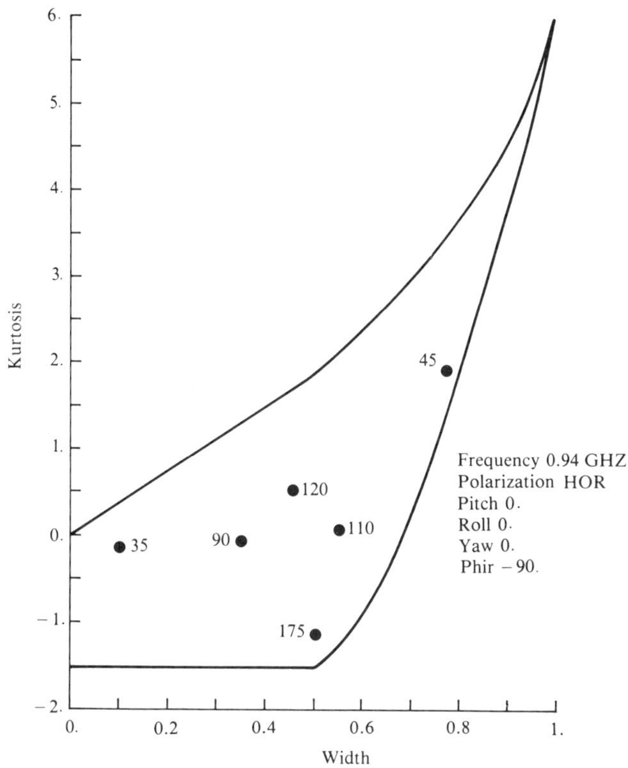

Figure 13-7 Selected width-kurtosis points for the Boeing 727–100C.

and extending roughly two-thirds of the way to the lower right corner. For points below this level and toward the left border, the corresponding density functions begin to resemble a χ^2; and for points further toward the lower left corner, their appearance becomes Gaussian.

If component scatterers are assumed to have other than constant or slowly varying amplitudes, then the scintillation statistics may change their behavior. Figure 13-8 shows computations for a number of w, g points assuming nonvarying or slowly varying component RCS. All of these computed points fall within the allowable region. When the computations are redone for 30 and 100° with the assumption of variable components (using Eqs. 12-30–12-33), we find that the w, g points have moved well outside of the allowable region. So amplitude components that are other than constant or slowly varying may indeed change the complex target scintillation statistics drastically.

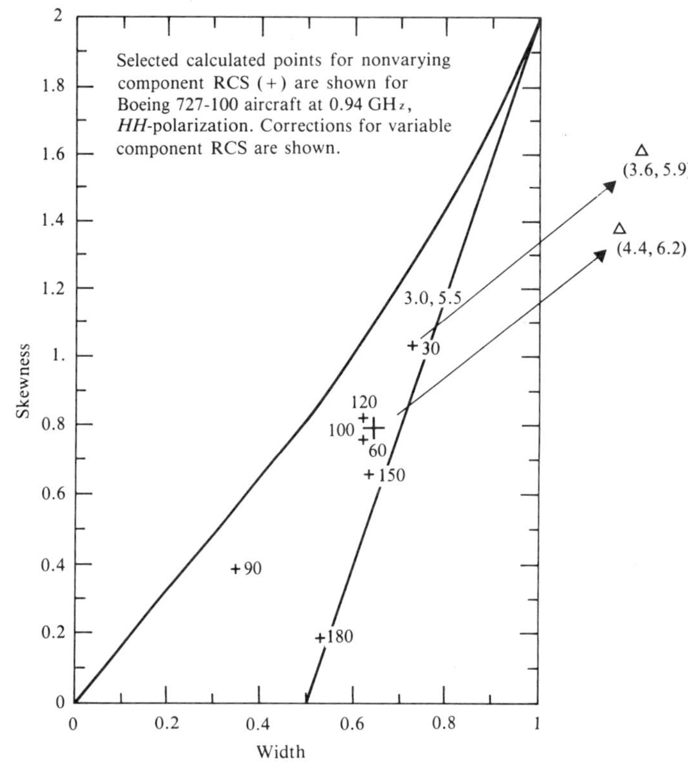

Figure 13-8 Allowable RCS scintillation region.

13-1-5 Distribution of RCS

For each (w, g) point of Fig. 13-6, and corresponding (w, K) point of Fig. 13-7, the β-parameters a and b (or n and p) have been computed and the appropriate β-density function plotted in Figs. 13-9–13-14. One sees at once that the selection of (w, g) and (w, K) points also illustrates the variety of different shapes that the RCS density may take: the resemblances range from Rayleigh, through χ^2, to Gaussian. However, as was remarked in the previous section, for most aspects in the horizontal plane the distributions of 727 RCS look very much like Rayleigh distributions. The very important difference, of course, is that the β-distributions have a finite tail, since the maximum RCS of the 727 is a finite number.

13-1-6 Glint

The 727–100C glint patterns versus observation direction have also been computed for α-values of 0 and 90°. Referring to the definition of the angle

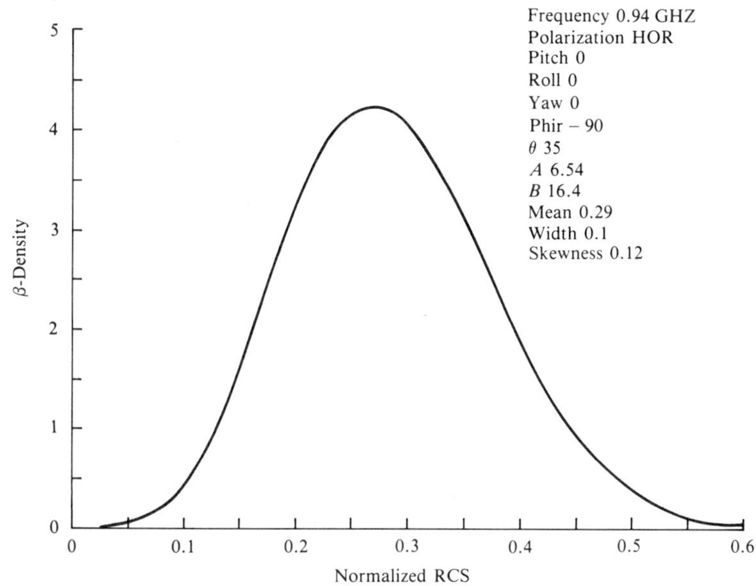

Figure 13-9 RCS distribution for Boeing 727–100C in the vicinity of 35° off-nose in the portside horizontal plane.

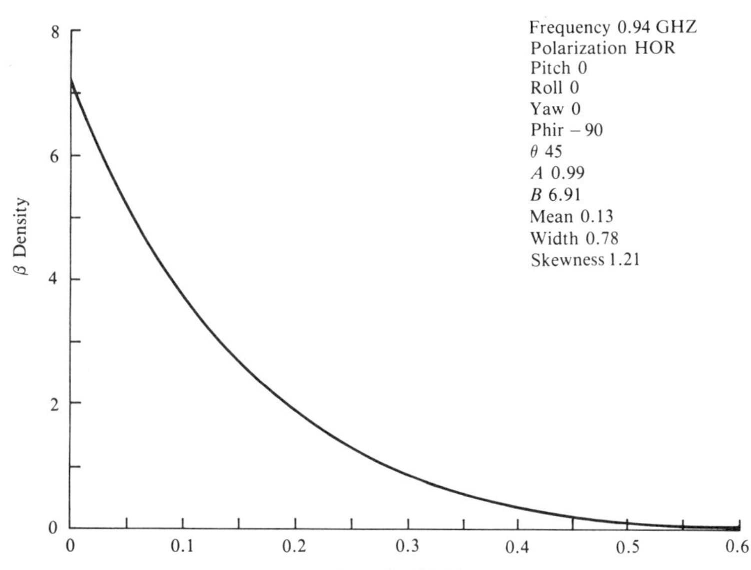

Figure 13-10 RCS distribution for Boeing 727–100C in the vicinity of 45° off-nose in the portside horizontal plane.

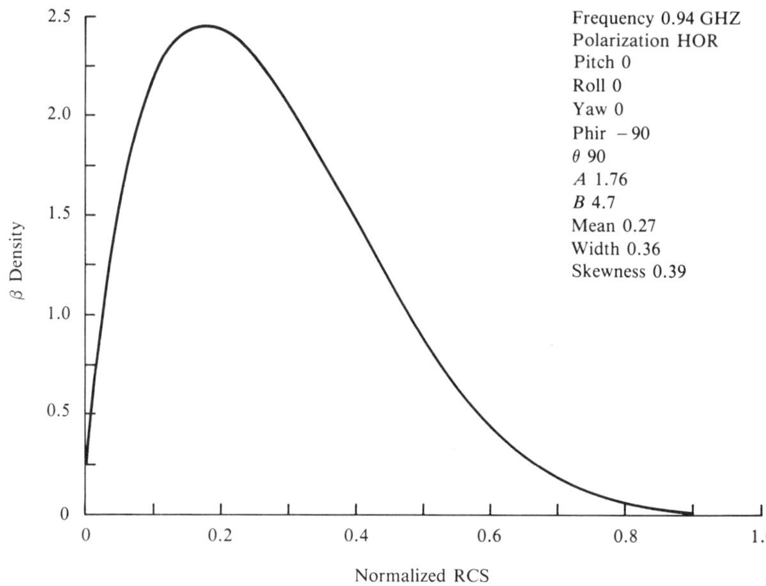

Figure 13-11 RCS distribution for Boeing 727–100C in the vicinity of 90° off-nose in the portside horizontal plane.

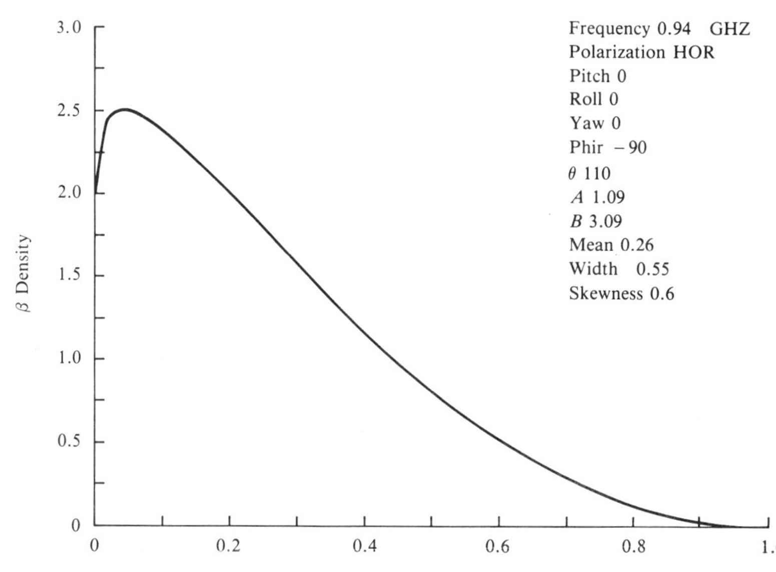

Figure 13-12 RCS distribution for Boeing 727–100C in the vicinity of 110° off-nose in the portside horizontal plane.

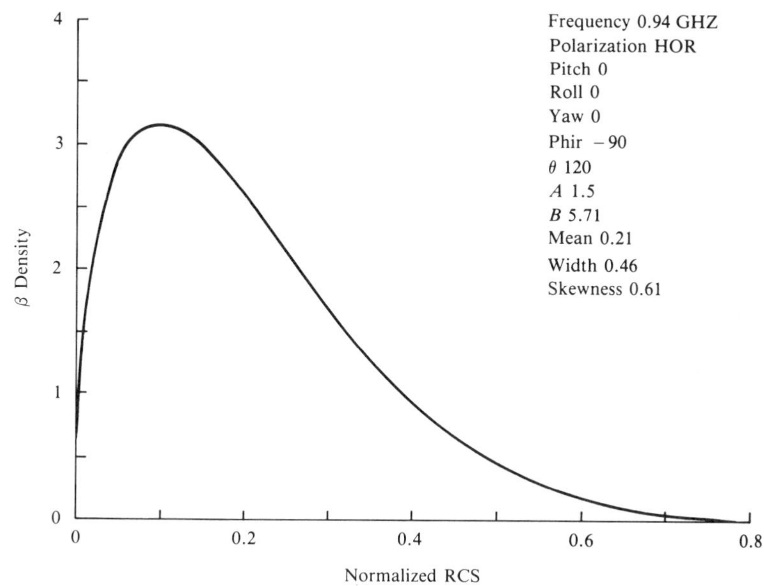

Figure 13-13 RCS distribution for Boeing 727–100C in the vicinity of 120° off-nose in the portside horizontal plane.

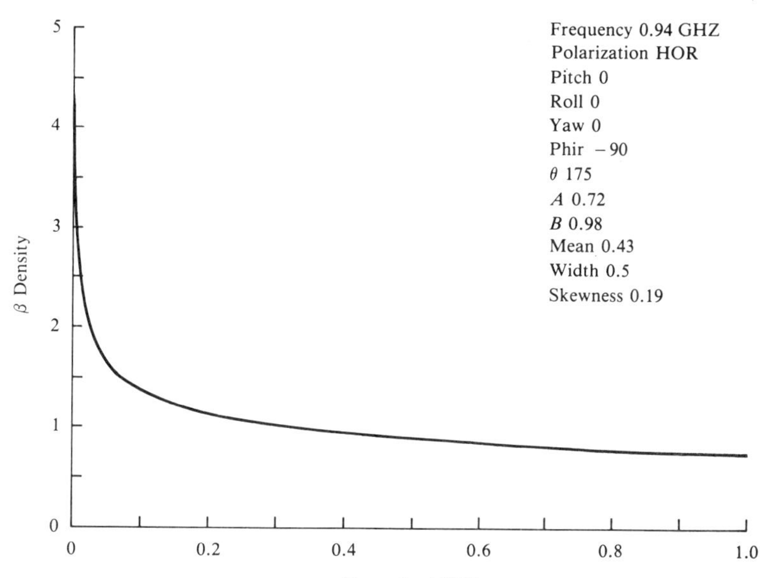

Figure 13-14 RCS distribution for Boeing 727–100C in the vicinity of 175° off-nose in the portside horizontal plane.

α, we see that in the monostatic case ($\beta = 0$) there is an ambiguous situation regarding the definition of the $\hat{\xi}$- and $\hat{\eta}$-directions. So we proceed as indicated in the footnote earlier: letting $\hat{\zeta}$ denote direction from receiver to target, define $\hat{\xi} = d\hat{\zeta}/d\theta$; then further fix $\hat{\eta}$ as $\hat{\eta} = \hat{\zeta} \times \hat{\xi}$. Under these circumstances the case $\alpha = 0°$ corresponds to finding the moment of RCS about an axis ($\hat{\xi}$) parallel to ground and normal to the incidence direction. The case $\alpha = 90°$ corresponds to the moment of RCS about an axis ($\hat{\eta}$) normal to both the incidence direction and the axis parallel to ground.

Figures 13-15 and 13-16 show, respectively, these results, which include not only the mean glint (central curve) but also the mean glint $\pm$ the standard deviation from the mean glint (upper/lower curves). The lobing characteristics of the glint patterns are also calculated and included on the figures. The reader notes at once that the lobing ovals (with the same interpretation as that given for the scintillation lobing) are essentially nonexistent, implying that the glint pattern is smooth at the aspects for which the computations are made.

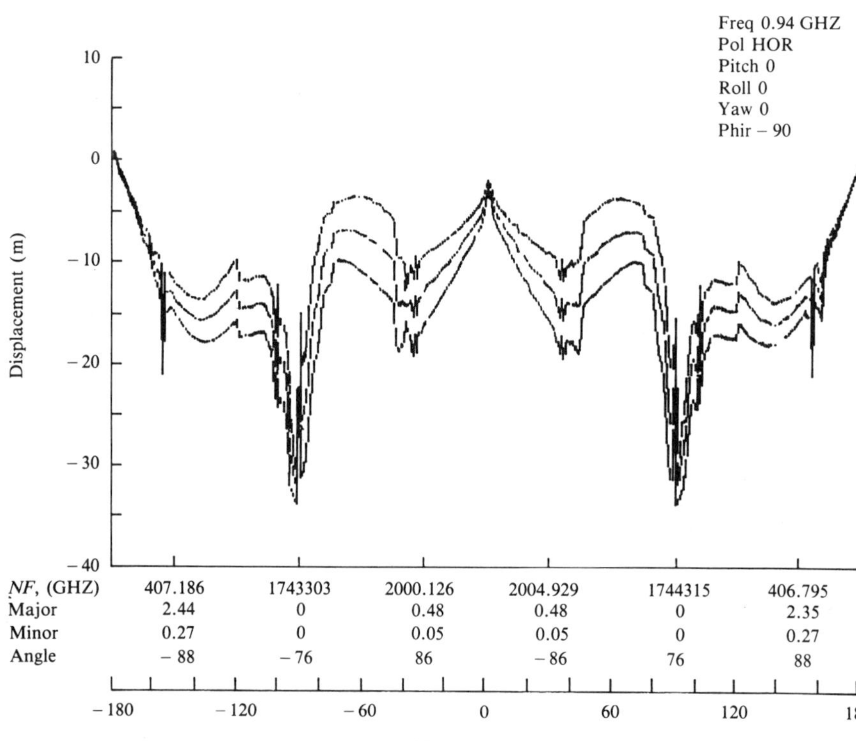

Figure 13-15 Monostatic glint for Boeing 727–100C (moment of RCS about axis parallel to ground and normal to incidence direction).

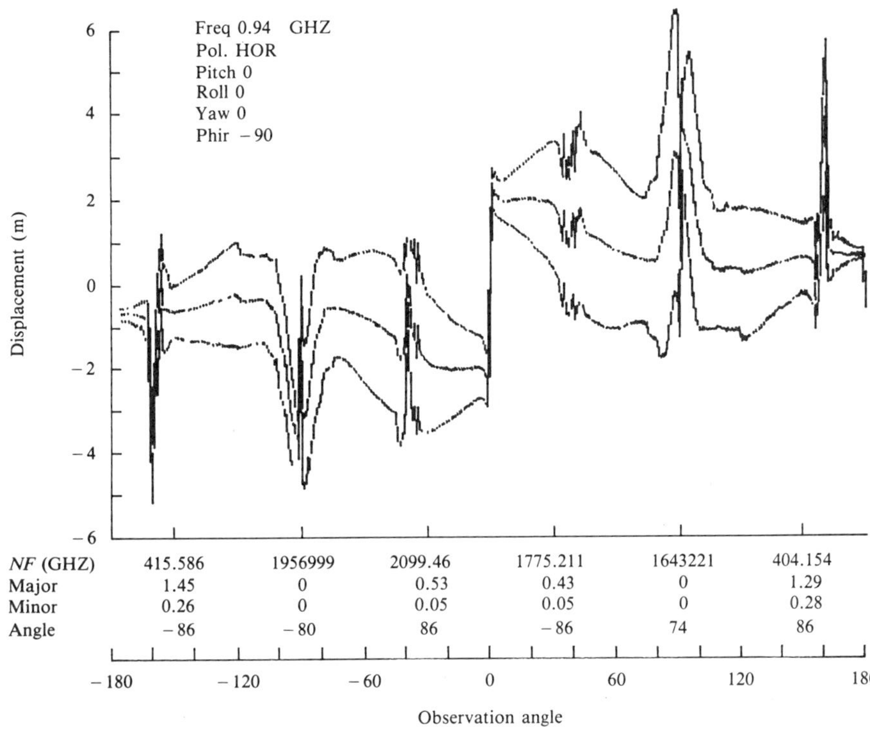

Figure 13-16 Monostatic glint for Boeing 727–100C (moment of RCS about axis normal to both incidence direction and axis parallel to ground).

13-1-7 727–100C Measured Data

For the 727–100C measured data the mean, width, skewness, and kurtosis are computed for 10° aspect sectors (in the horizontal plane). In general, the agreement is quite acceptable (Fig. 13-17). However, there are several sources of discrepancies that should be mentioned:

1. Correlation between RCS fluctuations: Since almost all components will have a linear variation with angle, both positive and negative correlation must exist and largely cancel.
2. The theory does a *two-dimensional averaging* about the aspect of interest, while the experimental data are generally determined from a single traverse. So components at the same distance from the axis of rotation cannot be treated as independent.
3. Errors result from sampling fractions of a lobe.

To evaluate the use of the β-distribution for this complex target, the

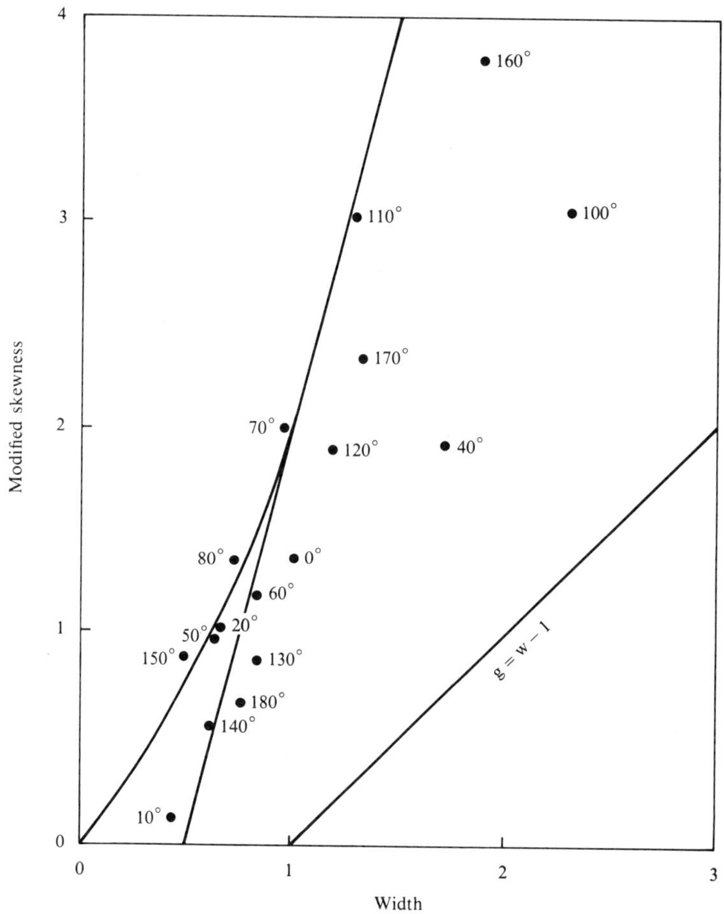

Figure 13-17. Modified skewness versus width using measured RCS data for Boeing 727–100C (10° windows; horizontal polarization).

following procedure is used:

The measured quantities w and g are used to compute the parameters a and b for a β-density; from this density function the kurtosis is determined. These kurtosis values K_β are then compared with the kurtosis K computed directly from the moments of the experimental data. A measure of how well the β-density is describing the complex target RCS is given by how closely K and K_β agree. Figures 13-18 and 13-19 show the results: In Fig. 13-18, K is plotted against K_β with acceptability judged by how near the points lie to a straight line at 45° with either axis. In Fig. 13-19, the two point sets are plotted in w, K space with reference to the allowable region. Roughly half of the points fall nicely within the region. (Vertical polarization results are quite similar.)

Cumulative distributions for experimental RCS data over 10° windows have also been determined. To each experimental distribution a β-distribution is fitted, using the (w, g) points. For comparison, a Rayleigh, a χ^2, and a lognormal are also shown. The RCS abscissa is normalized by the β-parameters:

$$\text{Abscissa} = \frac{a}{a+b} \times \frac{\text{RCS}}{\text{mean}}.$$

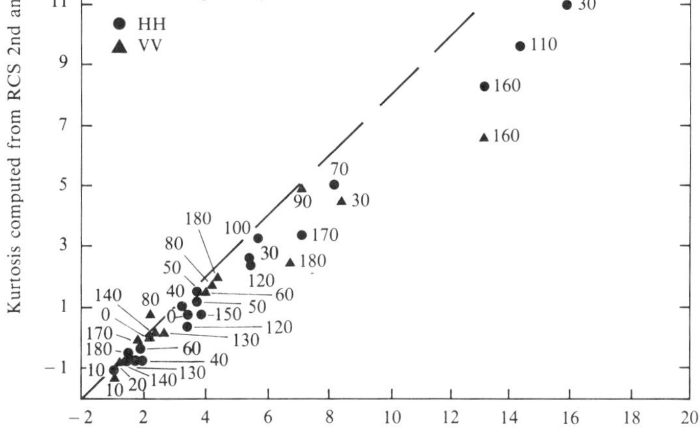

Figure 13-18. Kurtosis computed from second and fourth moments versus kurtosis computed from skewness-derived density for 727–100C measured data, using 10° windows.

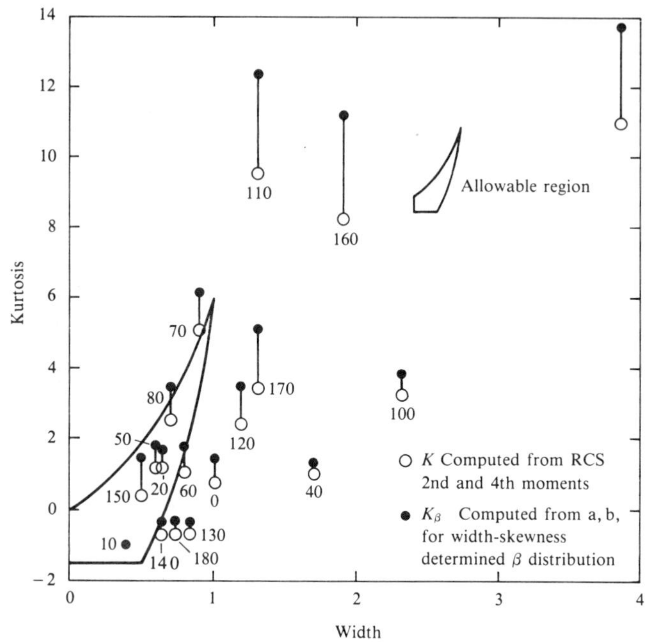

Figure 13-19 A different view of kurtosis computed from second and fourth moments versus kurtosis computed from skewness-derived density for 727–100C measured data, using 10° windows.

This normalization is chosen so as to make the mean of the measured data match the theoretical mean of the β-density; see Eq. 12-19. The data histogram (also drawn as a function of normalized RCS) is shown as an inset for each figure (Figs. 13-20–13-25).

Several comments can be made about these plots, and about the statistics of RCS scintillations:

1. The types of targets addressed are aircraft and missiles where the target RCS return must be distinguished from clutter and receiver noise. Here the aspect angles the target presents to a radar system can be inferred with reasonable accuracy, but in the case of ground targets, any aspect may be presented. Novak and Vote (1979) suggest that the distribution of RCS averaged over many aspect angles may be fitted by a lognormal distribution. We have the following observations:

   (a) Figure 13-26 shows log mean RCS versus probability for the 727–100C aircraft for the 19 data 10° windows covering the aspect

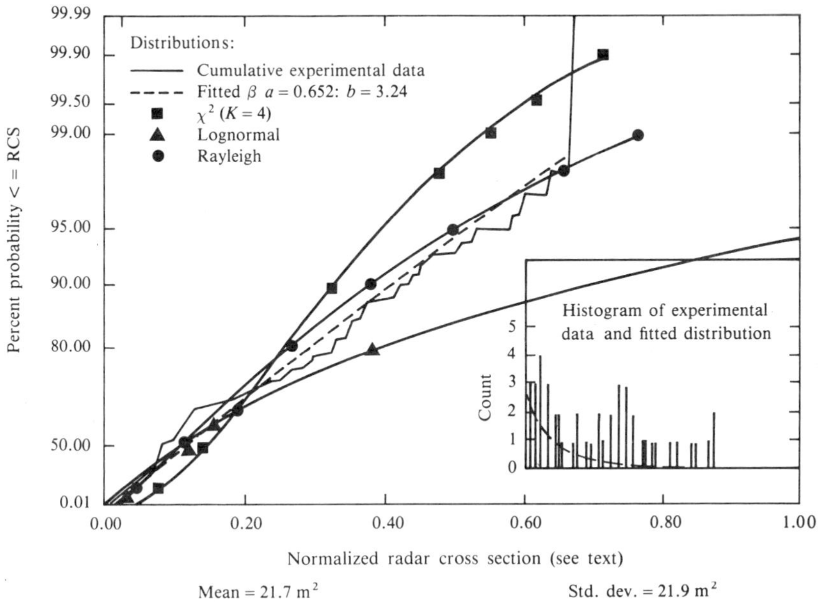

Figure 13-20 RCS distribution for Boeing 727–100C measured data window from −5° to 5° (0.94 GHz, horizontal polarization).

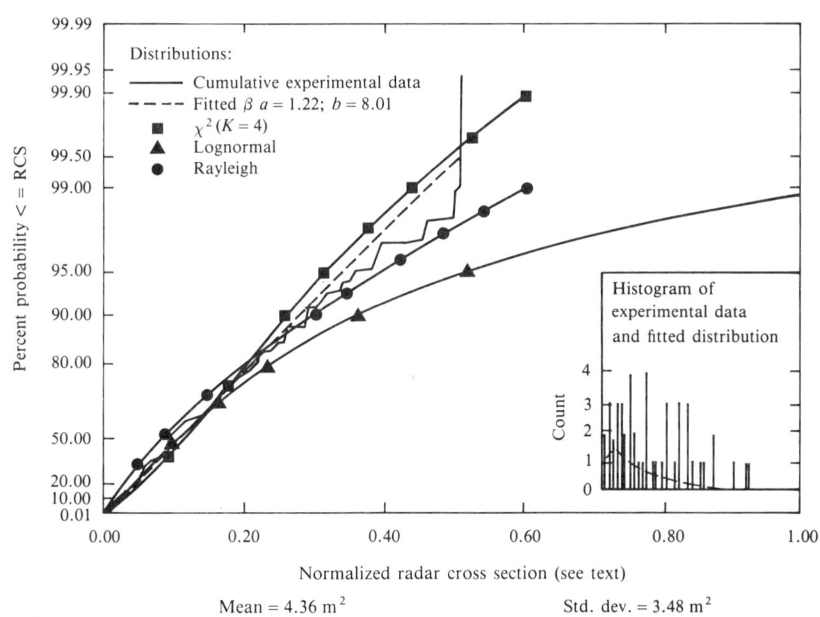

Figure 13-21 RCS distribution for Boeing 727–100C measured data window from 45° to 55° (0.94 GHz, horizontal polarization).

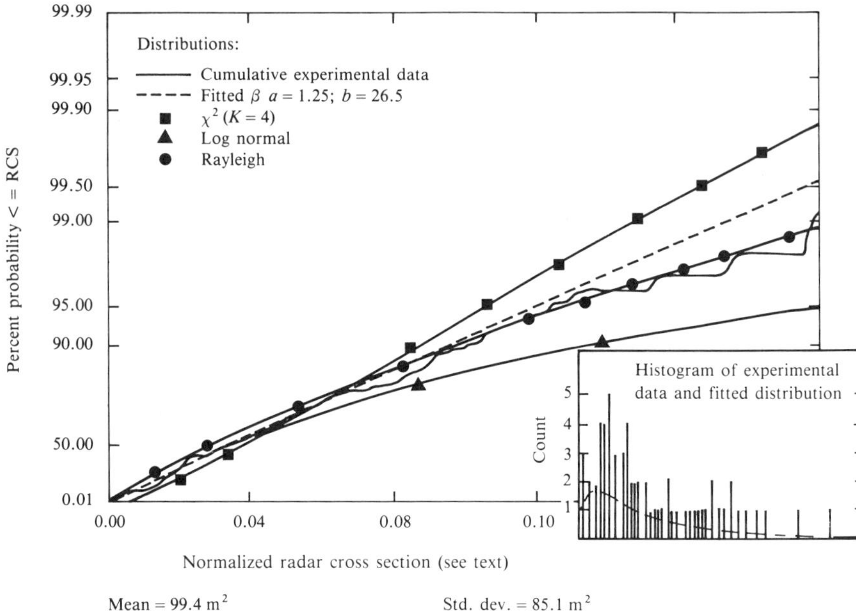

Figure 13-22 RCS distribution for Boeing 727–100C measured data window from 75° to 85° (0.94 GHz, horizontal polarization).

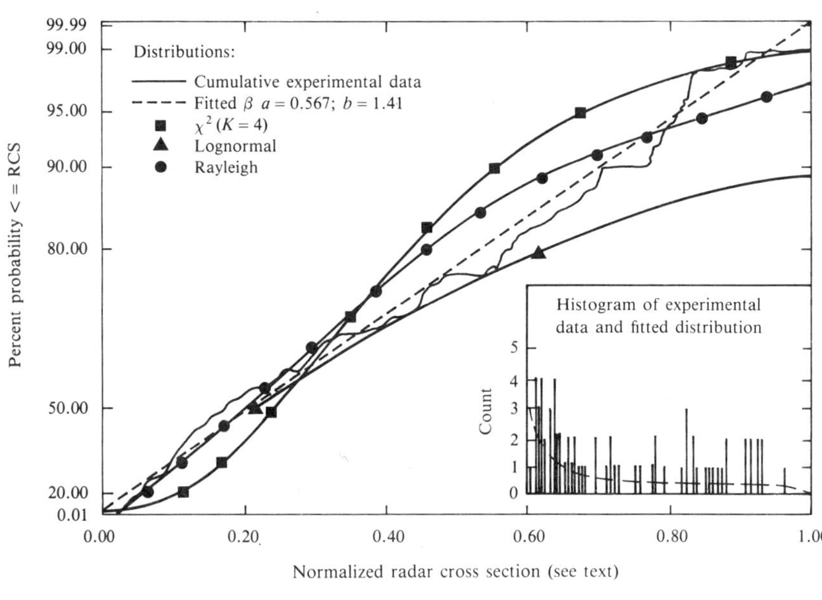

Figure 13-23 RCS distribution for Boeing 727–100C measured data window from 125° to 135° (0.94 GHz, horizontal polarization).

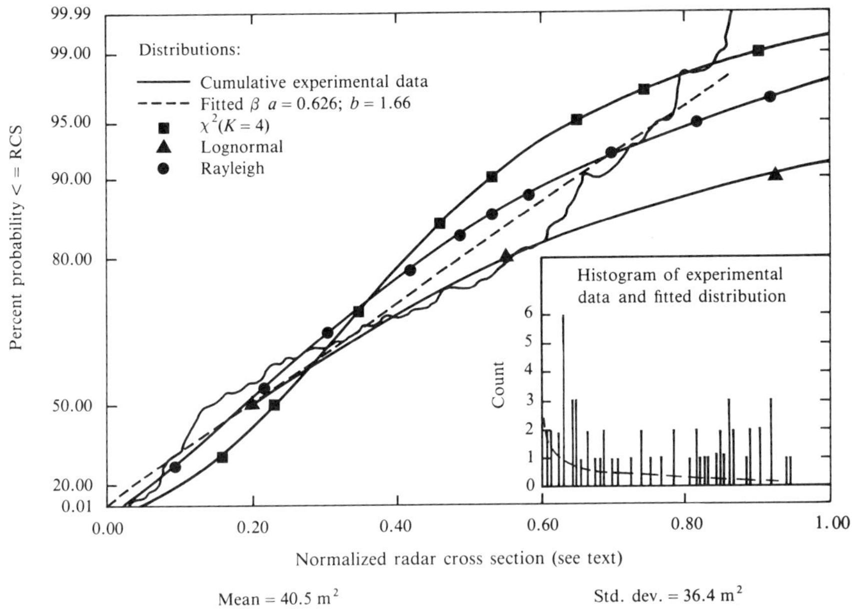

Figure 13-24 RCS distribution for Boeing 727–100C measured data window from 35° to 45° (0.94 GHz, vertical polarization).

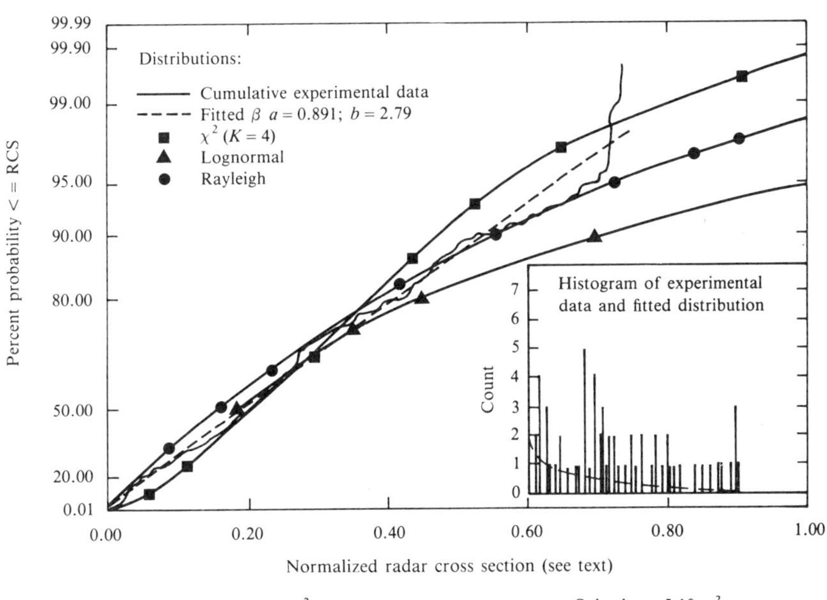

Figure 13-25 RCS distribution for Boeing 727–100C measured data window from 135° to 145° (0.94 GHz, vertical polarization).

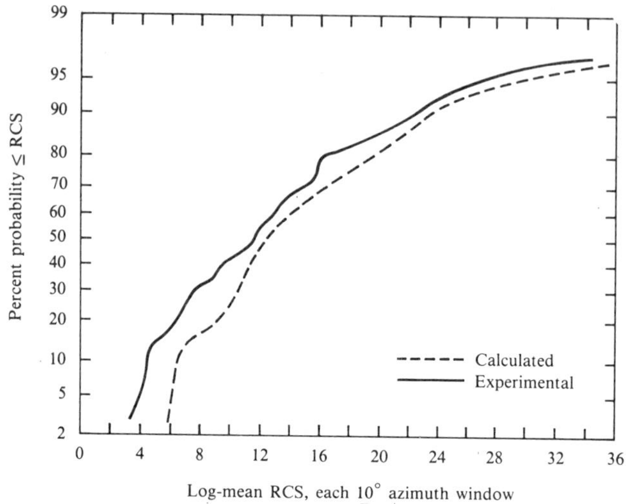

Figure 13-26 Log-mean RCS for each 10° Boeing 727–100C data set, nose to tail.

interval from 0–180°. The plot is not a straight line, indicating that lognormal is not a good approximation for such data.

(b) As Fig. 12-1 shows, the statistics of a series of 4π angular averages of cylinders of varying length are not well represented by lognormal. The same applies to the distribution of sector averages over several lobes of the scintillation pattern.

2. The cumulative distribution curves (see Fig. 14-3 below) provide a type of Kolmogoroff–Smirnov (K–S) test, that is, the difference between the cumulative distribution of the experimental data and the fitted β-distribution. An application of the K–S test shows that there is no portion of the RCS range where the K–S two-sided test at a 0.2 level of significance eliminates the β (and sometimes the Rayleigh) as a hypothetical distribution. But there are some ranges of RCS for which either the lognormal or the χ^2 must be rejected at that level of significance. (The K–S test is elaborated and used for an extension of this idea in Chapter 14 below.)
3. There is an argument that a two-parameter distribution is excessively complicated and is not needed to describe RCS scintillation statistics. However, in the process of fitting a distribution to any data set, the selection of a single-parameter fitting distribution from a candidate set is in effect equivalent to the introduction of a second parameter.
4. The data break from a fitted curve in the vicinity of a cumulative probability of 0.99 is appropriate to the number of data points used.

5. In general, the β-distribution fits the experimental data well but there is occasional data bunching, which implies that no smooth curve would give a good fit.
6. The Rayleigh distribution fits several experimental measurement sets but others require the greater flexibility of the β. The lognormal and the χ^2 are not good fits for any data set (cf. item 2 above).

13-2 THE FIREBEE

We are eager to examine RCS statistics of a complex target at a higher frequency. For our second example, we are fortunate to have obtained, from the RATSCAT† range, full-scale measurement data for the Teledyne Ryan Firebee at 9 GHz, HH.

Figures 13-27a and 13-27b show, respectively, a photograph and a

Figure 13-27a Photo of Teledyne Ryan Firebee.

† Radar Target Scatter Facility, Holloman AFB, New Mexico.

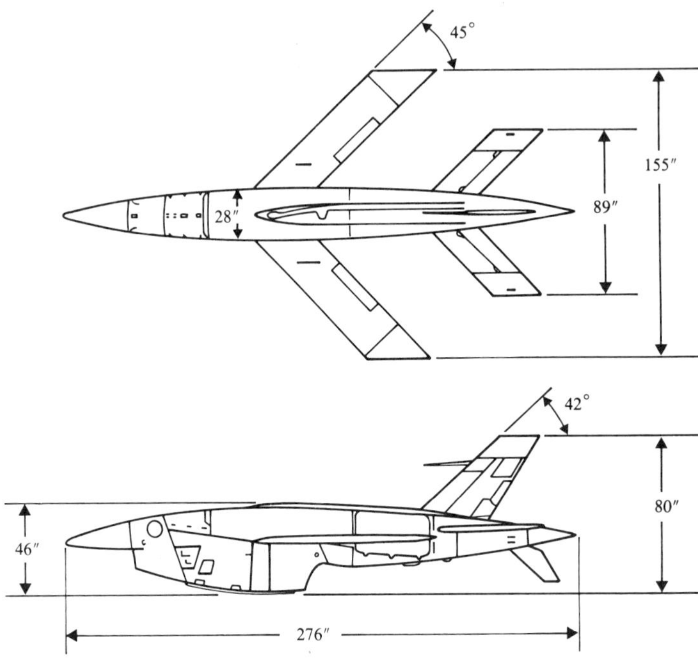

Figure 13-27b Drawing of Teledyne Ryan Firebee.

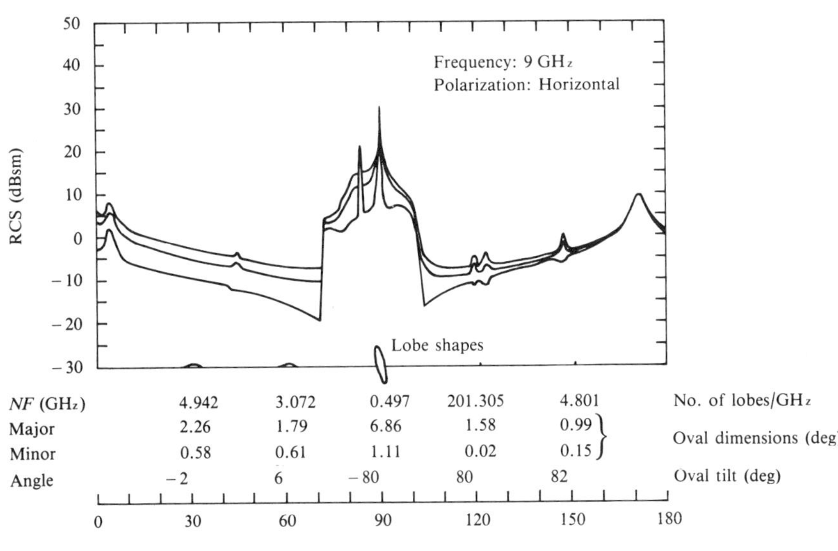

Figure 13-28 Computed monostatic RCS of Firebee.

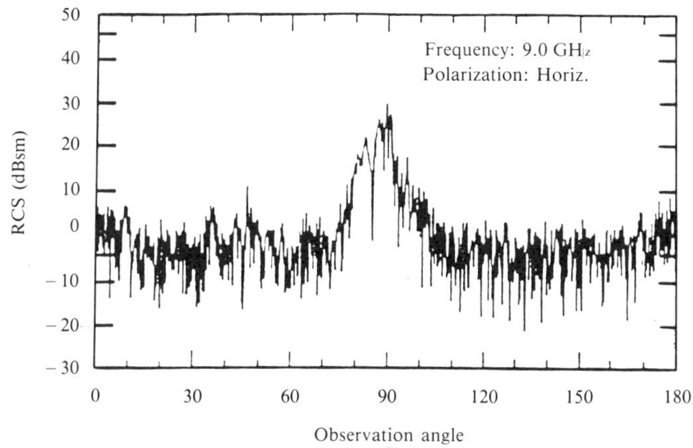

Figure 13-29 Measured monostatic RCS of Firebee.

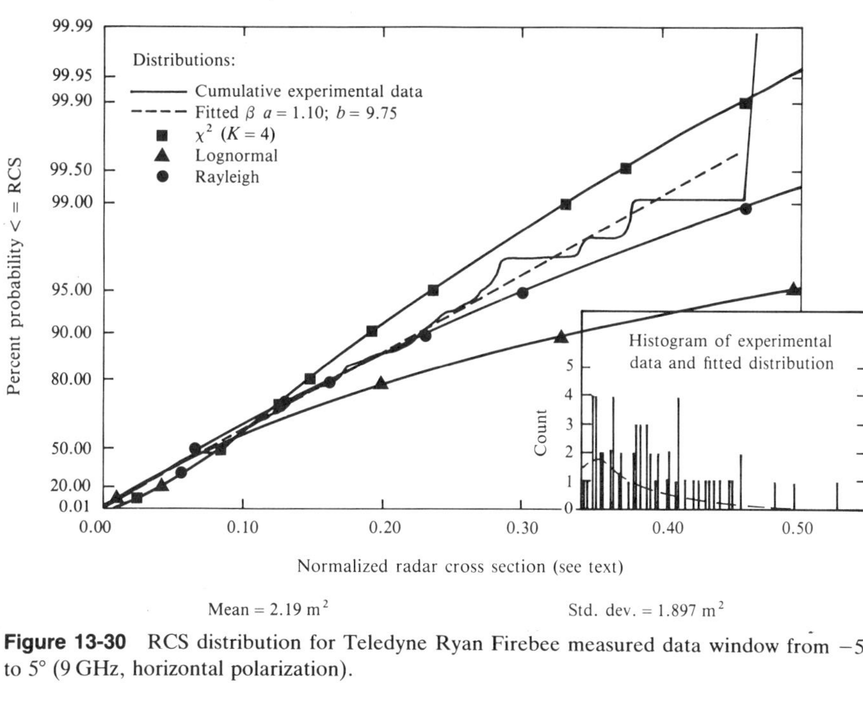

Figure 13-30 RCS distribution for Teledyne Ryan Firebee measured data window from $-5°$ to $5°$ (9 GHz, horizontal polarization).

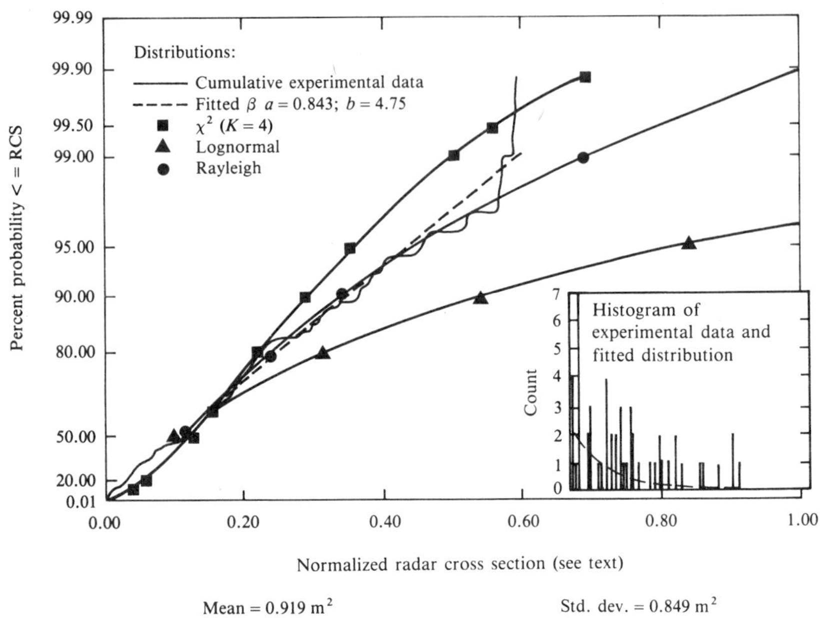

Figure 13-31 RCS distribution for Teledyne Ryan Firebee measured data window from 35° to 45° (9 GHz, horizontal polarization).

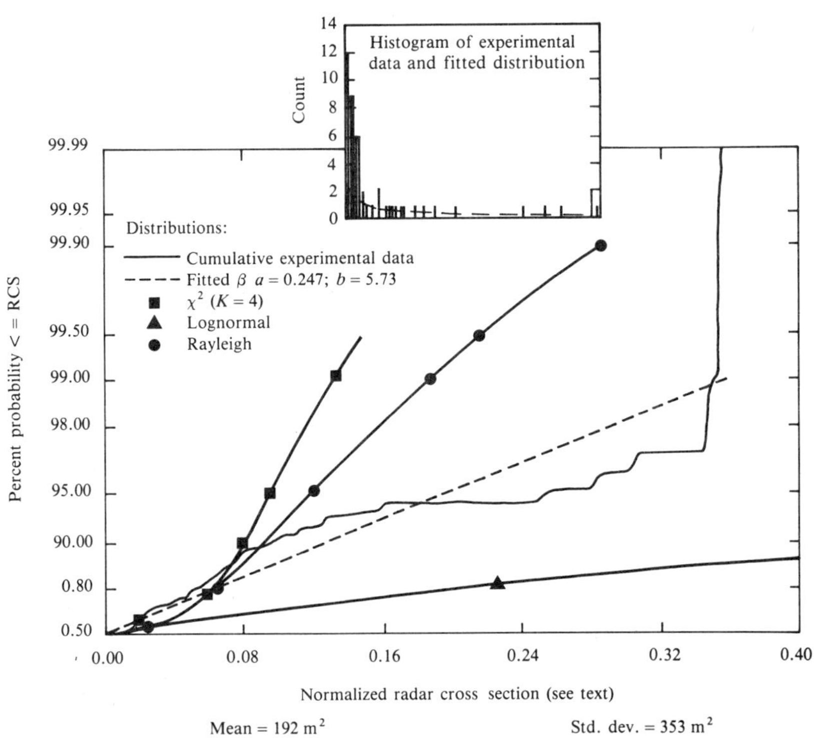

Figure 13-32 RCS distribution for Teledyne Ryan Firebee measured data window from 85° to 95° (9 GHz, horizontal polarization).

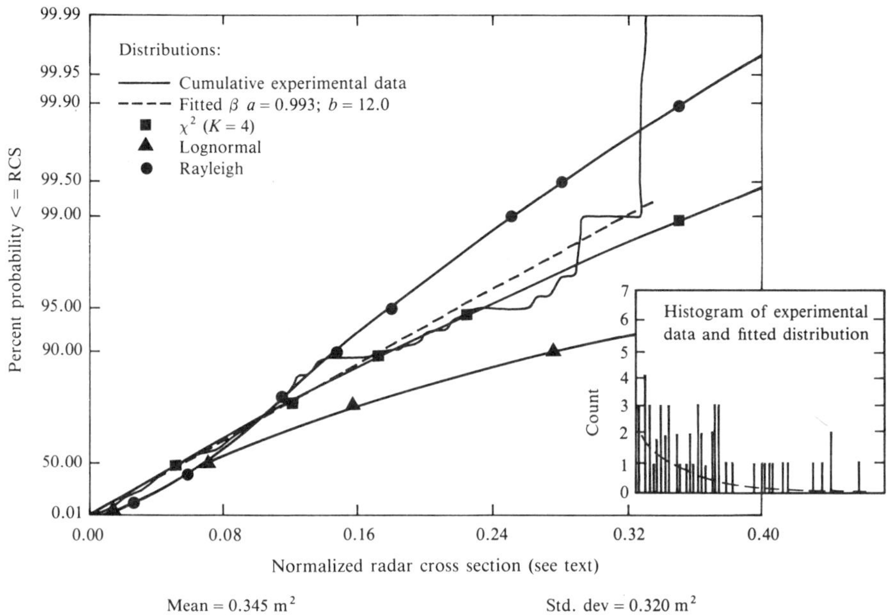

Figure 13-33 RCS distribution for Teledyne Ryan Firebee measured data window from 135° to 145° (9 GHz, horizontal polarization).

drawing† of the Teledyne Ryan Firebee which, for theoretical RCS analysis, we decompose into some 30 components. First order statistics of RCS, mean m and variance m_2, computed from this set of scatterers can be compared with RATSCAT-measured RCS patterns. Figure 13-28 shows the computed results, where the central curve represents m and the lower and upper curves represent $m - \sqrt{m_2}$ and $m + \sqrt{m_2}$, respectively. Clearly, all three curves are functions of aspect, frequency, and polarization. Figure 13-29 shows the RATSCAT RCS pattern for conditions corresponding to those of Fig. 13-28. Agreement between the two patterns is excellent.

Higher order RCS statistics of width, skewness, and kurtosis are computed, using 10° sectors of measured data. From these statistics we determine β-distribution fits for the data in 10° windows. Figures 13-30–13-33 show results for a selection of four windows. Each figure also includes Rayleigh, χ^2, and lognormal fits. We see, as we saw for the 727–100C, at a lower frequency, that the β- (and sometimes the Rayleigh) distribution offers a much better description of RCS distribution than either the lognormal or χ^2.

† Kindly furnished by G. D. Pugsley, Teledyne Ryan Aeronautical, 2701 Harbor Drive, San Diego, California 92138.

14

A NEW METHOD FOR COMPARING EXPERIMENTAL AND THEORETICAL DATA

Over the years since static radar ranges have been in operation there has been a contention that dynamically measured RCS data do not compare well with statically measured data. It is true that, as a complex flexible aircraft maneuvers, the relative motion of individual scatterers will affect the detailed constructive and destructive interference behavior of the RCS pattern. To observe the resulting modulation of the RCS pattern requires an appropriate bandwidth in the circuitry of tracking radars or intercepting missiles. Moreover, a flexible aircraft in flight has its wings deflected slightly upward whereas at rest its wings droop; on this basis alone, the dynamic and static RCS patterns would differ.

For stiff missiles, such as the early Polaris, static RCS data measured at RATSCAT have been shown to be equivalent to RCS data measured dynamically with Cape Kennedy instrumentation radars. Subsequently, there have been many attempts to correlate flight test data with static data (see ALBIS reference). But few of these attempts had instrumentation adequate to determine accurately the aircraft or missile attitude versus time. Further, few instrumentation radars are calibrated well enough to provide an accurate measurement of dynamic RCS.

Today, however, radars and evaluation ranges are much better equipped for such comparisons. And the development of statistical descriptions of RCS (described in Chapters 12 and 13), which is applicable to both experimental and theoretical methods, offers the opportunity to demonstrate whether or not there is agreement between dynamic data and static data or between either one and calculated data. This chapter is aimed at that end.

The new method for comparing RCS theory and experiment is based on the facts demonstrated in Chapters 12 and 13 that both theoretically determined and experimentally measured complex target RCS are β-distributed. On this basis we advocate that theoretical RCS calculations or statically measured RCS data be compared with each other, or with dynamically measured RCS data, by comparing their β-distributions.

The comparison of the two β-distributions might easily be based on an "eyeball" method (as were the comparisons of first order RCS statistics with experiment) because they look so good. But we propose a *firmer statistical basis,* adapted from a test of the Kolmogoroff–Smirnov type.

There is a Kolmogoroff test for deciding whether a particular measured data sample can be said to be drawn from a specified distribution. There is also a Smirnov test for deciding whether two samples are drawn from the same distribution (see, for example, Conover, 1980). We suggest a test of the latter type to decide whether the distributions derived from the theoretical and experimental results are essentially the same distribution.

These ideas are conveniently illustrated with results based on the experimental measurements for, and our theoretical model of, the 727 commercial circraft presented in Chapter 13. We extend these results to obtain our new method of comparison.

Figure 14-1 shows a histogram (the staircase curve) of measured data from the 135–145° interval in the horizontal plane. It also shows (lower

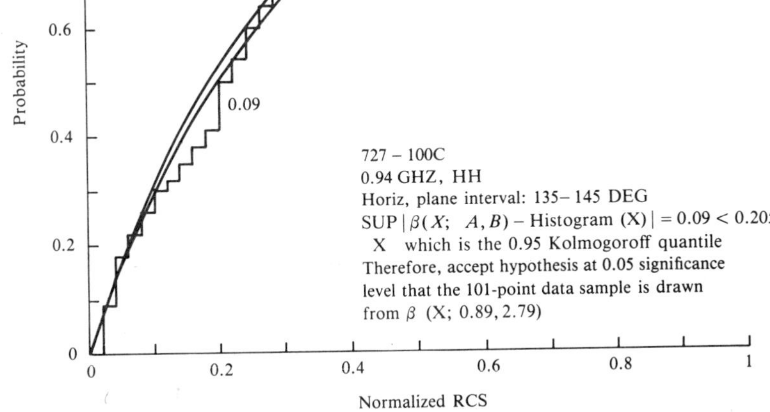

Figure 14-1 Kolmogoroff best-fit test for β-distributions.

smooth curve) a β-distribution of RCS computed from the special statistics of RCS. And finally, the figure also includes (upper smooth curve) a β-distribution of RCS computed from the special statistics derived from a theoretical model of the 727. Clearly, one could have a fine time here with an 'eyeball' comparison of the three curves. But it is possible to do better than that. By comparing the maximum vertical separation of the histogram from one of the smooth curves with the appropriate Kolmogoroff quantile, we can say that there is, say, a 95% probability that the data sample was drawn from the appropriate β-distribution. The matter is usually stated in the form of a hypothesis that the data are taken from the distribution in question. If the maximum vertical separation between histogram and distribution is less than the chosen Kolmogoroff quantile, then the hypothesis is accepted at the 5% level of significance, that is, there is at most a 5% chance that the hypothesis is rejected as false when in fact it is really true.

Figure 14-2 shows an automation of the differencing process discussed above. With such a graph in hand one can easily select (or have the computer select) the maximum separation. This separation can then be used to find the *critical* significance level, that is, the smallest significance level at which the hypothesis would be rejected, on the basis of the given sample. In

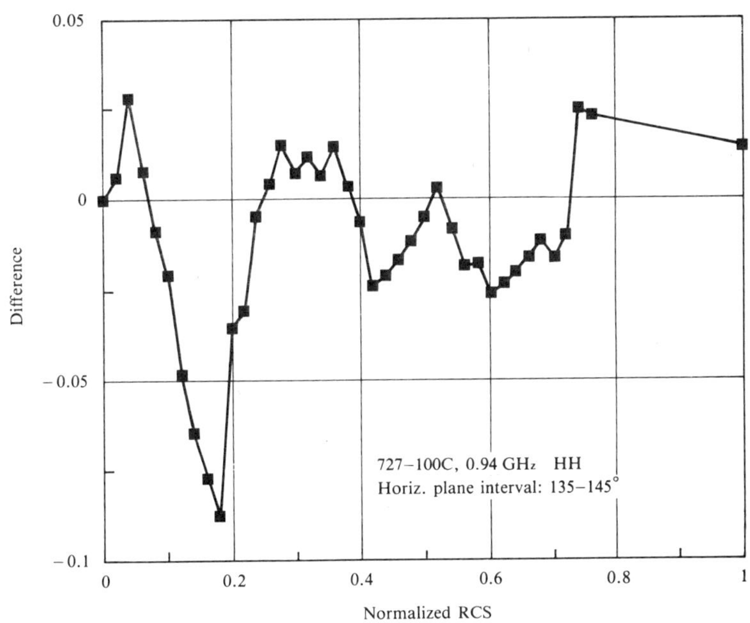

Figure 14-2 Kolmogoroff difference between measured 727–100C data histogram and theoretical β-model ($a = 0.88$, $b = 2.98$).

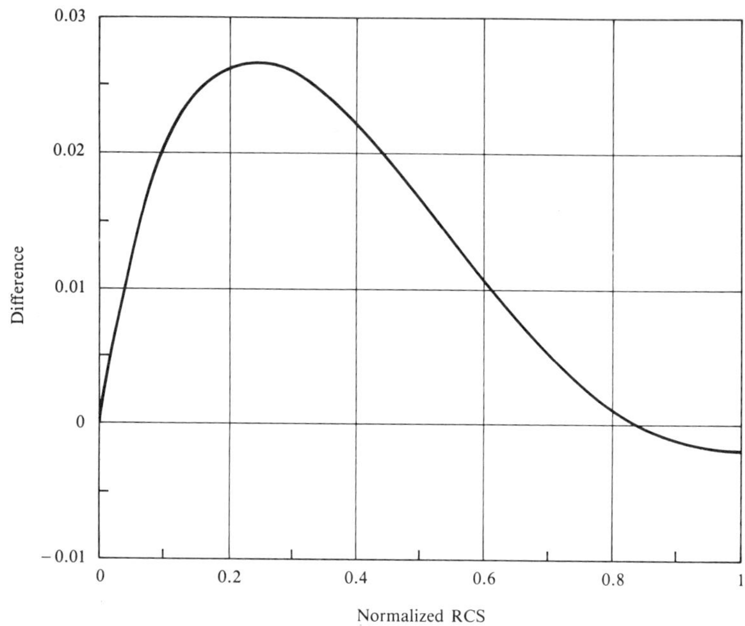

Figure 14-3 Kolmogoroff difference for two β-distributions for measured ($a = 0.89,\ b = 2.79$) and theoretical data ($a = 0.88,\ b = 2.98$).

the example of Fig. 14-2 one sees a maximum absolute difference of 0.08, which corresponds to a critical level somewhat in excess of 20% (actually outside the extent of the tables given by Conover, 1980).

As we mentioned above, there is a Smirnov test that allows one to say whether two (or more) samples are drawn from the same distribution. We propose that this test be adapted to the question of deciding whether two distributions are essentially alike. Such a test then gives us the new method of comparing theory and experiment. Figure 14-3 shows the difference between the β-distribution derived from our theoretical model and the β-distribution computed to fit the measured data. The maximum absolute difference is approximately 0.027. If these two β-distribution curves were considered to be 10,000-point samples, then the Smirnov test would declare them to be from the same distribution for a critical significance level of less than 0.5%. We would interpret that result as saying that the β-distributions calculated from theoretical model and measured data are so nearly alike as to make the theory and experiment in *very good* agreement in the present case. Another way of looking at this result would be to say that if the probability of detection were to be computed, one could obtain essentially the same results by using either the theoretical model or the experimentally measured data.

15

ELEMENTS OF DETECTION THEORY

The principles of detection theory are based on the statistical treatment of hypothesis testing. Hypotheses are statements of possible decisions being considered in the light of certain facts. For example, in a radar target detection problem we can select two hypotheses: a target is present, or no target is present. Corresponding to each hypothesis there is a probabilistic statement of possible outcomes. We then couple this probabilistic description with a criterion, or measure of goodness, that the decision will satisfy on the average. This coupling dictates a dichotomy, for the two hypotheses, of the sample space over which the outcome of the experiment is defined. The dichotomy represents the best decision rule subject to the goodness criterion.

15-1 EXAMPLE OF HYPOTHESIS TESTING

Consider a simple example. Assume a binary communication system that transmits a pulse $s(t)$ of unit amplitude, T seconds long, or no signal at all. Noise $n(t)$ is added at the receiver. We must decide, on a single observation, whether or not a signal is present. The observation is based on

$$r(t) = s(t) + n(t).$$

We form

H_0: no signal is present $[r(t) = n(t)]$, called the null hypothesis. (15-1)

H_1: a signal is present $[r(t) = 1 + n(t)$, assuming a 1-volt signal], called the alternative hypothesis. (15-2)

On the basis of a single observation $y = r(t_0)$, we must choose between H_0 and H_1. One reasonable criterion is to choose the hypothesis that is most likely to have occurred on the basis of a single observation. Let the probabilities that H_0 is true or that H_1 is true, given that $r(t_0) = y$, be, respectively,

$$P(H_0 \mid y) \qquad \text{and} \qquad P(H_1 \mid y).$$

These are called *a posteriori* probabilities. We choose H_0 if

$$P(H_0 \mid y) > P(H_1 \mid y) \qquad \text{or} \qquad P(H_0 \mid y)/P(H_1 \mid y) > 1. \tag{15-3}$$

Choose H_1 otherwise. This is called the maximum a posteriori probability criterion. The reasoning here is from effect to cause or from the individual to the general.

Stated in terms of probability densities, we have: choose H_0 if

$$P(H_0 \mid y \le Y \le y + dy) > P(H_1 \mid y \le Y \le y + dy); \tag{15-4}$$

choose H_1 otherwise; or, using the definition of conditional probability,

$$P(H_0 \mid y \le Y \le y + dy) = P(y \le Y \le y + dy \mid H_0) \\ \times P(H_0)/P(y \le Y \le y + dy), \tag{15-5}$$

where $P(H_0)$ is the probability that H_0 is true. The probability $P(H_1)$ that H_1 is true is then $1 - P(H_0)$. These are called *a priori* probabilities. The reasoning here is from cause to effect, or from the general to the individual.

15-2 LIKELIHOOD RATIO

Assign to the random variable Y a probability density function $p(y)$. Then

$$P(y \le Y \le y + dy) = p(y)\, dy. \tag{15-6}$$

Similarly, we set $P(y \le Y \le y + dy \mid H_0) = p_0(y)\, dy$, the subscript indicating that the probability is conditioned on H_0 being true. Thus,

$$P(H_0 \mid y \le Y \le y + dy) = p_0(y)\, dy\, P(H_0)/p(y)\, dy, \tag{15-7}$$

or, in the limit $dy \to 0$,

$$P(H_0 \mid y) = p_0(y)p(H_0)/p(y). \tag{15-8}$$

Similarly,

$$P(H_1 \mid y) = p_1(y)[1 - P(H_0)]/p(y). \tag{15-9}$$

So our decision rule may now be written:

- Choose H_0 if $p_0(y)P(H_0)/p_1(y)[1 - P(H_0)] > 1$ or if $p_0(y)/p_1(y) > [1 - P(H_0)]/P(H_0)$; (15-10)
- Choose H_1 if $p_1(y)/p_0(y) \ge P(H_0)/[1 - p(H_0)]$. (15-11)

The ratio $p_1(y)/p_0(y)$ is called the *likelihood ratio*; $p_1(y)$ and $p_0(y)$ are called *likelihood functions.*

15-3 LIKELIHOOD RATIO EXAMPLE

Assume $n(t)$ to be Gaussian with zero mean and unit variance. If H_0 is true, then with no signal present, y has the same density function as the noise:

$$p_0(y) = \frac{1}{\sqrt{2\pi}} \exp(-y^2/2). \tag{15-12}$$

If H_1 is true, with a 1-volt signal present, then y is still Gaussian, but with mean 1:

$$p_1(y) = \frac{1}{\sqrt{2\pi}} \exp[-(y-1)^2/2]. \tag{15-13}$$

The likelihood ratio $\lambda(y)$ is

$$\lambda(y) = p_1(y)/p_0(y) = \exp[-(y-1)^2/2 - y^2/2] = e^{y-\frac{1}{2}}. \tag{15-14}$$

So our rule is:

$$\text{Choose } H_1 \text{ if } e^{y-\frac{1}{2}} \geq P(H_0)/[1-P(H_0)]. \tag{15-15}$$

Sometimes a log-likelihood ratio is useful. Since $\ln x$ is a monotonically increasing function of x, inequality still holds. So

$$\text{Choose } H_1 \text{ if } y - \tfrac{1}{2} \geq \ln\{P(H_0)/[1-P(H_0)]\}; \tag{15-16}$$

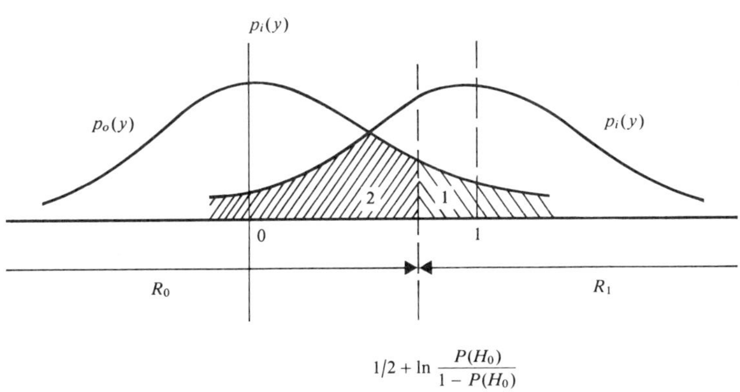

Figure 15-1 Dichotomy of the real line defined by a log likelihood ratio.

y is called the test statistic. The receiver that is optimum in the maximum a posteriori sense samples the signal and compares it to a threshold. If the sample is greater than

$$\tfrac{1}{2} + \ln\{P(H_0)/[1 - P(H_0)]\},$$

it is assumed that a "1" was sent. Otherwise, it is assumed that a "0" was sent.

Since y may take any value between $-\infty$ and ∞, the entire real line makes up the sample space. So we have a dichotomy of the real line with (Fig. 15-1)

$$\tfrac{1}{2} + \ln\{P(H_0)/[1 - P(H_0)]\}$$

as the dividing point. If y falls in region R_0, H_0 is chosen; if it falls in region R_1, H_1 is chosen. The decisions can be called, respectively, D_0 and D_1.

15-4 ERROR TYPES: PROBABILITIES OF FALSE ALARM (PFA) AND DETECTION (PD)

We can make *two kinds of errors*:

First kind: Choose H_1 when H_0 is true. Call the probability of so doing $P(D_1 \mid H_0)$. This probability corresponds to area 1 and is called *probability of false alarm.* We say there is a target present when there is none:

$$P(D_1 \mid H_0) = \text{PFA}. \tag{15-17}$$

Second kind: Choose H_0 when H_1 is true. Call the probability of so doing $P(D_0 \mid H_1)$. This probability corresponds to area 2. $P(D_1 \mid H_1)$ is the probability of choosing H_1 when H_1 is true, a correct decision:

$$P(D_1 \mid H_1) = 1 - P(D_0 \mid H_1) = PD. \tag{15-18}$$

This probability is the *probability of detection*; it is the probability of *correctly* deciding that a target is present. This probability corresponds to the *complement,* with respect to the curve $p_1(y)$, of the area 2.

We have

$$\text{PFA} = P(y \geq \tfrac{1}{2} + \ln\{P(H_0)/[1 - P(H_0)]\} \mid H_0) \tag{15-19}$$

If H_0 is true, then $p_0(y)$ applies, so that

$$\text{PFA} = \int_{R_1} p_0(y)\, dy = \frac{1}{\sqrt{2\pi}} \int_{Y_T} \exp(-y^2/2)\, dy \tag{15-20}$$

where $Y_T = \frac{1}{2} + \ln\{P(H_0)/[1 - P(H_0)]\}$. Similarly, when H_1 is true,

$$P(D_0 \mid H_1) = \int_{R_0} p_1(y)\, dy = \frac{1}{\sqrt{2\pi}} \int_{-\infty}^{Y_T} \exp[-(y-1)^2/2]\, dy \qquad (15\text{-}21)$$

and

$$PD = 1 - P(D_0 \mid H_1). \qquad (15\text{-}22)$$

The complement PD of the risk $P(D_0 \mid H_1)$ of choosing H_0 when in fact H_1 is true, that is, Eq. 15-22, is called the power of the test (Dixon and Massey, 1969). The greater the power PD, the more powerful the test.

15-5 TEST CRITERIA

The criterion used above is the *maximum a posteriori probability*. It assigns no particular weighting to the two error types, assuming that each is equally harmful. There are, of course, other criteria that can be used and some of these assign weights, or costs, to each error type. One such is *Bayes criterion* based on the minimization of an average cost; the resulting minimum cost is called Bayes risk.

In certain communication systems it is useful to minimize average error probability. But in a radar system a priori probabilities and error costs are difficult to determine. So here one usually turns to the *Neyman–Pearson criterion*: to maximize the probability of detection for a given, or fixed, false alarm probability. This objective can be accomplished by using a *likelihood ratio test*. Specifically, there exists some η, $\eta > 0$, such that if H_1 is chosen (Fig. 15-2) when

$$\lambda(y) = p_1(y)/p_0(y) \geq \eta \qquad (15\text{-}23)$$

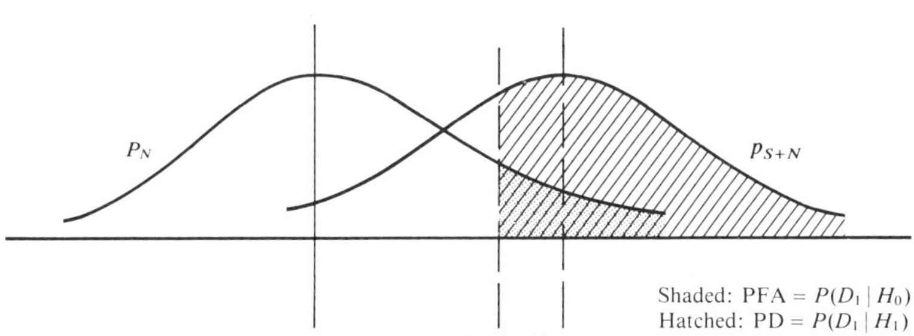

Figure 15-2 Definition of a threshold.

and H_0 is chosen otherwise, then this rule yields the maximum $P(D_1 \mid H_1)$ for all tests subject to the constraint that $P(D_1 \mid H_0)$ is less than some predetermined constant.

We look not at the proof, but the idea of the proof, of this statement. We must maximize $P(D_1 \mid H_1)$, subject to $P(D_1 \mid H_0) = \alpha$. Since $P(D_1 \mid H_1) = 1 - P(D_0 \mid H_1)$, it is equivalent to minimize $P(D_0 \mid H_1)$. And since $P(D_1 \mid H_0)$ is constant ($= \alpha$), adding it to $P(D_0 \mid H_1)$ does not influence the minimization. Consequently, maximizing $P(D_1 \mid H_1)$ is equivalent to minimizing

$$Q = P(D_0 \mid H_1) + \mu P(D_1 \mid H_0), \tag{15-24}$$

where μ is an arbitrary constant (Lagrange) multiplier. This last is identical with a Bayes minimum cost statement (cf. Whalen, 1971, p. 130) which is achieved if H_1 is chosen when

$$\mu \le p_1(y)/p_0(y). \tag{15-25}$$

Thus, a likelihood ratio test will maximize $P(D_1 \mid H_1)$ for a given $P(D_1 \mid H_0)$.

Example: Let the random variable y be

$$y = s + n,$$

where n is Gaussian with mean 0 and variance 2, and s is a constant that is 0 or 1.

On the basis of a *single sample y,* determine an optimum decision rule to choose between

$$H_0\colon s = 0;\ H_1\colon s = 1 \tag{15-26}$$

using a Neyman–Pearson (N-P) likelihood ratio test with $P(D_1 \mid H_0) = 0.1$.

The likelihood functions $p_0(y)$ and $p_1(y)$ are Gaussian since n is Gaussian:

$$p_0(y) = \frac{1}{\sqrt{2\pi}\sqrt{2}} \exp(-y^2/4), \quad p_1(y) = \frac{1}{\sqrt{2\pi}\sqrt{2}} \exp[-(y-1)^2/4]; \tag{15-27}$$

and the likelihood ratio is

$$\lambda(y) = p_1(y)/p_0(y) = e^{(y/2)-(1/4)}. \tag{15-28}$$

The N-P test is:

$$\text{Choose } H_1 \text{ if } e^{(y/2)-1} \ge \lambda_0 \text{ (threshold)}, \tag{15-29}$$

where the threshold λ_0 is chosen to satisfy the false alarm probability constraint. Since the exponential term is monotonically increasing with y, an equivalent test is

$$\text{Choose } H_1 \text{ if } y \ge \gamma, \tag{15-30}$$

where

$$P(D_1 \mid H_0) = \int_{R_1} p_0(y)\,dy = \int_{\gamma}^{\infty} \frac{1}{\sqrt{2\pi}\sqrt{2}} \exp(-y^2/4)\,dy = 0.1;$$
$$= \alpha; \tag{15-31}$$

this is the PFA constraint.

Changing variables $[x = y/\sqrt{2}]$ yields

$$P(D_1 \mid H_0) = \int_{\gamma/\sqrt{2}} \frac{1}{\sqrt{2\pi}} \exp(-x^2/2)\,dx = 0.1, \tag{15-32}$$

whence

$$\gamma = 1.8. \tag{15-33}$$

Therefore, our decision rule is

$$\text{Choose } H_1 \text{ if } y \geq 1.8;\ \text{choose } H_0 \text{ otherwise.} \tag{15-34}$$

The probability of detection, $PD = P(D_1 \mid H_1)$, based on a single observation y, is

$$P(D_1 \mid H_1) = \int_{R_1} p_1(y)\,dy = \int_{1.8}^{\infty} \frac{1}{\sqrt{2\pi}\sqrt{2}} \exp[-(y-1)^2/4]\,dy$$
$$= 0.285. \tag{15-35}$$

We can express this decision rule in terms of $\lambda(y)$ and λ_0 by observing that

$$\lambda(\gamma) = \lambda_0,$$

whence

$$\lambda(\gamma) = \frac{p_1(y)}{p_0(y)} = \frac{\dfrac{1}{\sqrt{2\pi}\sqrt{2}} \exp[-(0.8)^2/4]}{\dfrac{1}{\sqrt{2\pi}\sqrt{2}} \exp[-(1.8)^2/4]} = \lambda_0 \approx 1.9. \tag{15-36}$$

Finally, then the *decision rule* is

$$\text{Choose } H_1 \text{ if } \lambda(y) \geq 1.9;\ \text{choose } H_0 \text{ otherwise.} \tag{15-37}$$

Extension of Example. Suppose we have a priori probabilities for the signal s. If we assume $P(H_0) = 1/2$, then

$$\lambda(y) = p_1(y)/p_0(y) \geq P(H_0)/[1 - P(H_0)] = 1 \tag{15-38}$$

(see Section 15-1). So our decision rule is

$$\text{Choose } H_1 \text{ if } \lambda(y) \geq 1;\ \text{choose } H_0 \text{ otherwise.} \tag{15-39}$$

In terms of likelihood functions,

$$\text{Choose } H_1 \text{ if } \exp[-(y-1)^2/4]/\exp(-y^2/4) \geq 1, \tag{15-40}$$

or

$$\text{Choose } H_1 \text{ if } y \geq 1/2\text{; choose } H_0 \text{ otherwise.} \tag{15-41}$$

Here

$$\begin{aligned} \text{PFA} = P(D_1 \mid H_0) &= \int_{R_1} p_0(y)\,dy = \int_{1/2}^{\infty} \frac{dy}{\sqrt{2\pi}\sqrt{2}} \exp(-y^2/4) \\ &= 0.362. \end{aligned} \tag{15-42}$$

We can then show that $P(D_0 \mid H_1) = P(D_1 \mid H_0) = 0.362$ so that

$$\text{PD} = P(D_1 \mid H_1) = 0.638. \tag{15-43}$$

Note that although PD is higher here, so is PFA.

15-6 MULTIPLE SAMPLES

We need now to make a few remarks about extending our single sample case above to a set of N samples

$$y_1, y_2, \ldots, y_N$$

which may be represented as a vector $\mathbf{y}$. Our likelihood ratio is now

$$\lambda(\mathbf{y}) = p_1(\mathbf{y})/p_0(\mathbf{y}). \tag{15-44}$$

We can still frame our criterion in terms of a likelihood ratio test:

$$\text{Choose } H_1 \text{ if } \lambda(\mathbf{y}) \geq \lambda_0\text{; choose } H_0 \text{ if } \lambda(\mathbf{y}) < \lambda_0, \tag{15-45}$$

where, as before, λ_0 is a threshold that depends on the criterion and available a priori knowledge.

Example 1. On the basis of N statistically independent Gaussian samples with variance σ^2, determine a likelihood ratio test to choose between the hypotheses that their mean is zero (H_0) or that their mean is unity (H_1):

$$H_0\text{: mean} = 0;\ H_1\text{: mean} = 1 \tag{15-46}$$

with the decision rule:

$$\text{Choose } H_1 \text{ if } \lambda(\mathbf{y}) \geq \lambda_0\text{; choose } H_0 \text{ otherwise.} \tag{15-47}$$

Suppose

$$p_1(\mathbf{y}) = \prod_{i=1}^{N} \frac{\exp[-(y_i - 1)^2/2\sigma^2]}{(\sigma\sqrt{2\pi})^N} = (1/2\pi\sigma^2)^{N/2} \exp\left[-\tfrac{1}{2} \sum_{i=1}^{N} (y_i - 1)^2/\sigma^2\right] \tag{15-48}$$

and

$$p_0(\mathbf{y}) = \prod_{i=1}^{N} \frac{\exp(-y_i/2\sigma^2)^N}{(\sigma\sqrt{2\pi})^N} = (1/2\pi\sigma^2)^{N/2} \exp\left[-\tfrac{1}{2}\sum_{i=1}^{N}(y_i/\sigma)^2\right]. \tag{15-49}$$

The likelihood ratio is

$$\lambda(\mathbf{y}) = \exp\left[\tfrac{1}{2}\sum_{i=1}^{N}(2y_i - 1)/\sigma^2\right]. \tag{15-50}$$

Our decision rule is

$$\text{Choose } H_1 \text{ if } \lambda(\mathbf{y}) \geq \lambda_0; \text{ choose } H_0 \text{ otherwise.} \tag{15-51}$$

If we use the log-likelihood ratio (and some algebra), we can restate our rule as:

$$\text{Choose } H_1 \text{ if } \frac{1}{N}\sum_{i=1}^{N} y_i \geq \lambda_0', \tag{15-52}$$

$$\text{where } \lambda_0' = \frac{1}{2} + \frac{\sigma^2}{N}\ln\lambda_0.$$

Now

$$\frac{1}{N}\sum_{i=1}^{N} y_i = \hat{m}_y, \text{ the sample mean.} \tag{15-53}$$

The receiver operation is therefore to take N independent samples, calculate their mean, and compare the result to a threshold λ_0'. If the result is above the threshhold, we decide that the mean of the Gaussian distribution is unity.

The probability of incorrectly choosing H_1 is

$$P(D_1 \mid H_0) = \int_{R_1} p_0(\mathbf{y})\, d\mathbf{y} \tag{15-54}$$

where R_1 is the N-dimensional volume for which $\hat{m}_y$ is greater than some constant and in which H_1 is chosen. However, the error probability can be directly determined from the probability density for $\hat{m}_y$, since it too is Gaussian:

$$\hat{m}_y \text{ is Gaussian with variance } \sigma^2/N; \tag{15-55}$$

its mean depends on which hypothesis is assumed true.

If H_0 is true, $\hat{m}_y$ has mean 0; if H_1 is true, $\hat{m}_y$ has mean 1. For the density of the first, use $p_0(\hat{m}_y)$; for the second, $p_1(\hat{m}_y)$. We then have the following error statements:

$$\begin{aligned}\text{PFA} = P(D_1 \mid H_0) &= \int_{\lambda_0'}^{\infty} p_0(\hat{m}_y)\, d\hat{m}_y \\ &= \int_{\lambda_0'}^{\infty} (N/2\pi\sigma^2)^{1/2} \exp(-Nz^2/2\sigma^2)\, dz \end{aligned} \tag{15-56}$$

and

$$1 - \mathrm{PD} = P(D_0 \mid H_1) = \int_{-\infty}^{\lambda'_0} p_1(\hat{m}_y)\, d\hat{m}_y$$

$$= \int_{-\infty}^{\lambda'_0} (N/2\pi\sigma^2)^{1/2} \exp[-N(z-1)^2/2\sigma^2]\, dz. \qquad (15\text{-}57)$$

15-7 RADAR DETECTION OF TARGETS

Here our studies of probabilities become additionally complicated by the characteristics we expect a target to have, that is, whether it may be steady or may fluctuate; and we must take account of receiver characteristics, that is, what is the nature of the threshold criterion under which it operates and how it treats its received signals (incoherently or coherently). Nevertheless, the principles of detection are still the ones described above. We shall still have to choose between two hypotheses—H_0: no signal is present; and H_1: a signal is present. The signal may now be only describable by a probability distribution. And so we must be able to find not only a probability density $p_0(y)$, where the random variable y is noise only, but also a probability density $p_1(y)$, where y describes signal plus noise, each of which has its own probability density description.

The choice of a signal-to-noise ratio (S/N) essentially fixes the positions of the density functions $p_0(y)$ and $p_1(y)$ and, as a consequence, the probability of detection PD may be calculated for some fixed, but arbitrary, false alarm probability PFA. S/N may be specified as the ratio of average signal power to average noise power or as peak signal power to average noise power. In any case, some assumption must be made regarding the target signal, that is, whether it is steady or fluctuating, and if the latter, whether the fluctuations are slow or fast.† These assumptions lead to the classical detection theory of Marcum (1960) and Swerling (1960, 1970).

15-8 DETECTION THEORY USING THE β-DISTRIBUTION OF TARGET RCS

These models assume that target RCS is randomly distributed according to either a negative exponential distribution or a χ^2-distribution with 4 degrees of freedom.‡ Lognormal and Rice distributions were later included and Swerling (1970) extended his χ^2 model to a general χ^2-distribution.

† Slow pertains to scan-to-scan; fast to pulse-to-pulse.

‡ The number of degrees of freedom may be thought of as the rank of an orthogonal transformation which takes a bilinear form into a special quadratic form. The term was originated by Fisher (1925) in his study of sampling distributions.

However, these models do not bracket, either individually or as a group, the statistics of a real complex target, as we have already seen by examining our allowable region of width-skewness (or width-kurtosis) space. The Marcum and Swerling models are represented by just three points in this plane. The locus of points for each of several other distributions (general χ^2, Rice, and lognormal) is a line. For example, as we have seen, the general χ^2 model is represented by one boundary of our allowable region and contains the classical Swerling models (see Fig. 12-1).

Consequently, all of these models lack the necessary flexibility, possessed by the β-distribution, to span our allowable region. And the β-distribution is relatively easy to deal with computationally. Kulp (1984) has given a sampling of curves of detection probability versus average signal-to-noise ratio for a fixed false alarm probability; Kulp has also created a computer code for the computation of such curves (which we include, with his permission, as an appendix to this section). We shall consider some of his results.

For the illustration of β-detection curves, PFA $= 10^{-6}$ for all five of the curves we wish to discuss.

Figure 15-3 illustrates the flexibility of the β-target model for two widely separated a, b pairs:

Case 1	Case 2
$a = b = 0.5$	$a = 1.866,\ b = 40.0$
$w = 0.5,\ g = 0$	$w = 0.480,\ g = 0.857$
$N = 1$ (no. of pulses)	$N = 1.$

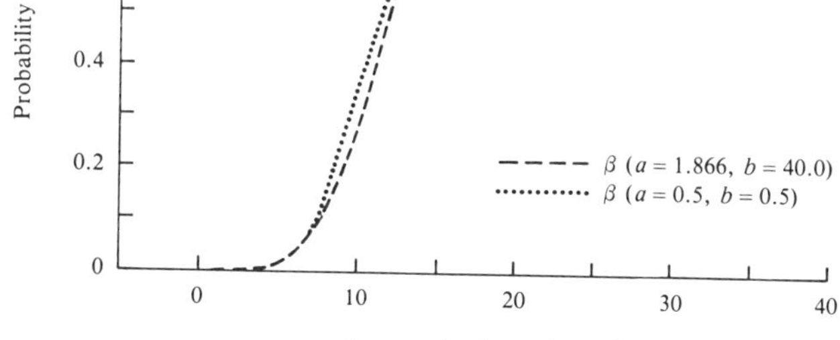

Figure 15-3 Comparison of detection probabilities for two choices of the shape parameters of the β-distribution for $N = 1$ and PFA $= 10^{-6}$.

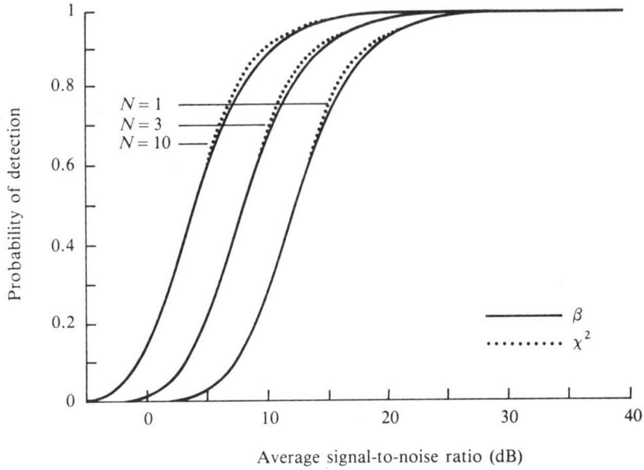

Figure 15-4 Detection probabilities for β ($a = 1.3$, $b = 13.0$) and χ^2 ($K = 1.53$) targets for slow-fluctuation case (PFA $= 10^{-6}$).

We note that, although the first two moments for these cases are very similar, the detection probabilities differ considerably, especially for large values of average signal-to-noise ratio. This difference is caused by the large disparity in third moments of the two distributions that is reflected in the disparity between the modified skewness (g) values. In the width–modified-skewness statistical space, the above two illustrations yield points so widely separated in this space that they represent essentially the total range of variation of detection curves over the allowable region for $w = 0.5$.

The locus of the Swerling II target model in the allowable w, g region is near that for choice 2 above. The β-distribution cannot represent Swerling II exactly but the first three moments of the β-distribution approach those of Swerling II (χ^2† with 4 degrees of freedom) in the limit as $b \to \infty$ and $a = 2$. Our choice 2 above then yields detection curves that are virtually indistinguishable from those of Swerling II (Fig. 15-4).

Figures 15-4 and 15-5 compare detection probabilities for the β- and χ^2-distributions with these statistics:

β	χ^2
$a = 1.3$	$k = 1.53$ (slow)
$b = 13$	$k = 1.53N$ (fast)
$w = 0.654$	$w = 0.654$
$g = 1.104$	$g = 1.308$

† Swerling's definition: Let v be a random variable with mean $\bar{v}$. Then v has a χ^2 distribution with $2k$ degrees of freedom if the probability density of v is

$$p_k(v, \bar{v}) = \frac{1}{(k-1)!}(k/\bar{v})(kv/\bar{v})^{k-1}\exp(-kv/\bar{v}),\ v > 0,$$

$$= 0,\quad v \le 0;\ k \text{ need not be an integer.}$$

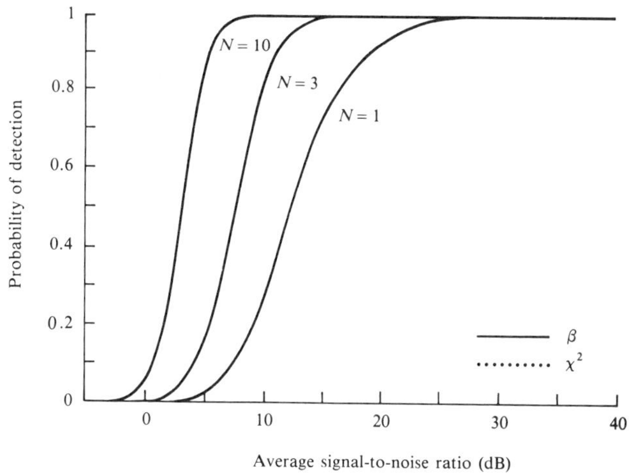

Figure 15-5 Detection probabilities for β ($a = 1.3$, $b = 13.0$) and χ^2 ($K = 1.53N$) targets for fast-fluctuation case (PFA $= 10^{-6}$).

The curves agree well, particularly as N increases (as N increases, all distributions become Gaussian). So the β-distribution handles these cases essentially as well as the χ^2. The point in w, g space corresponding to this choice of χ^2-target model can be reached by the β-distribution in the limit as $b \to \infty$ and $a = 1.529$.

Generally, the statistics of the β-distribution approach those for the χ^2 with $2k$ degrees of freedom as b becomes large and $a = k/2$. As both a and b become large, with $a < b$, the statistics of the β-distribution approach those for a nonfluctuating target, that is, $w = 0$, $g = 0$. Figures 15-6 and 15-7 show detection probability curves for $a = 12$, $b = 13$ (for which $w = 0.042$,

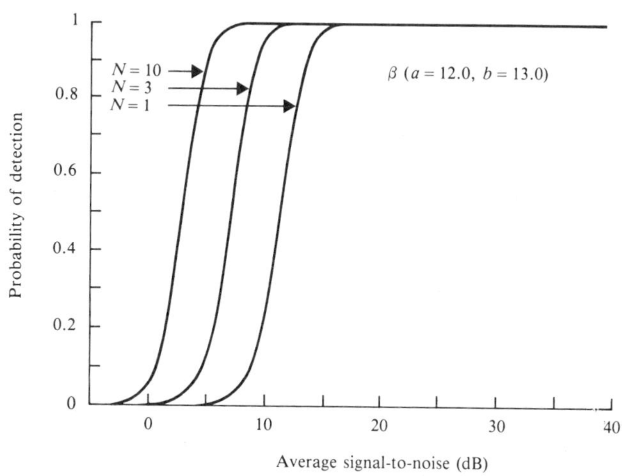

Figure 15-6 Detection probabilities for β ($a = 12.0$, $b = 13.0$) target for slow-fluctuation case (PFA $= 10^{-6}$).

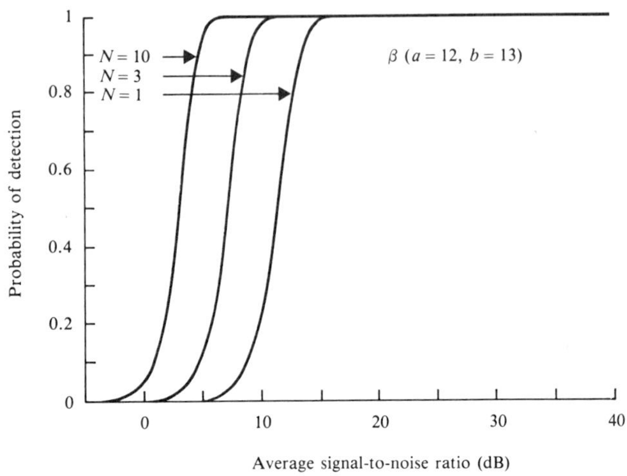

Figure 15-7 Detection probabilities for β ($a = 12.0$, $b = 13.0$) target for fast-fluctuation case ($\text{PFA} = 10^{-6}$).

$g = 0.0062$). Both fluctuating cases are very close to the nonfluctuating target models (i.e., detection probability curves for slow, fast, and nonfluctuating cases are nearly indistinguishable from one another on the scales of the graphs; Kulp (1984) gives a complete discussion of how the fluctuation rate, slow or fast, enters into the curve calculations).

15-9 EXAMPLE: 727 OVER A NORTH CAROLINA FOREST

Consider the simple hypothetical example of our 727–100C flying over a North Carolina forest. Suppose we wish to compute the PD of the 727 for a look-down viewing angle of 52° from vertical at a portside angle of 30° off-nose. The target signal is obscured by clutter background from the forest. Forest clutter signal can be modeled, for two- to three-look sampling (Ulaby et al., 1985), by an inverse Gaussian distribution (Tweedie, 1957; Chhikara and Folks, 1974) whose parameters are obtained as maximum likelihood estimates from a 1600-point sample of measured data (SAR imagery) at L-band. The distribution of RCS of the 727–100C in a neighborhood of the $\theta = 52°$, $\phi = 30°$ aspect is estimated from the methods and data of Chapter 13. The probability density functions for both the NC forest (30-year pine) and the 727–100C are shown in Fig. 15-8.

A threshold is chosen (arbitrarily) so that the $\text{PFA} = 0.05$ (made purposely modest so that the appropriate area can be easily identified in the figure). For this PFA the PD can be computed and is shown in Fig. 15-9. It is, approximately, 0.95 and corresponds to an average SNR (signal-to-noise ratio) of 18 dB, as Fig. 15-9 shows. The 18 dB is essentially the horizontal distance between peaks of the density functions shown Fig. 15-8.

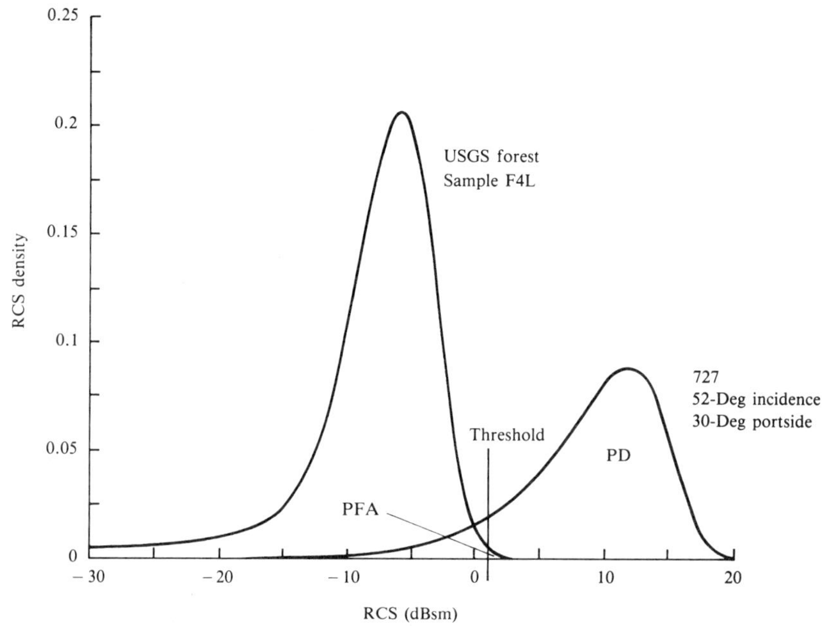

Figure 15-8 RCS densities of North Carolina forest clutter and 727–100C RCS at 1.3 GHz.

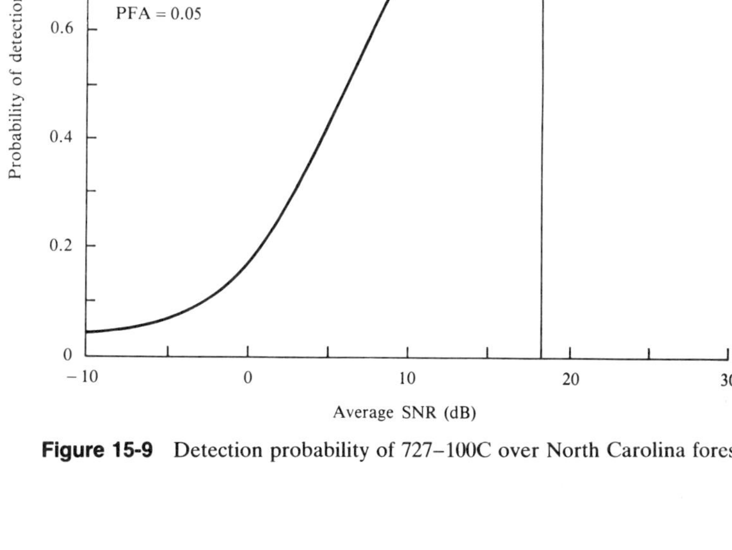

Figure 15-9 Detection probability of 727–100C over North Carolina forest.

PART IV

CONSTITUTIVE PARAMETERS

We have met with constitutive parameters earlier (Chapter 2) and have observed that permittivity and permeability may require not just a complex scalar form but, in certain cases, a more general complex tensor (matrix) representation.

For convenience, we note again that the most general form describing constitutive relations connects electric flux density **D** not only with **E** but also with magnetic flux density **B**; and magnetic flux density not only with **H** but also with **E**:

$$c\mathbf{D} = \mathfrak{P} \cdot \mathbf{E} + \mathfrak{L} \cdot c\mathbf{B}; \tag{1}$$

$$\mathbf{H} = \mathfrak{M} \cdot \mathbf{E} + \mathfrak{Q} \cdot c\mathbf{B}. \tag{2}$$

The relations are written in this interesting form because, so stated, they can be shown to be invariant under a Lorentz transformation. This transformation relates the properties of two media, one moving uniformly with respect to the other; its consequences are outside of the scope of our discussion here. The interested reader might consult the second chapter of Kong (1975). (Kong observes that in 1908 Minkowski postulated the invariance of the macroscopic Maxwell equations under a Lorentz transformation.)

Equations 1 and 2 are representable as matrix statements:

$$\begin{pmatrix} c\mathbf{D} \\ \mathbf{H} \end{pmatrix} = \mathfrak{C} \begin{pmatrix} \mathbf{E} \\ c\mathbf{B} \end{pmatrix} \tag{3}$$

where $\mathfrak{C}$ is a 6×6 constitutive matrix:

$$\mathfrak{C} = \begin{pmatrix} \mathfrak{P} & \mathfrak{L} \\ \mathfrak{M} & \mathfrak{Q} \end{pmatrix}. \tag{4}$$

Or, in a somewhat more satisfying form:

$$\begin{pmatrix} \mathbf{D} \\ \mathbf{B} \end{pmatrix} = \mathfrak{C}_{\mathbf{E}\ \mathbf{H}} \begin{pmatrix} \mathbf{E} \\ \mathbf{H} \end{pmatrix} \tag{5}$$

where

$$\mathfrak{C}_{\mathbf{E}\ \mathbf{H}} = \begin{pmatrix} \boldsymbol{\epsilon} & \boldsymbol{\xi} \\ \boldsymbol{\zeta} & \boldsymbol{\mu} \end{pmatrix} = \frac{1}{c} \begin{pmatrix} \mathfrak{P} - \mathfrak{L} \cdot \mathfrak{Q}^{-1} \cdot \mathfrak{M} & \mathfrak{L} \cdot \mathfrak{Q}^{-1} \\ -\mathfrak{Q}^{-1} \cdot \mathfrak{M} & \mathfrak{Q}^{-1} \end{pmatrix} \tag{6}$$

since we can readily reduce this latter to familiar special cases. The inverse here may be written as

$$\mathfrak{C}_{\mathbf{D}\ \mathbf{B}} = \begin{pmatrix} \boldsymbol{\kappa} & \boldsymbol{\chi} \\ \boldsymbol{\gamma} & \boldsymbol{\nu} \end{pmatrix}. \tag{7}$$

Now we amplify somewhat our earlier classification of media.

Classification of Media

In the main we shall assume that $\boldsymbol{\xi} = \boldsymbol{\zeta} = 0$ but that $\boldsymbol{\epsilon} \neq 0$, $\boldsymbol{\mu} \neq 0$ simultaneously.

Anisotropic. If a medium is described by a permittivity tensor $\boldsymbol{\epsilon}$, it is called *electrically anisotropic.* If it is described by a permeability tensor $\boldsymbol{\mu}$, it is called *magnetically anisotropic.* If the medium requires both $\boldsymbol{\epsilon}$ and $\boldsymbol{\mu}$ to describe it, then it is called electromagnetically anisotropic.

If the tensor is symmetric, then there always exists a coordinate transformation such that the tensor can be diagonalized (see Section 19-5-3). The system in which the tensor is diagonal is called the principal system. Suppose this is the case and we have

$$\boldsymbol{\epsilon} = \begin{pmatrix} \epsilon_x & 0 & 0 \\ 0 & \epsilon_y & 0 \\ 0 & 0 & \epsilon_z \end{pmatrix}; \tag{8}$$

then the medium can be further categorized by the nature of these entries on the diagonal.

Uniaxial $\epsilon_x = \epsilon_y \neq \epsilon_z$ (tetragonal, hexagonal, rhombohedral crystals);

Biaxial $\epsilon_x \neq \epsilon_y \neq \epsilon_z$ (orthorhombic, monoclinic, triclinic crystals);

Isotropic $\epsilon_x = \epsilon_y = \epsilon_z$.

If the tensor is of the special Hermitian form (a matrix is Hermitian if entries below the main diagonal are the complex conjugates of those above):

$$\mathfrak{A} = \begin{pmatrix} a & j\alpha & 0 \\ -j\alpha & a & 0 \\ 0 & 0 & a_z \end{pmatrix}, \tag{9}$$

then the medium is *gyroelectric* or *gyromagnetic* depending on whether $\mathfrak{A}$

stands for $\boldsymbol{\epsilon}$ or $\boldsymbol{\mu}$. If both tensors are of the form 9, the medium is *gyroelectromagnetic*.

If the off-diagonal tensors $\boldsymbol{\xi}$ and $\boldsymbol{\zeta}$ are not zero, then the medium is said to be *bianisotropic*. Such properties have been observed in antiferromagnetic chromium oxide [antiferromagnetic materials are ones in which it is energetically favorable for neighboring dipoles to take an antiparallel orientation; see Ramo, Whinnery, and Van Duzer (1965)]. For such media Maxwell's curl equations would read:

$$A\text{: } \nabla \times \mathbf{H} = j\omega(\boldsymbol{\epsilon} \cdot \mathbf{E} + \boldsymbol{\xi} \cdot \mathbf{H}), \tag{10}$$

$$F\text{: } \nabla \times \mathbf{E} = -j\omega(\boldsymbol{\zeta} \cdot \mathbf{E} + \boldsymbol{\mu} \cdot \mathbf{H}). \tag{11}$$

Using these equations we can derive the relations

$$\nabla \cdot (\mathbf{E} \times \mathbf{H}^*) = j\omega(\mathbf{E} \cdot \boldsymbol{\epsilon}^* \cdot \mathbf{E}^* + \mathbf{E} \cdot \boldsymbol{\xi}^* \cdot \mathbf{H}^* - \mathbf{H}^* \cdot \boldsymbol{\zeta} \cdot \mathbf{E} - \mathbf{H}^* \cdot \boldsymbol{\mu} \cdot \mathbf{H}) \tag{12}$$

and

$$\nabla \cdot (\mathbf{E}^* \times \mathbf{H}) = -j\omega(\mathbf{E}^* \cdot \boldsymbol{\epsilon} \cdot \mathbf{E} + \mathbf{E}^* \cdot \boldsymbol{\xi} \cdot \mathbf{H} - \mathbf{H} \cdot \boldsymbol{\zeta}^* \cdot \mathbf{E}^* - \mathbf{H} \cdot \boldsymbol{\mu}^* \cdot \mathbf{H}^*), \tag{13}$$

where * means complex conjugate. But

$$\langle \mathbf{S} \rangle = \frac{1}{2} \operatorname{Re}(\mathbf{E} \times \mathbf{H}^*) = \frac{1}{4} [\mathbf{E} \times \mathbf{H}^* + (\mathbf{E} \times \mathbf{H}^*)^*] \tag{14}$$

and so adding Eqs. 12 and 13, and assuming that the operations $\langle \cdot \rangle$ and $\nabla \cdot$ can be interchanged, we find that

$$\langle \nabla \cdot \mathbf{S} \rangle = \frac{-j\omega}{4} [\mathbf{E}^* \cdot (\boldsymbol{\epsilon} - \tilde{\boldsymbol{\epsilon}}^*) \cdot \mathbf{E} + \mathbf{H}^* \cdot (\boldsymbol{\mu} - \tilde{\boldsymbol{\mu}}^*) \cdot \mathbf{H} + \mathbf{H}^* \cdot (\boldsymbol{\zeta} - \boldsymbol{\zeta}^*) \cdot \mathbf{E} \\ + \mathbf{E}^*(\boldsymbol{\xi} - \boldsymbol{\xi}^*) \cdot \mathbf{H}]. \tag{15}$$

With relation 15 we can classify the medium further as

(a) passive if $\langle \nabla \cdot \mathbf{S} \rangle < 0$;
(b) active if $\langle \nabla \cdot \mathbf{S} \rangle > 0$; and
(c) lossless if $\langle \nabla \cdot \mathbf{S} \rangle = 0$.

For the latter it follows from above that

$$\boldsymbol{\epsilon} = \tilde{\boldsymbol{\epsilon}}^*, \qquad \boldsymbol{\mu} = \tilde{\boldsymbol{\mu}}^*, \qquad \boldsymbol{\xi} = \tilde{\boldsymbol{\xi}}^*, \qquad \boldsymbol{\zeta} = \tilde{\boldsymbol{\zeta}}^*,$$

where ˜ means transpose and * means complex conjugate.

We return to further discussion of gyroelectromagnetic materials in Chapters 18 and 19. But first, in Chapter 16, we examine some of the basic properties of radar-absorbing materials in the light of the impedance boundary condition. Then, in Chapter 17, we discuss some representations for the concepts of reflection and transmission at the boundaries between material layers. Finally, in Chapter 20, we examine the practical problem of determining constitutive parameters from a set of measurements.

16

IMPEDANCE BOUNDARY CONDITION

If the complex permittivity or permeability of a scattering body has a large modulus, then inside the body the wavelength (and the depth of penetration of the field into the body) will be small and so, inside the body, the conditions for geometrical optics obtain. If, in addition, the field varies slowly from point to point on the body surface (on the scale of a wavelength inside the body), and if there exist no sources inside the body, then the field near the surface (inside the body) will consist of a wave that is propagated and attenuated in the direction of the normal to the surface (into the interior) of the body.

Generally, this interior wave is not plane but its radius of curvature is large in comparison to the wavelength in the body and the depth of penetration. Therefore, to a first approximation, the electric and magnetic vectors in the body are parallel to the wave surface; they lie in a plane that is tangent to the body surface and are related to one another in the same way that electric and magnetic vectors are related for a plane wave, that is,

$$\mathbf{E} = \eta \hat{n} \times \mathbf{H}, \tag{16-1}$$

where $\hat{n}$ is the external normal to the body surface and η is impedance, $=\sqrt{\mu/\epsilon}$, ϵ and μ being (complex) permittivity and permeability of the body.

Since the tangential components of **E** and **H** are continuous, the tangential components of the *exterior* fields are related by the same expression on the surface. Thus, we obtain the so-called impedance boundary condition (IBC) ($e^{j\omega t}$ time dependence):

$$\hat{n} \times \mathbf{E} = \eta[\hat{n} \times (\hat{n} \times \mathbf{H})]. \tag{16-2}$$

If an x,y,z-system is introduced so that $\hat{x}$ and $\hat{y}$ form the tangent plane at the body surface and $\hat{z}$ points out of the body, then the IBC may be written $E_x = \eta H_y$, $E_y = \eta H_x$.

The above discussion follows closely a development given in Fock (1957) by Leontovich and the conditions are usually named after Leontovich. In Section 16-5 below we examine their application to a situation for which PO is appropriate.

16-1 THICK, THIN, AND RESONANT ABSORBERS

Consider a single layer of material with characteristics ϵ and μ over a perfect conductor (Fig. 16-1). The impedance at the layer surface can be written, taking account of polarization, as follows:

$$\eta_{\perp} = \frac{j\mu \tan(Nkd \cos \theta'')}{N \cos \theta''}, \tag{16-3}$$

$$\eta_{\parallel} = \frac{j\mu}{N} \cos \theta'' \tan(Nkd \cos \theta''), \tag{16-4}$$

where $N = \sqrt{\epsilon\mu}$ (ϵ and μ relative) and Snell's law has the form $\sin \theta = N \sin \theta''$. [Equations. 16-3 and 16-4 follow easily from the multiple layer expressions, given below in Section 16-2, by specifying the interior face to be metallic ($\eta_m = 0$). The form of the wave number and the polarization dependence of the layer impedances follow from Section 3-2.] We can write the corresponding reflection coefficients, respectively, as

$$\Gamma_{\perp} = \frac{\eta_{\perp} \cos \theta - 1}{\eta_{\perp} \cos \theta + 1}, \tag{16-5}$$

$$\Gamma_{\parallel} = \frac{\eta_{\parallel} - \cos \theta}{\eta_{\parallel} + \cos \theta}. \tag{16-6}$$

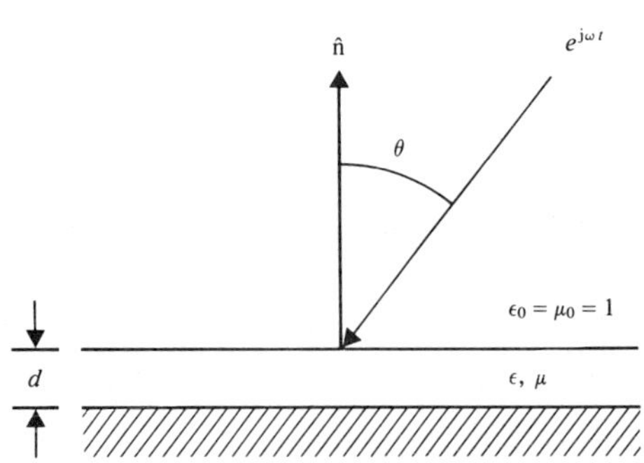

Figure 16-1 A single-layer absorber.

At normal incidence, of course, the polarization dependence disappears to give

$$\eta = j\frac{\mu}{N}\tan(Nkd), \qquad \Gamma = \frac{\eta - 1}{\eta + 1}. \tag{16-7}$$

The power reflection coefficient is $\Gamma\Gamma^*$, where * means complex conjugate.

However, the polarization dependence is also removable with the assumptions

$$|N| \gg 1, \qquad kd \ll |N|, \tag{16-8}$$

since

$$\eta_\perp \approx \eta_\parallel \approx j\frac{\mu}{N}\tan(Nkd)[1 + O(kd/N) + O(1/N^2)]. \tag{16-9}$$

The first of these assumptions is essentially that of Leontovich and permits the use of his IBC, which may also be written as (cf. previous section)

$$\mathbf{E} - (\hat{n} \cdot \mathbf{E})\hat{n} = \eta(\hat{n} \times \mathbf{H}). \tag{16-10}$$

Now we can look at the implications of some further assumptions that we can make regarding Nkd (i.e., the argument of the tangent function).

16-1-1 Thick Absorbers; $\mathrm{Im}(Nkd) \gg 1$

This condition means that the layer thickness is much much greater than several skin depths [or, that $d \gg 1/\mathrm{Im}(Nk)$, which means that thickness d is much greater than the wavelength in the material]; which implies that the metallic half-plane has no effect. The condition implies

$$\tan(Nkd) \approx j \tag{16-11}$$

which further implies

$$\eta \approx \frac{\mu}{N} = \sqrt{\frac{\mu}{\epsilon}}. \tag{16-12}$$

Therefore, in order to achieve zero reflectivity, we must have a material in which $\mu = \epsilon$.

16-1-2 Thin Absorbers; $|Nkd| \ll 1$

This condition implies that

$$\tan(Nkd) \approx Nkd. \tag{16-13}$$

So

$$\eta \approx j\mu kd \tag{16-14}$$

which, we note, is independent of permittivity (although permittivity continues to play its role through the original assumption of $|N| \gg 1$). Now if we wish to have a zero reflection coefficient, we must have $\eta = 1$ and so we must then have

$$kd = \frac{-j}{\mu}. \tag{16-15}$$

But kd is real and therefore μ must have a negligible real part and a large imaginary part.

16-1-3 Resonant Absorbers

These can be classified as two types:

1. *Quarterwave*: $\text{Re}(Nkd) = N'kd = \left(n - \frac{1}{2}\right)\pi$, $n = 1, 2, \ldots,$ so that

$$\eta = \frac{\mu}{N} \coth[\text{Im}(Nkd)]. \tag{16-16}$$

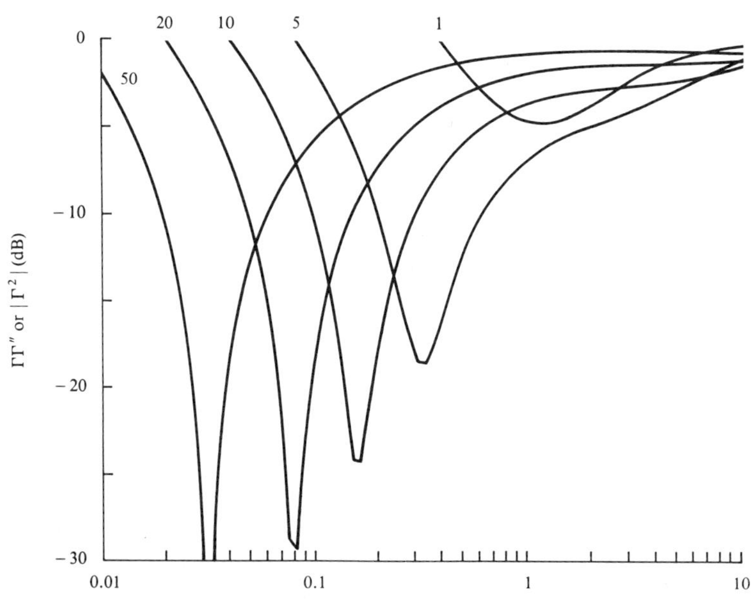

Figure 16-2 Power reflection coefficient versus $N''kd$ for a quarterwave absorber with $N'kd = \pi/2$ ($\epsilon'kd$ values as indicated for each curve).

2. *Halfwave*: $\mathrm{Im}(Nkd) = N'kd = n\pi$, $n = 1, 2, \ldots$, so that

$$\eta = \frac{\mu}{N}\tanh[\mathrm{Im}(Nkd)]. \tag{16-17}$$

In Eqs. 16-16 and 16-17 the trigonometric functions are hyperbolic and "Im" means "imaginary part of."

Figure 16-2 shows the power reflection coefficient $\Gamma\Gamma^*$ versus $N''kd$ for a quarterwave absorber ($N'kd = \pi/2$) for a sequence of values of $\mathrm{Re}(\epsilon kd) = \epsilon'kd$ ranging from unity to 50 with $\mathrm{Im}(\epsilon kd) = \epsilon''kd = 0$. We may then solve for μ' and μ'' in terms of $\mathrm{Re}(Nkd)$, $\mathrm{Im}(Nkd)$, $\mathrm{Re}(\epsilon kd)$, and $\mathrm{Im}(\epsilon kd)$. For example, $N''kd = 0.079$ on the $\epsilon'kd = 20$ curve (Fig. 16-2) yields a power reflection coefficient of -30 dB. Corresponding to these values, $\mu'kd = 0.123$ and $\mu''kd = 0.012$; whence $\eta = 0.996 - 0.050j$. Figure 16-3 gives additional information about $N''kd$ as a function of $\mu'kd$ for fixed ϵ and a range of $\mu''kd$ values. Figures 16-2 and 16-3 can be used together to determine power reflection coefficient as a function of μ for fixed ϵ.

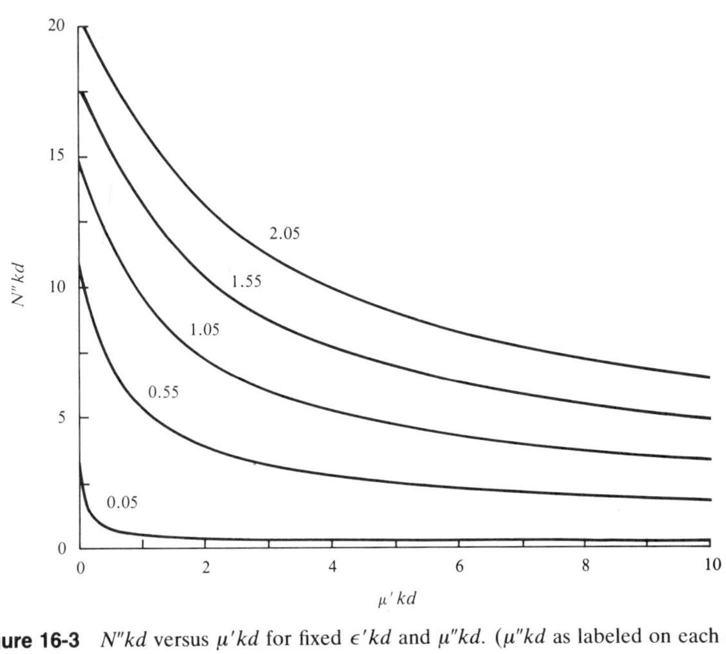

Figure 16-3 $N''kd$ versus $\mu'kd$ for fixed $\epsilon'kd$ and $\mu''kd$. ($\mu''kd$ as labeled on each curve.)

16-2 MULTIPLE LAYERS OVER AN ARBITRARY GROUND PLANE

The concept of impedance for plane layers (over an arbitrary ground plane) illuminated by normally incident energy can be phrased in terms of a (suitably normalized) layer-face impedance η and a layer impedance Z. For example, consider the layer pair shown in Fig. 16-4. Then we can move our position of interest from right to left as follows:

$$\frac{\eta_m}{Z_0} = \frac{(\eta_{m-1}/Z_0)\coth(\gamma_0 s) + 1}{\coth(\gamma_0 s) + (\eta_{m-1}/Z_0)}; \tag{16-18}$$

renormalize:

$$\frac{\eta_m}{Z} = \frac{\eta_m/Z_0}{Z/Z_0}; \tag{16-19}$$

$$\frac{\eta_{m+1}}{Z} = \frac{(\eta_m/Z)\coth(\gamma t) + 1}{\coth(\gamma t) + \eta_m/Z}; \tag{16-20}$$

renormalize:

$$\frac{\eta_{m+1}}{Z_0} = \frac{\eta_{m+1}/Z}{Z_0/Z}. \tag{16-21}$$

We have set $\gamma_0 = jk_0$ and find γ from $N\gamma_0$. The end result for one layer over metal is given by Eq. 16-7. For two layers over metal, it is

$$\eta = j\left[\frac{\eta_1 \tan(k_1 d_1) + \eta_2 \tan(k_2 d_2)}{1 - (\eta_1/\eta_2)\tan(k_1 d_1)\tan(k_2 d_2)}\right].$$

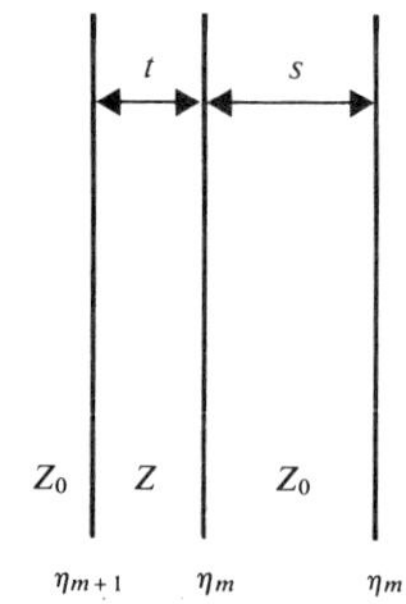

Figure 16-4 Impedance transfer from layer face to layer face.

16-3 SALISBURY SCREEN

Take the right-hand layer in Fig. 16-4 to be free space with thickness $s = \lambda_0/4$:

$$\gamma_0 = jk_0 = j(2\pi/\lambda_0). \tag{16-22}$$

Let the left-hand layer be a thin ($t \ll \lambda_0$) layer with unit permeability:

$$\gamma = j\omega\sqrt{\epsilon\mu} = jk_0\sqrt{\epsilon\mu/\epsilon_0\mu_0} = jk_0\sqrt{\epsilon/\epsilon_0}\,. \tag{16-23}$$

Then

$$\sqrt{\epsilon/\epsilon_0}\tanh(\gamma t) = j\sqrt{\epsilon/\epsilon_0}\tan(k_0\sqrt{\epsilon/\epsilon_0 t}) \approx jk_0 t\epsilon/\epsilon_0. \tag{16-24}$$

Let the thin layer have high conductivity σ:

$$\begin{aligned} jk_0 t\epsilon/\epsilon_0 &\approx jk_0 t(j\sigma/\omega\epsilon_0). \\ &= k_0\sigma t/\omega\epsilon_0. \end{aligned} \tag{16-25}$$

If the free-space layer is over metal, then $\eta_{m-1} = 0$ and $\eta/Z_0 = \infty$, and so

$$\frac{\eta_{m+1}}{Z_0} = \frac{Z}{Z_0}\coth(\gamma t) = (-1/-j\sqrt{\epsilon/\epsilon_0})\tan(\gamma t)\ \mathit{must} = 1 \tag{16-26}$$

for zero reflectivity. Therefore,

$$1/\sigma t = k_0/\omega\epsilon_0 = \sqrt{\mu_0/\epsilon_0}. \tag{16-27}$$

Restated in terms of resistivity ρ rather than conductivity σ, we have

$$\rho/t = 1/\sigma t = \sqrt{\mu_0/\epsilon_0} = Z_0\ (= 120\pi \text{ ohms}). \tag{16-28}$$

So to achieve a Salisbury screen absorber structure, we must place a thin sheet of 377 Ω/sq material at a distance of a quarter of a free-space wavelength in front of a metallic backing. The Salisbury screen structure gives perfect absorptivity only at that free-space wavelength.

16-4 TWO THEOREMS OF WESTON

The results given above for "thick" and "thin" absorber layers are closely connected with two basic theorems of absorber theory published in 1963 by Weston. Each theorem requires that the geometric shape and constitutive parameters of the scattering body be invariant under a rotation of $\pi/2$ about an axis parallel to the direction of incidence/observation.

Weston's first theorem states that the far backscattered field is zero if the scattering body is composed of material in which $\epsilon = \mu$. Note that if a metallic body (having the symmetry described above) is coated with a "thick" layer in which $\epsilon = \mu$, then the metal substrate matters not at all and, by the theorem, the far backscattered field is zero.

Weston's second theorem states that if the scattering body satisfies the Leontovich IBC, and if the impedance $\eta = 1$ at the surface, then the far backscattered field is zero. Here, the "thin"-layer coating can be used to ensure the Leontovich IBC. (The first theorem applied to the "thick" layer could actually have been included under the second. Thus, $\eta = \sqrt{\mu/\epsilon} = 1$ implies $\epsilon = \mu$; this is actually the case considered by Leontovich.)

Weston goes on to point out that his second theorem is frequency-independent and thus can be applied to the resonance and Rayleigh regions. But he does require that the radii of curvature of the surface be large compared with the interior wavelength [recall that $\mathrm{Im}(N) \gg 1$ must also be satisfied]; and, in addition, we have already seen that field variation outside the boundary must be small in a distance comparable with interior wavelength.

Weston extends his analysis to backscattering from arbitrary smooth convex bodies for which the IBC holds by separating the scattered field (for the high-frequency region) into its specular and diffracted parts. We examine his procedure and results briefly.

Since the specular return can be expressed in terms of the flat plate reflection coefficient $R = (\eta - 1)/(\eta + 1)$, it is clear that this return is zero for $\eta = 1$.

The diffracted part is harder.

Consider a ray traveling from a source Q to an observation point P by way of the path Q_1P_1 on the surface S (Fig. 16.5). To facilitate description of the field on the diffracted ray, introduce the unit tangent $\hat{d}_1$ to the ray, together with the unit outward normal $\hat{d}_2$ to S; put $\hat{d}_3 = \hat{d}_2 \times \hat{d}_1$. Since the diffracted ray is a geodesic, $\hat{d}_2$ lies along the principal normal to the ray and $\hat{d}_3$ along its binormal. At P the vectors $\hat{d}_1$, $\hat{d}_2$, $\hat{d}_3$ are obtained by translating those existing at the launch point P_1.

The basic assumption of Keller's GTD is that field components E_2 and E_3 propagate independently of each other and obey the principles of scalar fields, while $E_1 = 0$. Under these assumptions the Leontovich IBC becomes,

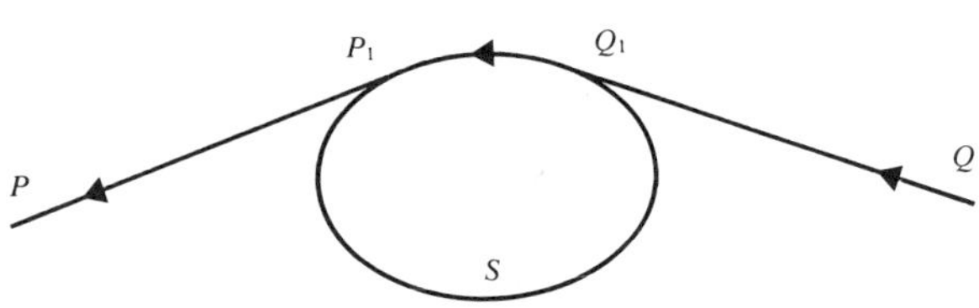

Figure 16-5 Diffracted ray path.

approximately,

$$\partial E_2/\partial n = jk\eta E_2,\ \partial E_3/\partial n = \frac{jk}{\eta} E_3,\ [e^{j\omega t}] \tag{16-29}$$

so that, for $\eta = 1$, E_2 and E_3 satisfy the same boundary condition. In contrast, for a perfectly conducting surface, $E_3 = 0$, $\partial E_2/\partial n = 0$.

From Keller and Levy (1959) the diffracted field $\mathbf{E}^d(P)$ is

$$\mathbf{E}^d(P) = E_2^i(Q_1)V(P, Q_1)\hat{a}_2(P_1) + E_3^i(Q_1)W(P, Q_1)\hat{a}_3(P_1), \tag{16-30}$$

where $E_2^i(Q_1)$, $E_3^i(Q_1)$ are components of the incident field at Q_1; V and W are functions of the diffraction coefficients at Q_1 and P_1, as well as arclength Q_1P_1 and ray length P_1P, and the principal radius of curvature at the launch point P_1. When $\eta = 1$, V and W are identical. But even so, $\mathbf{E}^d(P) \neq 0$. To make the total diffracted field zero one must take account of the $\pi/2$ rotational symmetry of the scatterer. Weston illustrates this situation using a body of revolution having symmetry axis along the z-axis. At the shadow boundary he observes the behavior of a quartet of symmetrically incident rays at Q_1, Q_2, Q_3, Q_4 with incident polarization as shown in Fig. 16-6 for the incidence direction in the $+\hat{z}$-direction and backscattering direction in the $-\hat{z}$-direction. As these rays travel geodesics in the shadowed region, they emerge and are launched at Q_3, Q_4, Q_1, Q_2, respectively, with polarizations as shown, $V(P, Q_1)$ being identical for each ray. Clearly, a pair of rays is in phase and a second pair is π out of phase with the first pair. Since their amplitudes are equal, their total combination is zero. The remaining set of rays can always be ordered by quartets such that this type of cancellation takes place.

Suppose now that the surface fields still obey the Leontovich IBC but that $\eta \neq 1$. Now there is no cancellation but a result for $\mathbf{E}^d$ is still available.

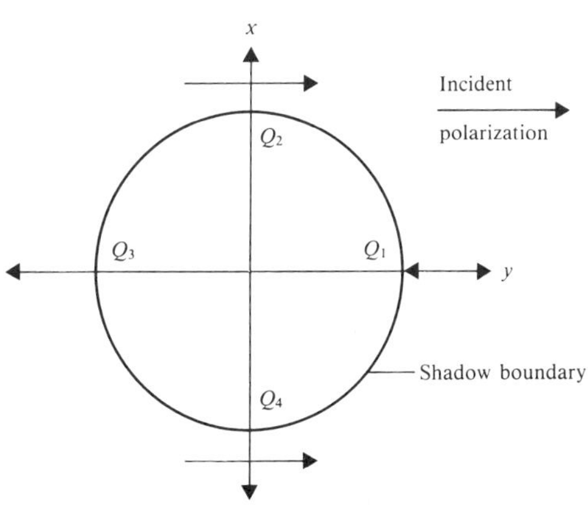

Figure 16-6 Ray paths for a body with axial symmetry.

Set up cylindrical coordinates r, ϕ, z so that $x = r\cos\phi$ and $y = r\sin\phi$; let a be the radius of the shadow boundary. A ray incident at Q_ϕ with coordinates $(a, \phi + \pi, 0)$ will travel a geodesic in the shadow to be launched in a backscattered direction at P_ϕ with coordinates $(a, \phi, 0)$; or, in an arbitrary direction after traveling a distance s from the shadow boundary. For the former one finds

$$\begin{aligned}
&\text{(a)}\quad E_2^i(Q_\phi) = -\sin\phi;\\
&\text{(b)}\quad E_3^i(Q_\phi) = \cos\phi;\\
&\text{(c)}\quad \hat{d}_2(P_\phi) = \hat{x}\cos\phi + \hat{y}\sin\phi;\\
&\text{(d)}\quad \hat{d}_3(P_\phi) = -\hat{x}\sin\phi + \hat{y}\cos\phi.
\end{aligned} \tag{16-31}$$

Weston shows that

$$W(P_\phi, Q_\phi, \eta) = V(P_\phi, Q_\phi, 1/\eta); \tag{16-32}$$

and then that an integration over ϕ from 0 to 2π yields, as the total contribution from all rays,

$$\mathbf{E}^d = \hat{y}\frac{\pi a \exp(+j\pi/4)}{(s\lambda)^{1/2}}[-V(P, Q, \eta) + V(P, Q, 1/\eta)]. \tag{16-33}$$

Again, it is clear that $\mathbf{E}^d = 0$ if $\eta = 1$. Otherwise, the behavior of V is such that $\|\mathbf{E}^d(\eta \neq 0)\| < \|\mathbf{E}^d(\eta = 0)\|$, that is, a coated body produces a smaller diffracted field amplitude than a metallic body does.

16-5 THE IMPEDANCE BOUNDARY CONDITION AND PHYSICAL OPTICS

In this section we show a particularly simple yet useful combination of the IBC with PO. For this combination, the effect of an absorber coat on the RCS of a perfectly conducting body can be approximated by an application of the flat-plate reflection coefficient for the absorber to the PO result.

16-5-1 Formulation of Approach

PO (also referred to as the Kirchoff approximation), which is developed in Section 8-1, can be revisited as a method for calculating the RCS of coated bodies. Coatings can include not only absorbing materials but also dielectric toppings over absorbing material as well. In the case of perfectly conducting bodies, the boundary condition $\hat{n} \times \mathbf{E} = 0$ on the surface considerably simplifies the surface integral representation for the scattered field (Eqs. 4-27 and 4-28). In the case of coated bodies the above condition is altered in form so as to continue to portray the condition that the tangential component of the electric field be continuous across the surface of the coat. At this point, then, we are able to see that either we cannot use a Kirchhoff (or PO) approximation of the surface integral representation, or we must make some further approximation that will yield a surface current density analogous to the one employed for a perfect conductor. That is, the boundary conditions at the surface of the coat require the use of fields both inside and outside the coat and in such circumstances we might even have to take account of the propagation of fields inside the coat.

Fortunately, there is a way around this dilemma, in the form of the so-called impedance boundary condition (IBC). As we have seen above, this condition requires either a strongly refracting or absorbing medium, by which we mean that the index of refraction of the coat has a very large magnitude or, equivalently, that the direction of propagation inside the coat is approximately normal to the surface of the coat. Under these circumstances we are then able to express the effects and behavior of the fields inside the coat entirely in terms of the impedance η, which relates the electric and magnetic fields on the surface of the coat. The outcome is the IBC.

As a result of the formulation of the IBC we are able to return to the concept of a surface current distribution for coated bodies, and so we can continue to make use of the representation for the scattered fields given by Eqs 4-27 and 4-28. Moreover, the natural opportunity still exists in this case for the application of a Kirchhoff approximation.

In the sections to follow we develop the steps outlined above. First we show how the IBC can be developed from the standard continuity boundary conditions when the coating has the required characteristics. Then we show how to develop the Kirchhoff approximation for the field scattered by such a coated body.

16-5-2 Surface Fields on a Coating

In this section we show in some detail the relations that can be obtained among the amplitudes of the incident, reflected, and refracted fields at the surface of a coated body. In anticipation of the application of these results to a PO approximation, we develop these relations for plane layers. Thus, we are able to extend and employ the results for plane layers given by Born and Wolf (1959) (see also Sections 4 and 5 below). We adapt a treatment used by Überall (1964) in another connection.

We assume that the field $\mathbf{E}$ outside the coating is made up of incident and reflected parts:

$$\mathbf{E} = \mathbf{E}_0 \exp(-j\mathbf{k}_0 \cdot \mathbf{r}) + \mathbf{E}_1 \exp(-j\mathbf{k}_1 \cdot \mathbf{r}) \tag{16-34}$$

and similarly, that the field $\mathbf{E}'$ inside the coating is made up of transmitted and reflected parts:

$$\mathbf{E}' = \mathbf{E}_0' \exp(-j\mathbf{k}_0' \cdot \mathbf{r}) + \mathbf{E}_1' \exp(-j\mathbf{k}_1' \cdot \mathbf{r}) \tag{16-35}$$

where the directions of propagation have been denoted by

$$\mathbf{k}_{0,1} = k\hat{k}_{0,1},\ \mathbf{k}'_{0,1} = k'\hat{k}'_{0,1},\ k'/k = N, \tag{16-36}$$

N being the complex index of refraction of the coating. The laws of reflection are

$$\hat{k}_1 = \hat{k}_0 - 2\hat{n}(\hat{n}\cdot\hat{k}_0),\ \hat{k}'_1 = \hat{k}'_0 - 2\hat{n}(\hat{n}\cdot\hat{k}'_0);$$

and of refraction are

$$k\hat{n}\times\hat{k}_0 = k'\hat{n}\times\hat{k}'_0,\ k\hat{n}\times\hat{k}_1 = k'\hat{n}\times\hat{k}'_1,$$

where $\hat{n}$ is the unit normal to the surface pointing from conductor to coat. Letting subscripts p and c denote the perfect conductor and coating, respectively, we note that, on the coat

$$\mathbf{r} = \mathbf{r}_c = \mathbf{r}_p + \boldsymbol{\delta},\ \boldsymbol{\delta} = \hat{n}\delta$$

($\hat{n}\cdot\mathbf{r}_p = 0$ if the origin is on the conductor), δ being the thickness of the coating.

The fields at the surface of the plane coating satisfy the boundary conditions

$$\begin{aligned} \hat{n}\times\mathbf{E}_c &= \hat{n}\times\mathbf{E}'_c, \\ \hat{n}\times\mathbf{H}_c &= \hat{n}\times\mathbf{H}'_c, \end{aligned} \tag{16-37}$$

and, on the conductor, they obey the conditions

$$\begin{aligned} \hat{n}\times\mathbf{E}'_p &= 0, \\ \hat{n}\cdot\mathbf{H}'_p &= 0. \end{aligned} \tag{16-38}$$

The latter are satisfied by taking

$$\mathbf{E}'_1 = 2\hat{n}(\mathbf{E}'_0\cdot\hat{n}) - \mathbf{E}'_0. \tag{16-39}$$

We represent the various fields in the form

$$\begin{aligned} \mathbf{E}_0 &= a_0[\hat{k}_0\times(\hat{n}\times\hat{k}_0)] + b_0[\hat{n}\times\hat{k}_0] \\ \mathbf{E}_1 &= a_1[\hat{k}_1\times(\hat{n}\times\hat{k}_1)] + b_1[\hat{n}\times\hat{k}_1] \\ \mathbf{E}'_0 &= a'_0[\hat{k}'_0 + (\hat{n}\times\hat{k}'_0)] + b'_0[\hat{n}\times\hat{k}'_0] \end{aligned} \tag{16-40}$$

and determine a_1, b_1, a'_0, and b'_0 in terms of a_0 and b_0 using Eqs. 16-37 through 16-39. Applying the first of Eqs. 16-37, we obtain

$$\begin{aligned} \hat{n}\times\mathbf{E}_c = {} & (\hat{n}\times\mathbf{E}_0)\exp(-j\mathbf{k}_0\cdot\mathbf{r}_c) + (\hat{n}\times\mathbf{E}_1)\exp(-j\mathbf{k}_1\cdot\mathbf{r}_c) \\ = {} & \{-a_0(\hat{n}\cdot\hat{k}_0)[\hat{n}\times\hat{k}_0] + b_0[\hat{n}\times(\hat{n}\times\hat{k}_0)]\}\exp(-j\mathbf{k}_0\cdot\mathbf{r}_c) \\ & + \{-a_1(\hat{n}\cdot\hat{k}_1)[\hat{n}\times\hat{k}_1] + b_1[\hat{n}\times(\hat{n}\times\hat{k}_1)]\}\exp(-j\mathbf{k}_1\cdot\mathbf{r}_c) \\ = {} & \{(\hat{n}\cdot\hat{k}_0)(-a_0\exp(-j\mathbf{k}_0\cdot\delta) + a_1\exp(j\mathbf{k}_0\cdot\delta))[\hat{n}\times\hat{k}_0] \\ & + (b_0\exp(-j\mathbf{k}_0\cdot\delta) + b_1\exp(j\mathbf{k}_0\cdot\delta))[\hat{n}\times(\hat{n}\times\hat{k}_0)]\}\exp(-j\mathbf{k}_0\cdot\mathbf{r}_p) \end{aligned}$$

$$= \hat{n} \times \mathbf{E}'_c = (\hat{n} \times \mathbf{E}'_0) \exp(-j\mathbf{k}'_0 \cdot \mathbf{r}_c) + (\hat{n} \times \mathbf{E}'_1) \exp(-j\mathbf{k}'_1 \cdot \mathbf{r}_c)$$
$$= (\exp(-j\mathbf{k}'_0 \cdot \mathbf{r}_c) - \exp(-j\mathbf{k}'_1 \cdot \mathbf{r}_c))(\hat{n} \times \mathbf{E}'_0) \quad \text{(by Eq. 16-39)}$$
$$= -2j \exp(-j\mathbf{k}'_0 \cdot \mathbf{r}_p) \sin(\mathbf{k}'_0 \cdot \delta)\{-a'_0(\hat{n} \cdot \hat{k}'_0)[\hat{n} \times \hat{k}'_0] + b'_0[\hat{n} \times (\hat{n} \times \hat{k}'_0)]\}$$
$$= -\frac{2j}{N} \exp(-j\mathbf{k}'_0 \cdot \mathbf{r}_p) \sin(\mathbf{k}'_0 \cdot \delta)\{-a'_0(\hat{n} \cdot \hat{k}'_0)[\hat{n} \times \hat{k}_0] + b'_0[\hat{n} \times (\hat{n} \times \hat{k}_0)]\}. \tag{16-41}$$

From this sequence of equalities we find the equations

$$(\hat{n} \cdot \hat{k}'_0) \frac{2j}{N} \exp(-j\mathbf{k}'_0 \cdot \mathbf{r}_p) \sin(\mathbf{k}'_0 \cdot \delta) a'_0 = \exp(-j\mathbf{k}_0 \cdot \mathbf{r}_p)(\hat{n} \cdot \hat{k}_0)(-a_0 \exp(-j\mathbf{k}_0 \cdot \delta) + a_1 \exp(j\mathbf{k}_0 \cdot \delta)),$$
$$-\frac{2j}{N} \exp(-j\mathbf{k}'_0 \cdot \mathbf{r}_p) \sin(\mathbf{k}'_0 \cdot \delta) b'_0 = \exp(-j\mathbf{k}_0 \cdot \mathbf{r}_p)(b_0 \exp(-j\mathbf{k}_0 \cdot \delta) + b_1 \exp(j\mathbf{k}_0 \cdot \delta)). \tag{16-42}$$

From the second of Eqs. 16-37 we derive, with the help of Maxwell's equation $-jk\mu\mathbf{H} = \nabla \times \mathbf{E}$:

$$\begin{aligned} \mathbf{H} &= (\hat{k}_0 \times \mathbf{E}_0) \exp(-j\mathbf{k}_0 \cdot \mathbf{r}) + (\hat{k}_1 \times \mathbf{E}_1) \exp(-j\mathbf{k}_1 \cdot \mathbf{r}) \\ \mathbf{H}' &= \frac{N}{\mu}\{(\hat{k}'_0 \times \mathbf{E}'_0) \exp(-j\mathbf{k}'_0 \cdot \mathbf{r}) + (\hat{k}'_1 \times \mathbf{E}'_1) \exp(-j\mathbf{k}'_1 \cdot \mathbf{r})\}, \end{aligned} \tag{16-43}$$

$$\hat{n} \times \mathbf{H}_c = \{-a_0[\hat{n} \times (\hat{n} \times \hat{k}_0)] - b_0(\hat{n} \cdot \hat{k}_0)[\hat{n} \times \hat{k}_0]\} \exp(-j\mathbf{k}_0 \cdot \mathbf{r}_c) + \{-a_1[\hat{n} \times (\hat{n} \times \hat{k}_1)] - b_1(\hat{n} \cdot \hat{k}_1)[\hat{n} \times \hat{k}_1]\} \exp(-\mathbf{j}\mathbf{k}_1 \cdot \mathbf{r}_c)$$
$$= \exp(-j\mathbf{k}_0 \cdot \mathbf{r}_p)\{(-a_0 \exp(-j\mathbf{k}_0 \cdot \delta) - a_1 \exp(j\mathbf{k}_0 \cdot \delta)[\hat{n} \times (\hat{n} \times \hat{k}_0)] + (\hat{n} \cdot \hat{k}_0)(-b_0 \exp(-j\mathbf{k}_0 \cdot \delta) + b_1 \exp(j\mathbf{k}_0 \cdot \delta)[\hat{n} \times \hat{k}_0]\} \tag{16-44}$$

$$= \hat{n} \times \mathbf{H}'_c = \frac{N}{\mu}\{-a'_0[\hat{n} \times (\hat{n} \times \hat{k}'_0)] - b'_0(\hat{n} \cdot \hat{k}'_0)[\hat{n} \times \hat{k}'_0]\} \times (\exp(-j\mathbf{k}'_0 \cdot \mathbf{r}_c) + \exp(-j\mathbf{k}'_1 \cdot \mathbf{r}_c))$$
$$= 2\mu^{-1} \exp(-j\mathbf{k}'_0 \cdot \mathbf{r}_p) \cos(\mathbf{k}'_0 \cdot \delta)\{-a'_0[\hat{n} \times (\hat{n} \times \hat{k}_0)] - b'_0(\hat{n} \cdot \hat{n}'_0)[\hat{n} \times \hat{k}_0]\}. \tag{16-45}$$

From this sequence of equalities we find two additional equations in the a's and the b's:

$$-2\mu^{-1} \exp(-j\mathbf{k}'_0 \cdot \mathbf{r}_p) \cos(\mathbf{k}'_0 \cdot \delta) a'_0 = \exp(-j\mathbf{k}_0 \cdot \mathbf{r}_p)(-a_0 \exp(-j\mathbf{k}_0 \cdot \delta) - a_1 \exp(j\mathbf{k}_0 \cdot \delta)$$

$$-2\mu^{-1}(\hat{n} \cdot \hat{k}'_0) \exp(-j\mathbf{k}'_0 \cdot \mathbf{r}_p) \cos(\mathbf{k}'_0 \cdot \boldsymbol{\delta}) b'_0 = (\hat{n} \cdot \hat{k}_0) \exp(-j\mathbf{k}_0 \cdot \mathbf{r}_p)(-b_0 \exp(-j\mathbf{k}_0 \cdot \delta) + b_1 \exp(j\mathbf{k}_0 \cdot \delta))$$

Equations 16-42 and 16-45 together now give

$$\begin{aligned} &a_0' \exp[-j(\mathbf{k}_0' - \mathbf{k}_0) \cdot \mathbf{r}_p] \\ &\quad = N(\hat{n} \cdot \hat{k}_0)(a_0 \exp(-j\mathbf{k}_0 \cdot \delta) + a_1 \exp(j\mathbf{k}_0 \cdot \delta))/2j(\hat{n} \cdot \hat{k}_0') \sin(\mathbf{k}_0' \cdot \delta) \\ &\quad = \mu[a_0 \exp(-j\mathbf{k}_0 \cdot \delta) + a_1 \exp(j\mathbf{k}_0 \cdot \delta)/2 \cos(\mathbf{k}_0' \cdot \delta)] \\ &b_0' \exp[-j(\mathbf{k}_0' - \mathbf{k}_0) \cdot \mathbf{r}_p] = N(b_0 \exp(-j\mathbf{k}_0 \cdot \delta) \\ &\qquad - b_1 \exp(j\mathbf{k}_0 \cdot \delta))/2j \sin(\mathbf{k}_0' \cdot \delta) = (\hat{n} \cdot \hat{k}_0)\mu(b_0 \exp(-j\mathbf{k}_0 \cdot \delta) \\ &\qquad - b_1 \exp(j\mathbf{k}_0 \cdot \delta))/2(\hat{n} \cdot \hat{k}_0') \cos(\mathbf{k}_0' \cdot \delta); \end{aligned} \tag{16-46}$$

whence, putting $\eta = (j\mu/N) \tan(\mathbf{k}_0' \cdot \delta)$ as before,

$$a_1 = a_0 \exp(-2j\mathbf{k}_0 \cdot \delta) \frac{1 + \eta(\hat{n} \cdot \hat{k}_0')/(\hat{n} \cdot \hat{k}_0)}{1 - \eta(\hat{n} \cdot \hat{k}_0')/(\hat{n} \cdot \hat{k}_0)} \tag{16-47}$$

$$b_1 = -b_0 \exp(-2j\mathbf{k}_0 \cdot \delta) \frac{1 + \eta(\hat{n} \cdot \hat{k}_0)/(\hat{n} \cdot \hat{k}_0')}{1 - \eta(\hat{n} \cdot \hat{k}_0)/(\hat{n} \cdot \hat{k}_0')} \tag{16-48}$$

Putting these results back into Eq. 16-46 gives

$$a_0' = a_0\mu \exp(j\mathbf{k}_0' \cdot \mathbf{r}_p - j\mathbf{k}_0 \cdot \mathbf{r}_c)/[1 - \eta(\hat{n} \cdot \hat{k}_0')/(\hat{n} \cdot \hat{k}_0)] \cos(\mathbf{k}_0' \cdot \delta) \tag{16-49}$$

and

$$\begin{aligned} b_0' = b_0\mu(\hat{n} \cdot \hat{k}_0) &\exp(j\mathbf{k}_0' \cdot \mathbf{r}_p - j\mathbf{k}_0 \cdot \mathbf{r}_c)/(\hat{n} \cdot \hat{k}_0') \\ &\times [1 - \eta(\hat{n} \cdot \hat{k}_0)/(\hat{n} \cdot \hat{k}_0')] \cos(\mathbf{k}_0' \cdot \delta). \end{aligned} \tag{16-50}$$

16-5-3 IBC

We now observe that if the coating is of a certain composition [e.g., highly absorbing; see Überall (1964) for other cases], we can establish a relation between the two boundary conditions given by Eq. 16-37. We form the expression

$$\begin{aligned} \hat{n} \times (\hat{n} \times \mathbf{E}_c) = \{&2a_0(\hat{n} \cdot \hat{k}_0')\eta[\hat{n} \times (\hat{n} \times \hat{k}_0)]/[1 - \eta(\hat{n} \cdot \hat{k}_0')/(\hat{n} \cdot \hat{k}_0)] \\ &+ 2b_0\eta(\hat{n} \cdot \hat{k}_0)(\hat{n} \times \hat{k}_0)/(\hat{n} \cdot \hat{k}_0')[1 - \eta(\hat{n} \cdot \hat{k}_0)/(\hat{n} \cdot \hat{k}_0')]\} \\ &\times \exp(-j\mathbf{k}_0 \cdot \mathbf{r}_c) \end{aligned} \tag{16-51}$$

and compare it with

$$\begin{aligned} \hat{n} \times \mathbf{H}_c = \exp(-j\mathbf{k}_0 \cdot \mathbf{r}_c)\{&-2a_0[\hat{n} \times (\hat{n} \times \hat{k}_0)]/[1 - \eta(\hat{n} \cdot \hat{k}_0')/(\hat{n} \cdot \hat{k}_0)] \\ &- 2b_0(\hat{n} \cdot \hat{k}_0)(\hat{n} \times \hat{k}_0)/[1 - \eta(\hat{n} \cdot \hat{k}_0)/(\hat{n} \cdot \hat{k}_0')]\}. \end{aligned} \tag{16-52}$$

We see that, provided $\hat{n} \cdot \hat{k}_0' \approx 1$ (that is, that $\hat{n} \approx \hat{k}_0'$), we can write

$$\hat{n} \times (\hat{n} \times \mathbf{E}_c) = -\eta\hat{n} \times \mathbf{H}_c, \tag{16-53}$$

or

$$\hat{n} \times \mathbf{E}_c = \eta[\hat{n} \times (\hat{n} \times \mathbf{H}_c)]$$

which is the IBC.† That is to say we can write conditions for the fields at the surface of the coating completely in terms of fields on the *outer* surface of the coating, and without reference to the fields inside the coating as in the previous section (Eqs. 16-37 and 16-38). The requirement for these circumstances to be possible is that the coating medium be either highly absorbing or highly refracting and hence that $\hat{n} \approx \hat{k}_0'$. Note that, in these circumstances, the impedance can be written in the form

$$\eta = \frac{j\mu}{N} \tan(Nk\delta)$$

as we saw in Eq. 16-7.

16-5-4 PO Based on an IBC

A PO, or Kirchhoff, approximation requires the radii of curvature in the neighborhood of every integration point on the scattering surface to be large enough to allow the surface fields to be approximated by those that would be present on an infinite plane. Thus, on the illuminated portion of the body we take the fields to be the geometric optics fields; on the shadowed portion of the body we take the surface fields to be zero.

The scattered field may be written (refer to Sections 8-1 and 11-6, particularly Eq. 11-66 in the latter), making use of the usual Green's function approximation

$$g(r, r') = \frac{\exp(-jk\,\|\mathbf{r} - \mathbf{r}'\|)}{4\pi\,\|\mathbf{r} - \mathbf{r}'\|} \approx \frac{\exp(-jkr)}{4\pi r'} \exp[2jk(\mathbf{r} \cdot \mathbf{r}')]$$

under the far field assumption that $\|\mathbf{r}\| \gg \|\mathbf{r}'\|$ and $\|\mathbf{r}'\| \gg \|\mathbf{r} - \mathbf{r}'\|$,

$$\mathbf{H}^s = \frac{-jk\exp(-jkr)}{4\pi r}\hat{k} \times \int_S \{[\hat{n} \times \mathbf{H}_c] + \eta[\hat{k} \times (\hat{n} \times \mathbf{E}_c)]\} \exp(jk\hat{k} \cdot \mathbf{r}')\, dS; \quad (16\text{-}54)$$

S represents the illuminated portion of the scattering body and $\hat{n}$ its unit outward normal. Since the unit vector $\hat{k}$ represents an arbitrary direction of incidence, Eq. 16-54 really provides a bistatic result. The cross-products $\hat{n} \times \mathbf{H}_c$ and $\hat{n} \times \mathbf{E}_c$ may be evaluated on the surface of the coating with the help of Eqs. 16-37 and 16-38 or by applying the IBC (16-53). Employing the

† In general, an IBC can be written down for each polarization separately (Überall, 1964). The impedance itself will then depend on polarization. The resulting polarization-dependent impedances can then be made independent of polarization by requiring a large magnitude for the index of refraction (refer to Section 16-2-1). The impedance then takes the form given above (see Eq. 16-9).

latter we find that Eq. 16-54 becomes

$$\mathbf{H}^s = \frac{-2jk\exp(-jkr)}{4\pi r}\int_S \{[a_0(\hat{n}\cdot\hat{k})u(\eta) + b_0\eta v(\eta)][\hat{n}\times\hat{k}] + [a_0(\hat{n}\cdot\hat{k})\eta u(\eta) - b_0 v(\eta)][\hat{k}\times(\hat{n}\times\hat{k})]\}\, dS \quad (16\text{-}55)$$

where

$$u(\eta) = 2(\hat{n}\cdot\hat{k})/[(\hat{n}\cdot\hat{k}) - \eta],$$
$$v(\eta) = 2(\hat{n}\cdot\hat{k})/[1 - \eta(\hat{n}\cdot\hat{k})].$$

If we put $\eta = 0$, that is, if we take the surface of the body to be perfectly conducting, we find that, since $u(0) = 2$ and $v(0) = 2(\hat{n}\cdot\hat{k})$, the brace-enclosed portion of the integrand of Eq. 16-55 takes the form:

$$2a_0(\hat{n}\cdot\hat{k})[\hat{n}\times\hat{k}] - 2b_0(\hat{n}\cdot\hat{k})[\hat{k}\times(\hat{n}\times\hat{k})].$$

This result of course checks with what would result from Eq. 16-54 if we used the fact that $\hat{n}\times\mathbf{E}_c = 0$ on a perfectly conducting surface.

It is useful to examine the form of the magnetic field itself on the surface of the coating. It is not simply a tangential field as it is for a perfect conductor:

$$\begin{aligned}
\mathbf{H}_c &= (\hat{k}_0\times\mathbf{E}_0)\exp(-j\mathbf{k}_0\cdot\mathbf{r}_c) + (\hat{k}_1\times\mathbf{E}_1)\exp(-j\mathbf{k}_1\cdot\mathbf{r}_c) \\
&\quad + (\hat{n}\times\mathbf{E}_0')\frac{N}{\mu}[\exp(-j\mathbf{k}_0'\cdot\mathbf{r}_c) + \exp(-j\mathbf{k}_1'\cdot\mathbf{r}_c)] \\
&\approx (\hat{k}_0\times\mathbf{E}_0)\exp(-j\mathbf{k}_0\cdot\mathbf{r}_c) + (\hat{k}_1\times\mathbf{E}_1)\exp(-j\mathbf{k}_1\cdot\mathbf{r}_c), \text{ if } \hat{n}\approx\hat{k}_1', \\
&= 2(\hat{n}\cdot\hat{k}_0)\exp(-j\mathbf{k}_0\cdot\mathbf{r}_c)\{-a_0(\hat{n}\times\hat{k}_0)/[(\hat{n}\cdot\hat{k}_0) - \eta] \\
&\quad + \eta b_0[\hat{k}_0\times(\hat{n}\times\hat{k}_0)]/[\eta(\hat{n}\cdot\hat{k}_0) - 1] \\
&\quad - b_0[\eta(\hat{n}\cdot\hat{k}_0) + 1][\hat{n}\times(\hat{n}\times\hat{k}_0)]/[\eta(\hat{n}\cdot\hat{k}_0) - 1]\}.
\end{aligned}$$

However, if we put $\eta = 0$, $\mathbf{H}_c$ reduces to a vector with only a tangential component at the surface (of what is now a perfect conductor since $\eta = 0$). In this case $\mathbf{r}_c$ signifies that $\mathbf{r}$ is evaluated on the surface of the perfect conductor, that is, that the layer thickness $\delta = 0$.

16-5-5 Example: Axial Backscattering for a body of Revolution

Consider the special case of backscattering, $\mathbf{k} = -\mathbf{k}_0$. In addition, choose a coordinate system so that the direction of incidence is $-\hat{z}$; thus, the illuminated portion S of the scattering surface can be projected onto an area A in the xy-plane. Then $\mathbf{k}_0 = -k\hat{z}$; $dS = (\hat{n}\cdot\hat{k}_0)^{-1}\,dx\,dy$; and $\mathbf{r}_c = x_c\hat{x} + y_c\hat{y} + z_c\hat{z}$ where $z_c = g(x, y)$ is the equation of S in the chosen coordinate system. With these conditions Eq. 16-55 becomes

$$\mathbf{H}^s = \frac{-2jk\exp(-jkr')}{4\pi r'} \int_A \{-a_0(\hat{n}\times\hat{k}_0)[1+\eta/(\hat{n}\cdot\hat{k}_0)]/[1-\eta/(\hat{n}\cdot\hat{k}_0)]$$
$$+ b_0[\hat{k}_0\times(\hat{n}\times\hat{k}_0)][1+\eta(\hat{n}\cdot\hat{k}_0)]/[1-\eta(\hat{n}\cdot\hat{k}_0)]\}$$
$$\times \exp(-2jkg(x, y))\,dx\,dy. \quad (16\text{-}56)$$

When the surface S is a surface of revolution we can obtain a further simplification for the nose-on direction of incidence. The equation of S can be written in the form $z = g(\sqrt{x^2+y^2})$ and a double stationary phase evaluation can be made on the basis of the fact that the stationary point (for the above choice of axes and direction of incidence) occurs for $x = y = 0$. Letting

$$r_0 = g_{xx}(0, 0),\ s_0 = g_{xy}(0, 0),\ \text{and}\ t_0 = g_{yy}(0, 0),$$

where the subscripts denote partial derivatives evaluated at the origin, and taking note of the fact that $\hat{n}\cdot\hat{k}_0 = -\cos\theta \rightarrow -1$, under these circumstances, at the stationary point, Eq. 16-56 reduces to

$$\mathbf{H}^s \approx \frac{\exp(-jkr')}{r'}\left(\frac{\eta-1}{\eta+1}\right)\mathbf{H}_0 \frac{\exp(-2jk\,g(0))}{(r_0t_0 - s_0^2)^{1/2}}. \quad (16\text{-}57)$$

This result is, as it should be, a normal incidence geometric optics result.

As a special example, consider a sphere of radius a, whence $z = \sqrt{a^2 - x^2 - y^2}$. We find $r_0 = t_0 = -2/a$ and $s_0 = 0$. Then

$$\mathbf{H}^s(\text{coated sphere}) \approx \frac{\exp(-jkr')}{r'}\left(\frac{\eta-1}{\eta+1}\right)\mathbf{H}_0 \frac{a}{2}\exp(-2jka).$$

RCS σ is then

$$\sigma(\text{coated sphere}) \approx |R|^2\pi a^2,$$

where R denotes the (complex) normal incidence flat plate reflection coefficient.

16-5-6 Discussion of Results

In order to connect further the results above with earlier sections, let us rephrase the expressions of the previous section in the language of reflection coefficients. Designate the plane determined by the vectors $\hat{n}$ and $\hat{k}_0$ as the plane of incidence. Then $\hat{n}\times\hat{k}_0$ is normal to the plane of incidence; $\hat{k}_0\times(\hat{n}\times\hat{k}_0)$ lies in the plane of incidence and is normal to both $\hat{k}_0$ and $\hat{n}\times\hat{k}_0$. When $\mathbf{E}_0$ has only a component along $\hat{n}\times\hat{k}_0$, i.e., when $b_0 = 0$, we call the incident field transverse electric. And when $a_0 = 0$, the incident magnetic field will lie along $\hat{n}\times\hat{k}_0$; so the incident field is transverse magnetic.

Since we may write $\hat{n} \cdot \hat{k}_0 = -\cos\theta$, θ being the acute angle between $\hat{n}$ and $\hat{k}_0$, we can rewrite the integrand of Eq. 16-55 in terms of the transverse electric reflection coefficient r_{TE} and the transverse magnetic reflection coefficient r_{TM} (Born and Wolf, 1959):

$$-\hat{k}_0 \times (\hat{n} \times \mathbf{H}_c) + \hat{k}_0 \times [\hat{k}_0 \times (\hat{n} \times \mathbf{E}_c)]$$
$$= -2\cos\theta\{r_{\mathrm{TE}} a_0[\hat{n} \times \hat{k}_0] + r_{\mathrm{TM}} b_0[\hat{k}_0 \times (\hat{n} \times \hat{k}_0)]\}\exp(-j\mathbf{k}_0 \cdot \mathbf{r}_c).$$

The IBC must of course hold to ensure the validity of Eq. 16-55; and, as was pointed out in Section 16-1, when $\hat{n} \approx \hat{k}_0'$, the impedance takes the form $\eta = (j\mu/N)\tan(Nk\delta)$.

We can apply the above results to multiple layer coatings provided we assume the validity of the IBC. In such cases the impedance η must be suitably adjusted to portray the characteristics of each and all of the layers (see Section 16-2).

16-6 TWO-DIMENSIONAL COATED SCATTERERS

We return now to the two-dimensional viewpoint introduced in Section 9-6 and discussed further with respect to its traveling wave and creeping wave

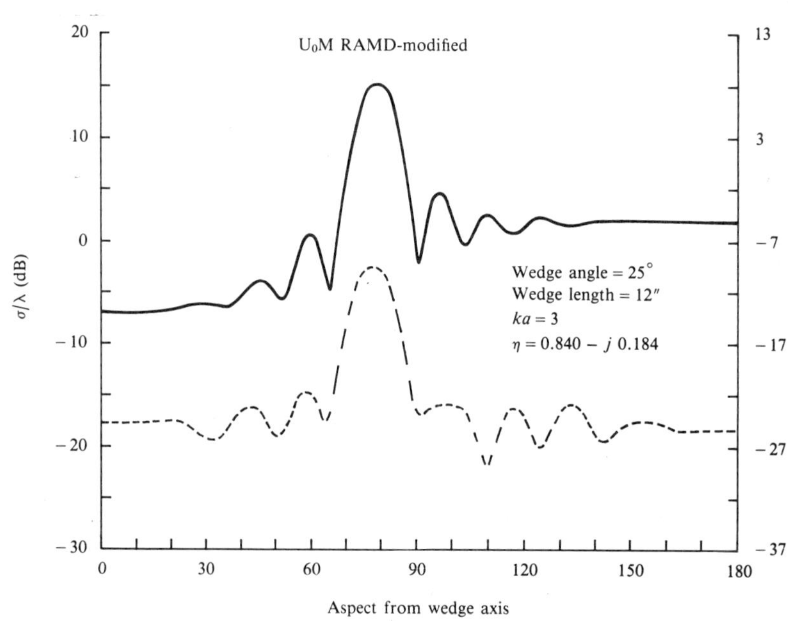

Figure 16-7 Backscattering RCS of bare and coated wedge cylinders for E-polarization.

ramifications in Section 11-9. Section 16-1 affords us a theoretical basis for, and a method by which to obtain, a nonzero surface impedance η for a metallic surface having one or more coating layers. Using such a nonzero impedance value in Eqs. 9-85 and 9-86 yields, for H-polarization and E-polarization, respectively, a surface current distribution from which RCW may easily be found. It is worth pointing out again the use of the duality relations. If we have the H-polarization RCW for a given impedance η, then the dual substitutions (Section 9-6) produce the E-polarization RCW for the impedance $1/\eta$. In particular, if $\eta = 1$ (perfect absorber), then the H-polarization and E-polarization RCWs are identical.

So we proceed to reexamine the wedge cylinder example given in Section 9-6, this time with an absorber coating. Consider the Emerson Cuming Eccosorb SFT-2.5 which, at a frequency of 3.0 GHz, has $\epsilon = 24 - j\,2.92$ and $\mu = 4.365 - j\,3.318$ [with time dependence $\exp(j\omega t)$]. For a thickness of 0.09 in., we find, from Eq. 16-7, a surface impedance $\eta = 0.840 - j\,0.184$. The corresponding normal incidence flat plate reflection coefficient $\Gamma\Gamma^*$ is approximately -18 dB.

Figures 16-7 and 16-8 show, for E- and H-polarizations, respectively, the same wedge cylinder considered in Section 9-6. The upper curve in each case is for the bare object, using $\eta = 0$, while the lower curve shows RCS

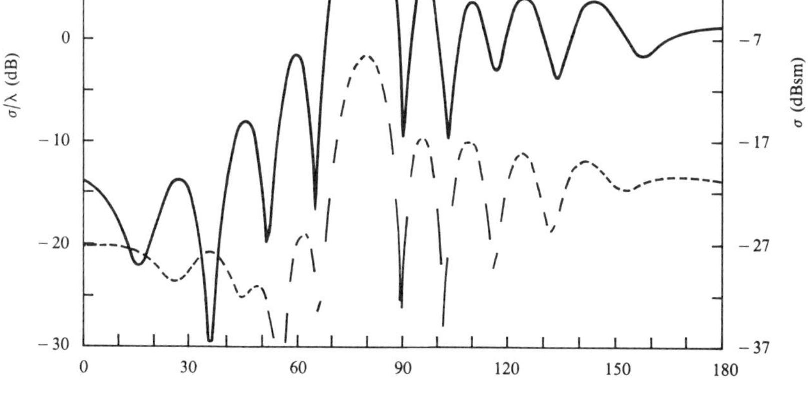

Figure 16-8 Backscattering RCS of bare and coated wedge cylinders for H-polarization.

for the coated object, using $\eta = 0.840 - 0.184j$. We note that the specular point at 77.5° shows a reduction in return essentially the same as the flat-plate reflection coefficient. At aspects which view the cylindrical surface specularly (160°–180°) we find increased reductions of some 2 dB. This gain in absorptivity may be due in part to phased creeping wave returns from the shadowed wedge edge.

17

REFLECTION AND TRANSMISSION: A DIFFERENT VIEWPOINT

In considering reflection and transmission of electromagnetic waves it is useful to distinguish between dispersion analysis and amplitude analysis. In the former we ask what types of waves are excited by an incident plane wave and what they look like. Phase matching offers a key here. In the latter, we calculate amplitudes of various waves in terms of incident wave amplitude. Here we introduce continued fractions (Chapter 18) and matrix methods (Chapter 19).

17-1 PHASE MATCHING

Let two homogeneous media be separated by a plane boundary at $x = 0$. Let an incident plane wave in region O be reflected at the boundary and transmitted into region t (Fig. 17-1). We have

$$\begin{aligned}
&\text{Incident:} && \exp(-j\mathbf{k}\cdot\mathbf{r} + j\omega t);\\
&\text{Reflected:} && \exp(-j\mathbf{k}_r\cdot\mathbf{r} + j\omega t);\\
&\text{Transmitted:} && \exp(-j\mathbf{k}_t\cdot\mathbf{r} + j\omega t).
\end{aligned} \tag{17-1}$$

The boundary condition at $x = 0$ requires tangential **E** and **H** to be continuous there for all y, z, and t. So (from Section 3-2)

$$\begin{aligned}
k_y &= k_{ry} = k_{ty};\\
k_z &= k_{rz} = k_{tz};\\
\omega &= \omega_r = \omega_t.
\end{aligned} \tag{17-2}$$

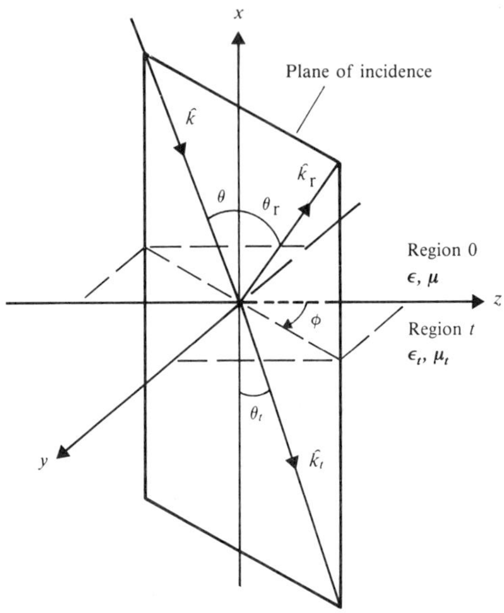

Figure 17-1 A plane wave incident on a plane boundary at $x = 0$.

Note that the last of these relations implies that there is no change of frequency for reflected and transmitted waves. The first two relations say that tangential **k** is continuous across $x = 0$. And out of these two come the laws of reflection and transmission:

1. Incident, reflected, and transmitted wave vectors all lie in an *incidence plane.*
2. Angle of reflection θ_r and of incidence θ are related by

$$k(\theta) \sin \theta = k_r(\theta_r) \sin \theta_r. \tag{17-3}$$

3. Angle of transmission (or refraction) θ_t and incidence angle are related by

$$k(\theta) \sin \theta = k_t(\theta_t) \sin \theta_t. \tag{17-4}$$

The sine of the angle is the ratio of tangential component to magnitude for all three waves. These results hold for general bianisotropic media. If both media are isotropic, then

$$\begin{aligned} k = k_r &= \omega\sqrt{\epsilon\mu} = \omega N/c \\ k_t &= \omega\sqrt{\epsilon_t\mu_t} = \omega N_t/c. \end{aligned} \tag{17-5}$$

The magnitudes of all three **k**-vectors are angle-independent and so Eq. 17-3 yields

$$\theta = \theta_r.$$

And because of 17-5, 17-4 becomes *Snell's law:*

$$N \sin \theta = N_t \sin \theta_t. \qquad (17\text{-}6)$$

These laws of reflection and transmission are therefore stated in terms of angles of incidence, reflection, and transmission and they are direct consequences of the condition that **k**-tangential be continuous across $x = 0$. Compare Section 3-2.

17-2 k-SURFACES

The dispersion relations provide a functional relation among components of **k** and angular frequency ω. In a three-dimensional space with axes k_x, k_y, k_z, a quadratic equation describes a two-dimensional surface called a wave surface or **k**-surface.

In the isotropic case the **k**-surface is a sphere with radius $\omega\sqrt{\epsilon\mu}$. In the uniaxial case, where the permittivity may be taken as the matrix

$$\mathrm{diag}(\epsilon, \epsilon, \epsilon_z) \qquad (17\text{-}7)$$

and μ as a scalar so that the dispersion relations are (see Section 19-2 for a development of dispersion relations for general anisotropic media)

$$k^2 = \omega^2 \mu\epsilon, \qquad (17\text{-}8a)$$

$$k_x^2 + k_y^2 + (\epsilon_z/\epsilon)k_z^2 = \omega^2 \mu\epsilon_z, \qquad (17\text{-}8b)$$

the **k**-surface for ordinary waves (17-8a) is a sphere, but for extraordinary waves (17-8b) is an ellipsoid (Figs. 17-2a and 17-2b).

The **k**-vector intersects the **k**-surface at a point and its magnitude in this direction is proportional to the length from the origin to this point. The phase velocity of a wave in the medium is, for the ith component,

$$v_{pi} = \frac{\omega}{k_i}; \qquad (17\text{-}9)$$

while the group velocity is

$$v_{gi} = \frac{\partial \omega}{\partial k_i} = \frac{\dfrac{\partial f}{\partial k_i}}{\dfrac{\partial f}{\partial \omega}}, \qquad (17\text{-}10)$$

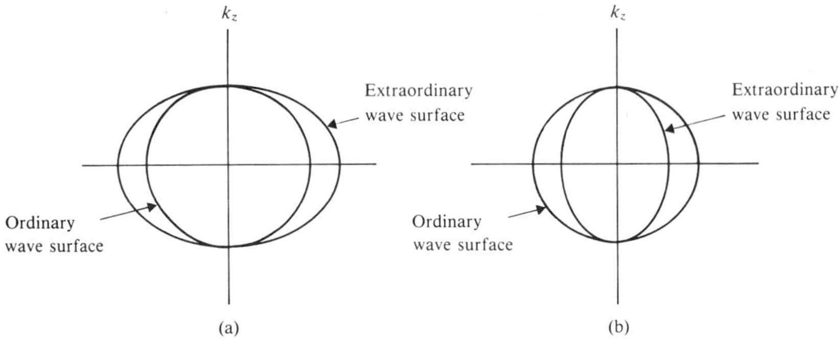

Figure 17-2 Ordinary and extraordinary wave surfaces for uniaxial media: (a) positive uniaxial media; (b) negative uniaxial media. Reprinted with permission from Kong (1975), Copyright © 1975, John Wiley & Sons.

where $f = \omega/2\pi$. Now, since a vector with components $\partial f/\partial k_i$ is normal to $f = \text{constant}$, we see that, as a wave propagates with phase velocity u along the direction of $\mathbf{k}$ which intersects the wave surface at a point, the group velocity is in a direction normal to the wave surface at this point.

Example 1. Using a positive uniaxial medium: (a) For the ordinary wave the phase and group velocities are in the same direction; whereas (b) for the extraordinary wave, they are in different directions.

The direction of group velocity is in the direction of the time-averaged Poynting vector: Take the differential of $\mathbf{k} \times \mathbf{E} = \omega\mathbf{B}$ and the complex conjugate of $\mathbf{k} \times \mathbf{H} = -\omega\mathbf{D}$:

$$\begin{aligned} \delta\mathbf{k} \times \mathbf{E} + \mathbf{k} \times \delta\mathbf{E} &= \omega\delta\mathbf{B}; \\ \delta\mathbf{k} \times \mathbf{H}* + \mathbf{k} \times \delta\mathbf{H}* &= -\omega\delta\mathbf{D}*. \end{aligned} \tag{17-11}$$

Dot-multiply the first of these by $\mathbf{H}*$ and the second by $\mathbf{E}$ and subtract the second from the first. The identitites $\mathbf{A} \cdot \mathbf{B} \times \mathbf{C} = \mathbf{B} \cdot \mathbf{C} \times \mathbf{A} = \mathbf{C} \cdot \mathbf{A} \times \mathbf{B}$ imply

$$2\delta\mathbf{k} \cdot \mathbf{E} \times \mathbf{H}* = \omega(\mathbf{H}* \cdot \delta\mathbf{B} + \mathbf{E} \cdot \delta\mathbf{D}*) + \delta\mathbf{E} \cdot \mathbf{k} \times \mathbf{H}* - \delta\mathbf{H}* \cdot \mathbf{k} \times \mathbf{E}. \tag{17-12}$$

Finally, use Eq. 17-11 to obtain from Eq. 17-12

$$2\delta\mathbf{k} \cdot \mathbf{E} \times \mathbf{H}* = \omega(\mathbf{H}* \cdot \delta\mathbf{B} + \mathbf{E} \cdot \delta\mathbf{D}* - \mathbf{D}* \cdot \delta\mathbf{E} - \mathbf{B} \cdot \delta\mathbf{H}*). \tag{17-13}$$

Using the lossless conditions for a bianisotropic medium (Chapter 16), the right-hand side of Eq. 17-13 becomes pure imaginary. But the time-averaged Poynting vector is the real part of $\mathbf{E} \times \mathbf{H}*$. So we must have

$$\delta\mathbf{k} \cdot \langle \mathbf{S} \rangle = 0. \tag{17-14}$$

Thus, the direction of the time-averaged Poynting vector is normal to the wave surface, since change of **k** is tangential to the wave surface, at the point in question. But group velocity is also normal to the **k**-surface. Therefore, the group velocity direction coincides with the Poynting vector flow.

We can define a dimensionless **k**-vector $\mathbf{k}/\omega\sqrt{\epsilon_0\mu_0} = \mathbf{n}$ which we can call an *index* vector. If we use **n** in Eq. 17-8, the resulting component relations are called *Fresnel equations.*

17-3 RAY VECTORS AND RAY SURFACES

Wave propagation direction becomes ambiguous in a complicated medium. We have power flow normal to both **E** and **H** (Poynting theorem); and direction of propagation of wavefronts is along **k** which is normal to both **D** and **B** (Section 1-1). And in a bianisotropic medium the directions of **k** and **s** do not in general coincide.

Poynting's power flow direction is characterized by a ray vector **s**, where we define the magnitude of **s** by (**s** has dimension of length)

$$\mathbf{s} \cdot \mathbf{k} = 1, \tag{17-15}$$

s being perpendicular to both **E** and **H**:

$$\mathbf{s} \cdot \mathbf{E} = 0, \qquad \mathbf{s} \cdot \mathbf{H} = 0. \tag{17-16}$$

We can use **s** to arrive at a set of equations similar to those obtained for **k** with a plane wave in Section 1-1. Cross-multiply $\mathbf{k} \times \mathbf{E} = -\omega\mathbf{B}$ by **s** and use the vector identity $\mathbf{s} \times (\mathbf{k} \times \mathbf{E}) = \mathbf{k}(\mathbf{s} \cdot \mathbf{E}) - (\mathbf{k} \cdot \mathbf{s})\mathbf{E}$ to find

$$\mathbf{s} \times \mathbf{B} = (1/\omega)\mathbf{E}. \tag{17-17}$$

Similarly, from $\mathbf{k} \times \mathbf{H} = -\omega\mathbf{D}$ we get

$$\mathbf{s} \times \mathbf{D} = -(1/\omega)\mathbf{H}. \tag{17-18}$$

We can also define a ray surface in a manner similar to the way we defined a wave surface. In fact, noting the duality between the $\mathbf{k}\times$ and $\mathbf{s}\times$ equations and between the constitutive relations in the **E**, **H** and **D**, **B** bases (Section 16-1) we can obtain the ray surfaces by making the substitutions

$$\boldsymbol{\epsilon} \to \boldsymbol{\kappa}, \quad \boldsymbol{\mu} \to \boldsymbol{\nu}, \quad \boldsymbol{\xi} \to \boldsymbol{\chi}, \quad \boldsymbol{\zeta} \to \boldsymbol{\gamma}, \quad \omega \to 1/\omega, \quad \text{and} \quad \mathbf{k} \to \mathbf{s}. \tag{17-19}$$

Example 2. For a uniaxial medium we find for the ordinary wave

$$s_x^2 + s_y^2 + s_z^2 = 1/\omega^2\epsilon\mu \tag{17-20}$$

and for the extraordinary wave

$$s_x^2 + s_y^2 + (\epsilon/\epsilon_z)s_z^2 = 1/\omega^2\epsilon_z\mu. \tag{17-21}$$

We showed above that $\mathbf{s} \cdot \delta\mathbf{k} = 0$, that is, the normal to the wave surface gives the direction of the corresponding ray vector (shades of geometric optics). Similarly, the normal to the ray surface gives the direction of the corresponding **k**-vector, which follows from

$$\mathbf{k} \cdot \mathbf{s} = 1. \tag{17-22}$$

Differentiation gives

$$\mathbf{k} \cdot \delta\mathbf{s} + \mathbf{s} \cdot \delta\mathbf{k} = 0, \tag{17-23}$$

whence

$$\mathbf{k} \cdot \delta\mathbf{s} = 0. \tag{17-24}$$

The ray surface describes constant phase, for the phase of a wave along a ray can be written

$$\psi = \int \mathbf{k} \cdot \mathbf{dc} = \int \mathbf{k} \cdot (\mathbf{s}/s)\, dc = c/s, \tag{17-25}$$

where c denotes length of the segment along the ray path. In geometric optics the quantity ψ is called the *eikonal* of the wave. When $c = s$ (constant), then the eikonal is ω/c (constant). The ray surface gives the magnitudes s in all directions. So the ray surface is also the surface of constant phase. Compare Section 7-2.

17-4 PHASE MATCHING WITH k-SURFACES

Let the incidence plane be the xz-plane. Then the continuity of tangential **k** can be illustrated by a diagram (Fig. 17-3) in the $k_x k_z$-plane. The k_z-axis has nothing to do with the physical z-axis, but it provides a feeling of a physical boundary. So dashed lines are used for the **k**-surface of the medium below the k_z-line and, similarly, for the **k**-surface of medium t above the k_z-line. Incident **k** is plotted toward the origin. Then, requiring that tangential k_z-components be equal, we construct reflected and transmitted **k** values.

However, it is not always possible to construct reflected and transmitted **k** values. Suppose medium 0 to be denser than medium t (i.e., $N > N_t$). Then the radius of the wave surface in medium t is shorter than that in medium 0 (Fig. 17-4), implying that as k_z of the incident wave becomes larger than k_t, there is no intersection with the small sphere, which would amount to requiring that one component of a vector be larger than its magnitude—an impossibility unless the vector is complex. The wave surface in medium t is

$$k_{tx}^2 + k_z^2 = k_t^2. \tag{17-26}$$

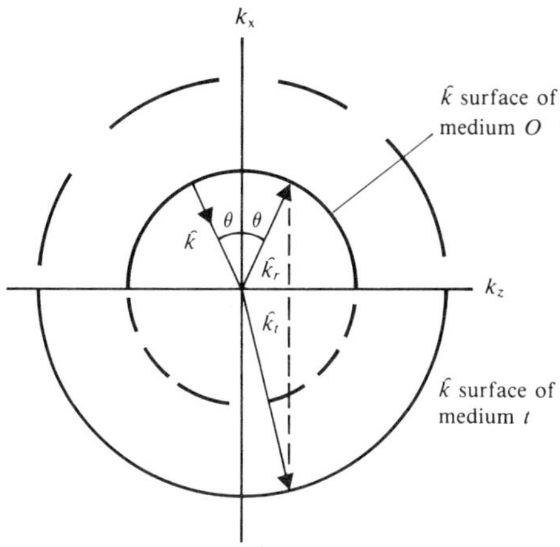

Figure 17-3 A $\hat{k}$-space representation of incident, reflected, and transmitted $\hat{k}$-vectors. (By phase-matching conditions, the k_z-components of the three vectors are equal. The incident $\hat{k}$-vector can also be drawn from the center toward the $\hat{k}$-surface.) Reprinted with permission from Kong (1975), Copyright © 1975, John Wiley & Sons.

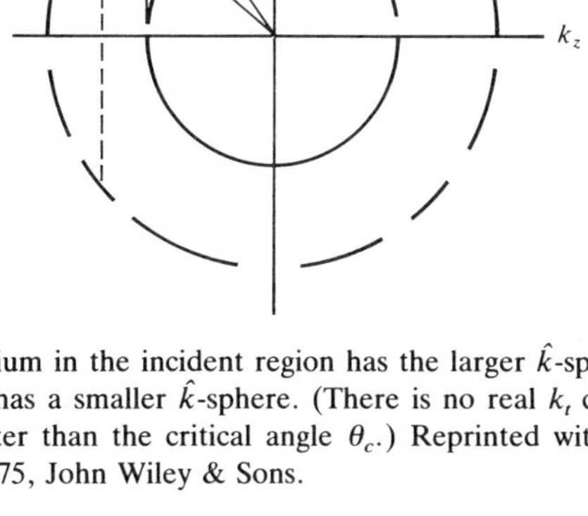

Figure 17-4 The medium in the incident region has the larger $\hat{k}$-sphere, and the medium in the transmitted region has a smaller $\hat{k}$-sphere. (There is no real k_t corresponding to k_i if the incidence angle is greater than the critical angle θ_c.) Reprinted with permission from Kong (1975), Copyright © 1975, John Wiley & Sons.

Since $k_z > k_t$,

$$k_{tx} = \pm\sqrt{k_t^2 - k_z^2} = \pm j\alpha_{tx}, \ \alpha \text{ real}. \tag{17-27}$$

But in medium t the wave is characterized by $\exp(-jk_{tx}x - jk_z z + j\omega t)$ which, for $k_z > k_t$, becomes

$$\exp(\alpha_{tx}x - jk_z z + j\omega t). \tag{17-28}$$

Thus, the transmitted wave decays exponentially in the negative $\hat{x}$-direction ($\alpha_{tx} > 0$) and propagates along the $\hat{z}$-direction with phase velocity ω/k_z. This wave is plane and nonuniform with varying amplitude on a constant phase front. Since it has maximum amplitude at the boundary surface and

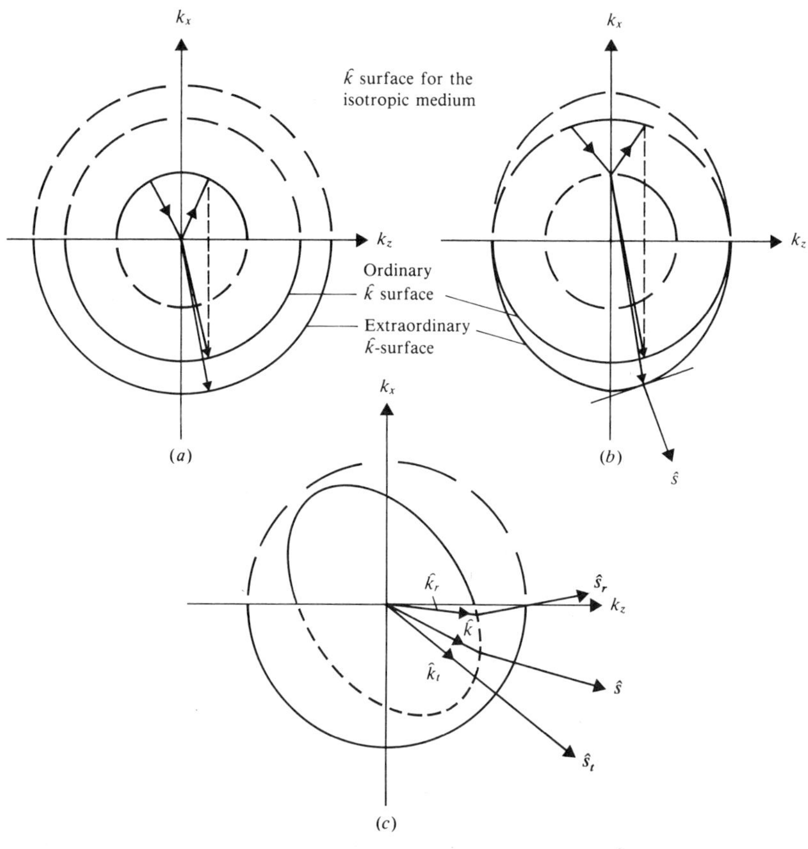

Figure 17-5 A plane wave incident upon a uniaxial medium: (a) phase matching for plane of incidence normal to the optic axis of a positive uniaxial medium; (b) phase matching for plane of incidence parallel to the optic axis of a positive uniaxial medium. The direction of $\hat{s}$ is different from that of $\hat{k}$ for the extraordinary wave; (c) an extraordinary wave incident from a uniaxial medium upon an isotropic medium. The reflected wave is a backward wave. Reprinted with permission from Kong (1975), Copyright © 1975, John Wiley & Sons.

decays exponentially away from the surface, it is called a surface wave. A wave having exponential decay in some direction is called evanescent in that direction. Ours is evanescent in the $-\hat{x}$-direction.

Evanescence in the transmitted wave begins when $k_t = k_z = k \sin \theta_c$; so θ_c is called the critical angle. It can be shown, by examining energy flow for the waves, that there is no time-averaged power penetrating medium t when $\theta > \theta_c$. So this phenomenon is considered to be one of total reflection.

Construction of **k**-surfaces permits us to visualize the **k**-vectors in reflected and transmitted regions.

The idea can also be applied to nonisotropic media. Let the optic axis of the medium be perpendicular to the incidence plane. See Fig. 17-5a. The power flow directions for ordinary and extraordinary waves are the same as the directions of the wave vectors. Next, let the optic axis be parallel to the incidence plane. See Fig. 17-5b. By the nature of the wave surface, the power flow direction $\hat{s}$ for the extraordinary wave is no longer the same as the direction $\hat{k}$.

Now let an extraordinary wave be excited in a uniaxial medium and be incident on a plane boundary with an isotropic medium. See Fig. 17-5c. Remember again that our plot is for **k**-space, not physical space. We draw **k** from center to surface. Now $\mathbf{k}_r$ points in $-k_x$ and $+k_z$ directions. But the power flow for the reflected wave *does* have $+k_x$ and $+k_z$ components. Therefore, the reflected wave carries energy away from the surface but has a phase front propagating into the surface. This phenomenon is called a *backward wave.*

18

STRATIFIED MEDIA [exp($j\omega t$)]

Consider a stratified medium as shown in Fig. 18-1. We concentrate on a TE incident wave $\mathbf{E} = \hat{y}E_y$. The TM case can then be gotten from the TE by duality: $\mathbf{E} \rightarrow \mathbf{H}$, $\mathbf{H} \rightarrow -\mathbf{E}$, $\boldsymbol{\epsilon} \rightarrow \boldsymbol{\mu}$, $\boldsymbol{\mu} \rightarrow \boldsymbol{\epsilon}$, where the tensor notation signifies biaxial (diagonal) layers.

Since $\mathbf{E} = \hat{y}E_y$, we can use the Maxwell–Faraday equations to find $\mathbf{H}$:

$$\begin{aligned} H_x &= (1/j\omega\mu_x)(\partial E_y/\partial z); \\ H_z &= (-1/j\omega\mu_z)(\partial E_y/\partial x). \end{aligned} \tag{18-1a}$$

Substitution of these expressions into the Maxwell–Ampère equation gives

$$[(\mu_y/\mu_z)(\partial^2/\partial x^2) + (\mu_y/\mu_x)(\partial^2/\partial z^2) + \omega^2\mu_y\epsilon_y)]E_y = 0, \tag{18-1b}$$

which is the wave equation, expressing the value of E_y in all regions of the stratified medium. A solution of Eq. 18-1b for any layer or, in particular, for the hth is

$$\begin{aligned} E_{hy} &= [A_h \exp(-jk_{hx}x + B_h \exp(jk_{hx}x)] \exp(-jk_z z) \\ H_{hx} &= -(k_z/\omega\mu_{hx})[A_h \exp(-jk_{hx}x) + B_h \exp(jk_{hx}x)] \exp(-jk_z z) \\ H_{hz} &= (k_{hx}/\omega\mu_{hz})[A_h \exp(-jk_{hx}x - B_h \exp(jk_{hx}x)] \exp(-jk_z z) \end{aligned} \tag{18-2}$$

where the subscript h for k_z is not needed since the phase-matching condition makes k_z the same for all layers. Substituting the first of Eqs. 18-2 into the wave equation above gives the dispersion relation

$$(\mu_{hy}/\mu_{hz})k_{hx}^2 + (\mu_{hy}/\mu_{hx})k_z^2 = \omega^2\mu_{hy}\epsilon_{hy}. \tag{18-3}$$

We must now find the wave amplitudes A_h and B_h from the boundary conditions at the layer faces.

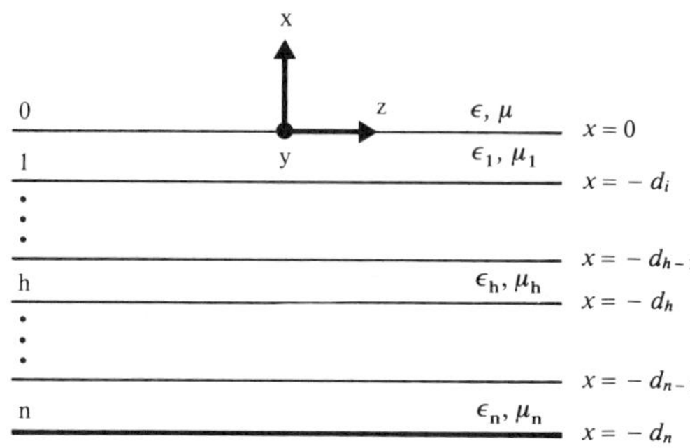

Figure 18-1 A stratified medium.

The boundary condition at $x = -d_h$ requires E_y and H_z to be continuous across the boundary:

$$\begin{aligned} &A_h \exp(jk_{hx}d_h) + B_h \exp(-jk_{hx}d_h) \\ &= A_{h+1} \exp(jk_{(h+1)x}d_h) + B_{h+1} \exp(-jk_{(h+1)x}d_h); \\ &(k_{hx}/\omega\mu_{hz})[A_h \exp(jk_{hx}d_h) - B_h \exp(-jk_{hx}d_h)] \\ &= (k_{(h+1)x}/\omega\mu_{(h+1)x})[A_{h+1} \exp(jk_{(h+1)x}d_h) \\ &\quad - B_{h+1} \exp(-jk_{(h+1)x}d_h)]. \end{aligned} \tag{18-4}$$

There are $2n+2$ unknowns [A_h and B_h, $h = 1, 2, \ldots, n$ plus R and T]. Solutions can be realized by an impedance concept which is represented in the form of a Smith chart (Kong, 1975, p. 116). Instead we give a continued fractions representation.

From Eqs. 18-4:

$$\begin{aligned} A_h \exp(jk_{hx}d_h) &= \tfrac{1}{2}[1 + (\mu_{hz}k_{(h+1)x}/\mu_{(h+1)z}k_{hx})] \\ &\quad \times [A_{h+1} \exp(jk_{(h+1)x}d_h) + R_{h(h+1)}B_{h+1} \exp(-jk_{(h+1)x}d_h)] \\ B_h \exp(-jk_{hx}d_h) &= \tfrac{1}{2}[1 + (\mu_{hz}k_{(h+1)x}/\mu_{(h+1)z}k_{hx})] \\ &\quad \times [R_{h(h+1)}A_{h+1} \exp(jk_{(h+1)x}d_h) + B_{h+1} \exp(-jk_{(h+1)x}d_h)] \end{aligned} \tag{18-5}$$

where

$$R_{h(h+1)} = \frac{1 - [\mu_{hz}k_{(h+1)x}/\mu_{(h+1)z}k_{hx}]}{1 + [\mu_{hz}k_{(h+1)x}/\mu_{(h+1)z}k_{hx}]} \tag{18-6}$$

is the reflection coefficient for TE waves in the hth region caused by the boundary between layers h and $h+1$. Forming the ratio of the first to the second of Eqs. 18-5 yields

$$\frac{A_h}{B_h} e^{jk_{hx}d_h} = \frac{1}{R_{h(h+1)}} + \frac{[1 - 1/R^2_{h(h+1)})] \exp[2jk_{(h+1)x}(d_{h+1} - d_h)]}{[1/R_{h(h+1)}] \exp[2jk_{(h+1)x}(d_{h+1} - d_h)] + [A_{h+1}/B_{h+1}] \exp(2jk_{(h+1)x}d_{h+1})} \tag{18-7}$$

Equation 18-7 offers the opportunity to form a continued fraction by introducing an expression similar to Eq. 18-7 for the fractional term

$$[A_{h+1}/B_{h+1}] \exp(2jk_{(h+1)x}d_{h+1})$$

in terms of

$$[A_{h+2}/B_{h+2}] \exp(2jk_{(h+2)x}d_{h+2}),$$

and so on. On the basis of this process we find

$$R = \frac{1}{R_{01}} + \cfrac{[1 - 1/R^2_{01}] \exp(2jk_{1x}d_1)}{\cfrac{\exp(2jk_{1x}d_1)}{R_{01}} + \cfrac{1}{R_{12}} + \cfrac{[1 - 1/R^2_{12}] \exp[2jk_{2x}(d_2 - d_1)]}{\cfrac{\exp[2jk_{2x}(d_2 - d_1)]}{R_{12}} + \cdots + \cfrac{1}{R_{(n-1)n}} + \cfrac{[1 - 1/R^2_{(n-1)n}] \exp[2jk_{nx}(d_n - d_{n-1})]}{\cfrac{\exp[2jk_{nx}(d_n - d_{n-1})]}{R_{(n-1)n}} + R_{nt}}}} \; . \tag{18-8}$$

Special Cases: One Layer $(n = 0)$:

$$R = R_{0t} = \frac{1 - (\mu_z k_{tx}/\mu_{tz} k_x)}{1 + (\mu_z k_{tx}/\mu_{tz} k_x)} . \tag{18-9}$$

For an isotropic medium, this reduces to the Fresnel reflection coefficient.

Two Layers $(n = 1)$

$$R = \frac{1}{R_{01}} + \frac{[1 - (1/R^2_{01})] \exp(2jk_{1x}d_1)}{(1/R_{01}) \exp(2jk_{1x}d_1) + R_{1t}} = \frac{R_{01} + R_{1t} \exp(-2jk_{1x}d_1)}{1 + R_{01}R_{1t} \exp(-2jk_{1x}d_1)} . \tag{18-10}$$

Note: Reflection coefficient for a TE-wave at a plane boundary: Consider a TE-wave (polarized perpendicular to the plane of incidence, with the

xz-plane the incidence plane), incident on a plane $x=0$ separating two media (compare Fig. 3-2):

$$\begin{aligned}
\mathbf{E}_i &= \hat{y}E_0 \exp(jk_x x - jk_z z + j\omega t) \\
\mathbf{E}_r &= \hat{y}R^{\mathrm{TE}}E_0 \exp(-jk_x x - jk_z z + j\omega t) \\
\mathbf{E}_t &= \hat{y}T^{\mathrm{TE}}E_0 \exp(jk_x x - jk_z z + j\omega t) \\
\mathbf{H}_i &= -(E_0/\omega\mu)(z\hat{k}_x + x\hat{k}_z) \exp(jk_x x - jk_z z + j\omega t) \\
\mathbf{H}_r &= -R^{\mathrm{TE}}(E_0/\omega\mu)(-\hat{z}k_x + \hat{x}k_z) \exp(-jk_x x - jk_z z + j\omega t) \\
\mathbf{H}_t &= -T^{\mathrm{TE}}(E_0/\omega\mu_t)(\hat{z}k_{tx} + \hat{x}k_z) \exp(jk_{tx} x - jk_z z + j\omega t).
\end{aligned} \tag{18-11}$$

Since E_y and H_z must be continuous at $x=0$, it follows that

$$\begin{aligned}
1 + R^{\mathrm{TE}} &= T^{\mathrm{TE}} \\
(k_x/\mu)(-1 + R^{\mathrm{TE}}) &= -(k_{tx}/\mu_t)T^{\mathrm{TE}}
\end{aligned} \tag{18-12}$$

whence

$$R^{\mathrm{TE}} = \frac{1 - (\mu k_{tx}/\mu_t k_x)}{1 + (\mu k_{tx}/\mu_t k_x)}, \qquad T^{\mathrm{TE}} = \frac{2}{1 + (\mu k_{tx}/\mu_t k_x)}. \tag{18-13}$$

If t denotes a perfect conductor, then $R^{\mathrm{TE}} = -1$. If t is a good conductor, then there exists a transmitted wave. Since $|k_{tx}| \gg |k_z|$, the last three equations of Eq. 18-11 show a very large z-component and a very small x-component for $\mathbf{H}_t$. Thus, $\mathbf{k}_t$ is almost perpendicular to the surface $x=0$, independent of the incidence angle. Compare this result with the remarks of Chapter 17 concerning the impedance boundary condition. Note that $R^{\mathrm{TE}} + T^{\mathrm{TE}} \neq 1$. However, the sum of the *power* reflection and transmission coefficient *is* equal to unity.

In using the propagation matrices we note that d_t is not a physical parameter and so we must set $d_t = 0$. For two layers:

$$\begin{aligned}
\begin{pmatrix} R \\ 1 \end{pmatrix} &= \frac{1}{4}\left(1 + \frac{\mu_z k_{1x}}{\mu_{1z} k_x}\right)\left(1 + \frac{\mu_{1z} k_{tx}}{\mu_{tz} k_{1x}}\right) \\
&\times \begin{pmatrix} \exp(-jk_{1x}d_1) & R_{01}\exp(jk_{1x}d_1) \\ R_{01}\exp(-jk_{1x}d_1) & \exp(jk_{1x}d_1) \end{pmatrix} \\
&\times \begin{pmatrix} \exp(jk_{tx}d_1) & R_{1t}\exp(-jk_{tx}d_1) \\ R_{1t}\exp(jk_{tx}d_1) & \exp(-jk_{tx}d_1) \end{pmatrix} \begin{pmatrix} 0 \\ T \end{pmatrix},
\end{aligned} \tag{18-14}$$

whence

$$T = \frac{4\exp[-j(k_{1x} - k_{tx})d_1]}{[1 + (\mu_z k_{1x}/\mu_{1z}k_x)][1 + (\mu_{1z}k_{tx}/\mu_{tz}k_{1x})][1 + R_{01}R_{1t}\exp(-2jk_{1x}d_1)]}. \tag{18-15}$$

18-1 REFLECTIVITY

Reflectivity, or power reflection coefficient, is defined as the ratio of the normal component of a reflected Poynting vector to the normal component of an incident Poynting vector, that is,

$$r^{\mathrm{TE}} = \frac{\hat{x} \cdot \mathbf{S}_r}{-\hat{x} \cdot \mathbf{S}_i}.$$

Similarly, the power transmission coefficient is

$$t^{\mathrm{TE}} = (-\hat{x} \cdot \mathbf{S}_t)/(-\hat{x} \cdot \mathbf{S}_i),$$

where

$$\mathbf{S}_i = [-\hat{x}(k_x/\omega\mu) + \hat{z}(k_z/\omega\mu)]\,|E_0|^2;$$

$$\mathbf{S}_r = [\hat{x}(k_x/\omega\mu) + \hat{z}(k_z/\omega\mu)]\,|R^{\mathrm{TE}}|^2\,|E_0|^2;$$

and

$$\mathbf{S}_t = [-\hat{x}(k_{tx}/\omega\mu_t) + \hat{z}(k_z/\omega\mu_t)]\,|T^{\mathrm{TE}}|^2\,|E_0|^2.$$

Thus,

$$r^{\mathrm{TE}} = |R^{\mathrm{TE}}|^2$$

and

$$t^{\mathrm{TE}} = (\mu k_{tx}/\mu_t k_x)\,|T^{\mathrm{TE}}|^2.$$

Using Eq. 18-13, we have

$$r^{\mathrm{TE}} + t^{\mathrm{TE}} = 1$$

for real k_{tx}.

19

GYROELECTROMAGNETIC LAYERS

In this chapter we examine the reflection from and/or transmission through a structure of plane layers having free space on one side and an isotropic medium (i.e., a dielectric or a metal) on the other. (These relationships are carefully defined in Section 19-1 below.) In general, each layer is to consist of a material that is linear and homogeneous, but anisotropic, exhibiting biaxial, gyroelectric, and gyromagnetic properties.

We look first at the nature of the dispersion relation for each layer; then at the conditions under which a material may be called biaxial; and finally, at the development of a chain matrix that connects the field behavior in any layer with that in any other layer via the boundary conditions at the interfaces.

The practical use of anisotropic materials has increased significantly in recent years. Aside from crystals, ferrites, and plasmas, which have been studied extensively in the past few decades, new materials have been developed for applications in fiberoptics, integrated optical devices, microstrip antennas, radar absorbers, and antenna radomes. These materials include certain ceramic compounds, honeycomb structures, and composites containing fibers with preferred orientation.

Maxwell's equations now involve tensor permittivity and permeability ($e^{j\omega t}$ time dependence suppressed):

$$\nabla \times \mathbf{H} = j\omega\epsilon_0 \boldsymbol{\epsilon}\mathbf{E}; \tag{19-1a}$$

$$\nabla \times \mathbf{E} = -j\omega\mu_0 \boldsymbol{\mu}\mathbf{H}; \tag{19-1b}$$

$$\boldsymbol{\epsilon} = \begin{pmatrix} \epsilon_1 & \epsilon_4 & 0 \\ \epsilon_5 & \epsilon_2 & 0 \\ 0 & 0 & \epsilon_3 \end{pmatrix}, \qquad \boldsymbol{\mu} = \begin{pmatrix} \mu_1 & \mu_4 & 0 \\ \mu_5 & \mu_2 & 0 \\ 0 & 0 & \mu_3 \end{pmatrix}; \tag{19-2}$$

ϵ_0 and μ_0 being free-space permittivity and permeability and ϵ_i, μ_i, $i = 1, \ldots, 5$ being dimensionless complex quantities that describe the anisotropic medium.

The propagation of electromagnetic waves in inhomogeneous or stratified anisotropic media has been studied by many authors. Low-frequency radiation by a horizontal dipole into an inhomogeneous anisotropic earth was first examined by Tikhonov (1959) and Chetaev (1963), and later by Wait (1966) to include the displacement current. Negi and Saraf (1973) carried out a similar study for a vertical dipole, whereas Daniels and Zich (1973) considered a formulation based on the operator theory of Bresler and Marcuvitz (1956a and b) combined with a Fourier integral representation. General studies of radiation from antennas embedded in anisotropic media were performed by Kong (1972), Tang (1979), and Ali and Mahmoud (1979); the problem of propagation in a uniaxial optical fiber has been reexamined recently by Paul and Shevgaonkar (1981) whose work contains an extensive bibliography. Among the books devoted to anisotropic media, the most comprehensive treatment is given by Kong (1981); specialized works on magnetic materials (Lax and Button, 1962) and gyrotropic waveguides (Hlawiczka, 1981) are also available. Finally, we remark that Graglia and Uslenghi (1987) extended the methods for planar layers, described in this chapter, to a treatment of layered absorbers on cylindrical structures. They have, in addition, developed, and give numerical examples for, an optimization procedure for the minimization of RCS over a desired frequency band.

19-1 GEOMETRY OF THE PROBLEM

The structure we wish to consider is composed of n layers parallel to the xy-plane, with the z-axis perpendicular to the interfaces (see Fig. 19-1). With respect to the x,y,z-axes, the tensor permittivity $\boldsymbol{\epsilon}_h$ and tensor permeability $\boldsymbol{\mu}_h$ for the hth layer are

$$\boldsymbol{\epsilon}_h = \begin{pmatrix} \epsilon_{h1} & \epsilon_{h4} & 0 \\ \epsilon_{h5} & \epsilon_{h2} & 0 \\ 0 & 0 & \epsilon_{h3} \end{pmatrix}, \qquad \boldsymbol{\mu}_h = \begin{pmatrix} \mu_{h1} & \mu_{h4} & 0 \\ \mu_{h5} & \mu_{h2} & 0 \\ 0 & 0 & \mu_{h3} \end{pmatrix}. \tag{19-3}$$

The interfaces between layers are the planes $z=0, -d_1, \ldots, -d_n$. The $\frac{1}{2}$-space $z>0$ is free-space while the $\frac{1}{2}$-space $z<-d_n$ can be either a homogeneous isotropic dielectric or a perfect conductor.

The incident plane wave is

$$\hat{k}_0 = \hat{x}\sin\gamma\cos\beta + \hat{y}\sin\gamma\sin\beta - \hat{z}\cos\gamma \tag{19-4}$$

whose projection on the xy-plane is in the $\hat{x}_0$ direction, which makes an angle β with $\hat{x}$ (Fig. 19-2); $\hat{k}_0$ makes an angle γ with the negative $\hat{z}$-axis, and

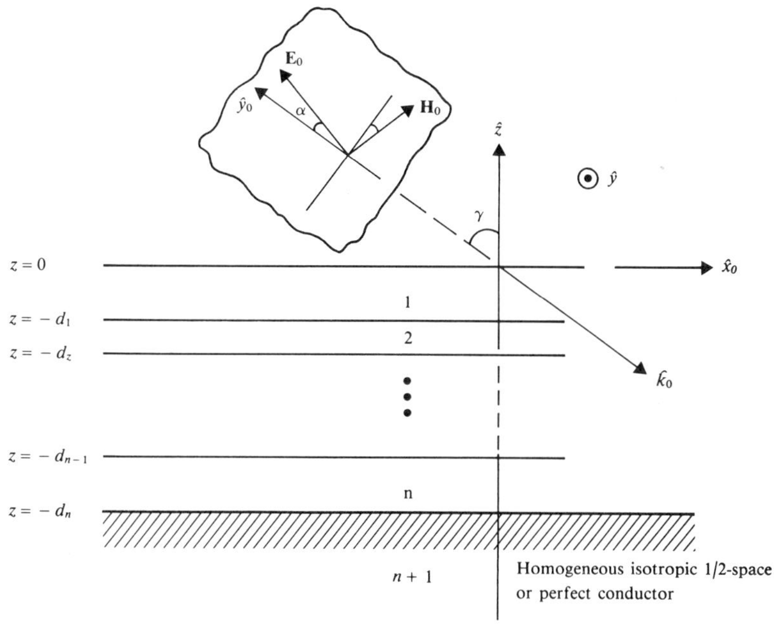

Figure 19-1 Plane wave incidence on n isotropic layers over a homogeneous isotropic half-space. Reprinted with permission from Damaskos, Maffett, and Uslenghi (1985). "Reflection and Transmission for Gyroelectromagnetic Biaxial Layered Media," *Journal of the Optical Society of America,* **2,** pp. 454–461.

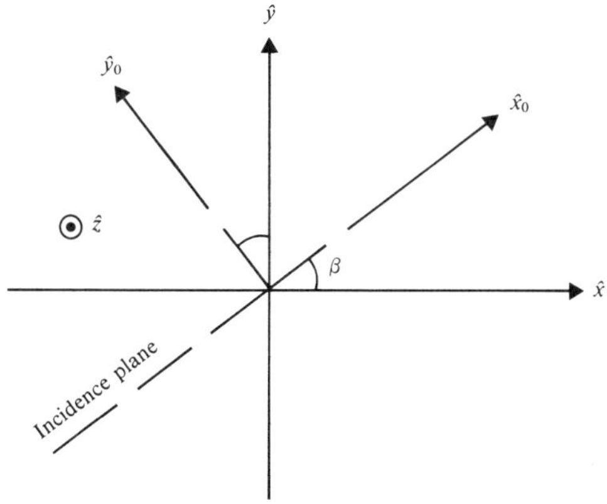

Figure 19-2 Orientation of plane wave direction. Reprinted with permission from Damaskos, Maffett, and Uslenghi (1985). "Reflection and Transmission for Gyroelectromagnetic Biaxial Layered Media," *Journal of the Optical Society of America,* **2,** pp. 454–461.

the incident electric field $\mathbf{E}_0$ makes an angle α with the $\hat{y}$-axis (which is normal to the incidence plane); $0 \le \alpha,\ \beta,\ \gamma \le \pi/2$. So we have for the incident field

$$\begin{aligned} \mathbf{E}_0 = [&\hat{x}(\sin\alpha\cos\beta\cos\gamma - \cos\alpha\sin\beta) \\ &+ \hat{y}(\sin\alpha\sin\beta\cos\gamma + \cos\alpha\cos\beta) \\ &+ \hat{z}(\sin\alpha\sin\gamma)]\exp(-jk_0\hat{k}_0\cdot\mathbf{r}); \end{aligned} \tag{19-5}$$

$$\begin{aligned} \mathbf{H}_0 = Y_0\hat{k}_0\times\mathbf{E}_0 = Y_0[&\hat{x}(\cos\alpha\cos\beta\cos\gamma + \sin\alpha\sin\beta) \\ &+ \hat{y}(\cos\alpha\sin\beta\cos\gamma - \sin\alpha\cos\beta) \\ &+ \hat{z}(\cos\alpha\sin\gamma)]\exp(-jk_0\hat{k}_0\cdot\mathbf{r})]; \end{aligned} \tag{19-6}$$

where

$$\mathbf{r} = x\hat{x} + y\hat{y} + z\hat{z} \tag{19-7}$$

locates the position of the observation point, and

$$Z_0 = Y_0^{-1} = \sqrt{\mu_0/\epsilon_0}, \qquad k_0 = \omega\sqrt{\epsilon_0\mu_0} = 2\pi/\lambda_0 \tag{19-8}$$

are the intrinsic impedance and the wave number of free space, respectively. The incident electric field has unit amplitude.

We are now ready to find the dispersion relation. To accomplish this task we adapt some results from Damaskos, Maffett, and Uslenghi (1982).†

19-2 THE DISPERSION RELATION FOR GENERAL ANISOTROPIC MEDIA

Consider a plane electromagnetic wave with space-time dependence as above:

$$\exp(-jk_0\hat{k}\cdot\mathbf{r} + j\omega t) \tag{19-9}$$

which propagates in an anisotropic medium with tensor permittivity $\boldsymbol{\epsilon}\epsilon_0$ and tensor permeability $\boldsymbol{\mu}\mu_0$; as above, $k_0 = \omega\sqrt{\epsilon_0\mu_0}$ is free-space wave number, $\mathbf{r}$ the position vector, ω the angular frequency, t the time, and $j = \sqrt{-1}$. We wish to find the functional dependence $\hat{k} = \hat{k}(\boldsymbol{\epsilon}, \boldsymbol{\mu})$ of the dimensionless propagation vector $\hat{k}$ upon the dimensionless tensors $\boldsymbol{\epsilon}$ and $\boldsymbol{\mu}$.

For plane waves such as Eq. 19-9, Maxwell's curl equations become (see Section 1-1)

$$\hat{k}\times\mathbf{H} = -\boldsymbol{\epsilon}\mathbf{E}; \qquad \hat{k}\times\mathbf{E} = \boldsymbol{\mu}\mathbf{H}; \tag{19-10}$$

where $\boldsymbol{\epsilon}$ and $\boldsymbol{\mu}$ may be written in rectangular x, y, z coordinates as

$$\boldsymbol{\epsilon} = \begin{pmatrix} \epsilon_{xx} & \epsilon_{xy} & \epsilon_{xz} \\ \epsilon_{yx} & \epsilon_{yy} & \epsilon_{yz} \\ \epsilon_{zx} & \epsilon_{zy} & \epsilon_{zz} \end{pmatrix}; \qquad \boldsymbol{\mu} = \begin{pmatrix} \mu_{xx} & \mu_{xy} & \mu_{xz} \\ \mu_{yx} & \mu_{yy} & \mu_{yz} \\ \mu_{zx} & \mu_{zy} & \mu_{zz} \end{pmatrix}. \tag{19-11}$$

† With permission of the Institute of Electrical and Electronics Engineers. © 1982 IEEE.

and **H** is the product of the magnetic field times $\sqrt{\mu_0}/\sqrt{\epsilon_0}$. Introduce the singular matrix (cf. Eq. 1-22)

$$\mathfrak{k} = \begin{pmatrix} 0 & -k_z & k_y \\ k_z & 0 & -k_x \\ -k_y & k_x & 0 \end{pmatrix} \tag{19-12}$$

so that the vector operator $\hat{k} \times$ ($\hat{k}$ "cross") can be written as the tensor $\mathfrak{k}$ (which, we observe again, is singular); then Eqs. 19-10 become

$$\mathfrak{k}\mathbf{H} = -\boldsymbol{\epsilon}\mathbf{E}, \qquad \mathfrak{k}\mathbf{E} = \boldsymbol{\mu}\mathbf{H}, \tag{19-13}$$

from which the wave equations

$$(\mathfrak{k}\boldsymbol{\mu}^{-1}\mathfrak{k} + \boldsymbol{\epsilon})\mathbf{E} = 0, \qquad (\mathfrak{k}\boldsymbol{\epsilon}^{-1}\mathfrak{k} + \boldsymbol{\mu})\mathbf{H} = 0, \tag{19-14}$$

follow at once. In Eqs. 19-14 we have assumed that $\boldsymbol{\epsilon}$ and $\boldsymbol{\mu}$ are nonsingular. For Eqs. 19-14 to have solutions, the relations

$$\det(\mathfrak{k}\boldsymbol{\mu}^{-1}\mathfrak{k} + \boldsymbol{\epsilon}) = 0, \qquad \det(\mathfrak{k}\boldsymbol{\epsilon}^{-1}\mathfrak{k} + \boldsymbol{\mu}) = 0 \tag{19-15}$$

must be satisfied, respectively. The fact that Eqs. 19-15 are identical follows at once from the symmetry of Maxwell's equations. Either of Eqs. 19-15 yields the following general dispersion relation:

$$\begin{aligned} &k_x^4\epsilon_{xx}\mu_{xx} + k_y^4\epsilon_{yy}\mu_{yy} + k_z^4\epsilon_{zz}\mu_{zz} + k_x^2k_y^2A_1 + k_x^2k_z^2A_2 \\ &\quad + k_y^2k_z^2A_3 + k_x^3k_yB_1 + k_x^3k_zB_2 + k_xk_y^3B_3 + k_y^3k_zB_4 + k_xk_z^3B_5 \\ &\quad + k_yk_z^3B_6 + k_x^2k_yk_zC_1 + k_xk_y^2k_zC_2 + k_xk_yk_z^2C_3 + k_x^2D_1 \\ &\quad + k_y^2D_2 + k_z^2D_3 + k_xk_yF_1 + k_xk_zF_2 + k_yk_zF_3 \\ &\quad + (\det\boldsymbol{\epsilon})(\det\boldsymbol{\mu}) = 0, \end{aligned} \tag{19-16}$$

where

$$\begin{aligned} A_1 &= \epsilon_{xx}\mu_{yy} + \epsilon_{yy}\mu_{xx} + (\epsilon_{xy} + \epsilon_{yx})(\mu_{xy} + \mu_{yx}), \\ A_2 &= \epsilon_{xx}\mu_{zz} + \epsilon_{zz}\mu_{xx} + (\epsilon_{xz} + \epsilon_{zx})(\mu_{xz} + \mu_{zx}), \\ A_3 &= \epsilon_{yy}\mu_{zz} + \epsilon_{zz}\mu_{yy} + (\epsilon_{yz} + \epsilon_{zy})(\mu_{yz} + \mu_{zy}), \\ B_1 &= \epsilon_{xx}(\mu_{xy} + \mu_{yx}) + \mu_{xx}(\epsilon_{xy} + \epsilon_{yx}), \\ B_2 &= \epsilon_{xx}(\mu_{xz} + \mu_{zx}) + \mu_{xx}(\epsilon_{xz} + \epsilon_{zx}), \\ B_3 &= \epsilon_{yy}(\mu_{xy} + \mu_{yx}) + \mu_{yy}(\epsilon_{xy} + \epsilon_{yx}), \\ B_4 &= \epsilon_{yy}(\mu_{yz} + \mu_{zy}) + \mu_{yy}(\epsilon_{yz} + \epsilon_{zy}), \\ B_5 &= \epsilon_{zz}(\mu_{xz} + \mu_{zx}) + \mu_{zz}(\epsilon_{xz} + \epsilon_{zx}), \\ B_6 &= \epsilon_{zz}(\mu_{yz} + \mu_{zy}) + \mu_{zz}(\epsilon_{yz} + \epsilon_{zy}), \\ C_1 &= \epsilon_{xx}(\mu_{yz} + \mu_{zy}) + \mu_{xx}(\epsilon_{yz} + \epsilon_{zy}) \\ &\quad + (\epsilon_{xy} + \epsilon_{yx})(\mu_{xz} + \mu_{zx}) + (\epsilon_{xz} + \epsilon_{zx})(\mu_{xy} + \mu_{yx}), \end{aligned} \tag{19-17}$$

$$
\begin{aligned}
C_2 &= \epsilon_{yy}(\mu_{xz} + \mu_{zx}) + \mu_{yy}(\epsilon_{xz} + \epsilon_{zx}) \\
&\quad + (\epsilon_{xy} + \epsilon_{yx})(\mu_{yz} + \mu_{zy}) + (\epsilon_{yz} + \epsilon_{zy})(\mu_{xy} + \mu_{yx}), \\
C_3 &= \epsilon_{zz}(\mu_{xy} + \mu_{yx}) + \mu_{zz}(\epsilon_{xy} + \epsilon_{yx}) \\
&\quad + (\epsilon_{xz} + \epsilon_{zx})(\mu_{yz} + \mu_{zy}) + (\epsilon_{yz} + \epsilon_{zy})(\mu_{xz} + \mu_{zx}), \\
D_1 &= M_{23}N_{23} + M_{32}N_{32} - M_{22}N_{33} - M_{33}N_{22}, \\
D_2 &= M_{13}N_{13} + M_{31}N_{31} - M_{11}N_{33} - M_{33}N_{11}, \\
D_3 &= M_{12}N_{12} + M_{21}N_{21} - M_{11}N_{22} - M_{22}N_{11}, \\
F_1 &= M_{33}(N_{12} + N_{21}) + N_{33}(M_{12} + M_{21}) - M_{13}N_{23} - M_{23}N_{13} \\
&\quad - M_{31}N_{32} - M_{32}N_{31}, \\
F_2 &= M_{22}(N_{13} + N_{31}) + N_{22}(M_{13} + M_{31}) - M_{12}N_{32} - M_{32}N_{12} \\
&\quad - M_{21}N_{23} - M_{23}N_{21}, \\
F_3 &= M_{11}(N_{23} + N_{32}) + N_{11}(M_{23} + M_{32}) - M_{12}N_{13} - M_{13}N_{12} \\
&\quad - M_{21}N_{31} - M_{31}N_{21},
\end{aligned}
$$

with

$$
\begin{aligned}
M_{11} &= \epsilon_{yy}\epsilon_{zz} - \epsilon_{yz}\epsilon_{zy}, \\
M_{22} &= \epsilon_{xx}\epsilon_{zz} - \epsilon_{xz}\epsilon_{zx}, \\
M_{33} &= \epsilon_{xx}\epsilon_{yy} - \epsilon_{xy}\epsilon_{yx}, \\
M_{12} &= \epsilon_{yz}\epsilon_{zx} - \epsilon_{yx}\epsilon_{zz}, \\
M_{21} &= \epsilon_{xz}\epsilon_{zy} - \epsilon_{xy}\epsilon_{zz}, \\
M_{13} &= \epsilon_{yx}\epsilon_{zy} - \epsilon_{yy}\epsilon_{zx}, \\
M_{31} &= \epsilon_{xy}\epsilon_{yz} - \epsilon_{xz}\epsilon_{yy}, \\
M_{23} &= \epsilon_{xy}\epsilon_{zx} - \epsilon_{xx}\epsilon_{zy}, \\
M_{32} &= \epsilon_{xy}\epsilon_{yx} - \epsilon_{xx}\epsilon_{yz};
\end{aligned}
\tag{19-18}
$$

and N_{ij}, $i, j = 1, 2, 3$ is obtained from M_{ij} by replacing ϵ with μ.

Observe that Eq. 19-16 is invariant under the interchange $\epsilon \rightarrow \mu$, $\mu \rightarrow \epsilon$; in addition, it is invariant when $\hat{k}$ is replaced by $-\hat{k}$. Equation 19-16 is the most general dispersion relation for a medium with electric and magnetic anisotropies.

In a boundary value problem involving plane wave incidence upon layered anisotropic media whose interfaces are planes $z = \text{constant}$, the transverse components k_x and k_y of the propagation vector are determined by the boundary conditions. Then Eq. 19-16 becomes a fourth degree equation for the unknown longitudinal wave number k_z; its four roots $k_z^{(i)}$ correspond to waves propagating in the $\pm z$ direction, depending on whether the real part of the root is, respectively, positive or negative. The total field in each layer may be written as a sum of hybrid modes, that is, E_z and

H_z are linear combinations of the four terms

$$\exp(-jk_0k_z^{(i)}z + j\omega t), \qquad i = 1, 2, 3, 4.$$

This aspect of the anisotropic layers problem is expanded in Section 19-6.

19-3 PARTICULAR CASES

The dispersion relation (Eq. 19-16) simplifies considerably in the antisymmetric case:

$$\begin{aligned} &\epsilon_{xy} = -\epsilon_{yx}, \qquad \epsilon_{xz} = -\epsilon_{zx}, \qquad \epsilon_{yz} = -\epsilon_{zy}, \\ &\mu_{xy} = -\mu_{xx}, \qquad \mu_{xz} = -\mu_{zx}, \qquad \mu_{yz} = -\mu_{zy}, \end{aligned} \tag{19-19}$$

which entails (the tilde means that Eqs. 19-19 are satisfied):

$$\begin{aligned}
\tilde{A}_1 &= \epsilon_{xx}\mu_{yy} + \epsilon_{yy}\mu_{xx}, \\
\tilde{A}_2 &= \epsilon_{xx}\mu_{zz} + \epsilon_{zz}\mu_{xx}, \\
\tilde{A}_3 &= \epsilon_{yy}\mu_{zz} + \epsilon_{zz}\mu_{yy}, \\
\tilde{B}_1 &= \tilde{B}_2 = \tilde{B}_3 = \tilde{B}_4 = \tilde{B}_5 = \tilde{B}_6 = 0, \\
\tilde{C}_1 &= \tilde{C}_2 = \tilde{C}_3 = 0, \\
\tilde{D}_1 &= 2\epsilon_{xx}\mu_{xx}\epsilon_{yz}\mu_{yz} + 2\epsilon_{xy}\mu_{xy} \\
&\quad - (\epsilon_{xx}\epsilon_{yy} + \epsilon_{xy}^2)(\mu_{xx}\mu_{zz} + \mu_{xz}^2) \\
&\quad - (\epsilon_{xx}\epsilon_{zz} + \epsilon_{xz}^2)(\mu_{xx}\mu_{yy} + \mu_{xy}^2), \\
\tilde{D}_2 &= 2\epsilon_{yy}\mu_{yy}\epsilon_{xz}\mu_{xz} + 2\epsilon_{xy}\mu_{xy}\epsilon_{yz}\mu_{yz} \\
&\quad - (\epsilon_{xx}\epsilon_{yy} + \epsilon_{xy}^2)(\mu_{yy}\mu_{zz} + \mu_{yz}^2) \\
&\quad - (\epsilon_{yy}\epsilon_{zz} + \epsilon_{yz}^2)(\mu_{xx}\mu_{yy} + \mu_{xy}^2), \\
\tilde{D}_3 &= 2\epsilon_{zz}\mu_{zz}\epsilon_{xy}\mu_{xy} + 2\epsilon_{xz}\mu_{xz}\epsilon_{yz}\mu_{yz} \\
&\quad - (\epsilon_{xx}\epsilon_{zz} + \epsilon_{xz}^2)(\mu_{yy}\mu_{zz} + \mu_{yz}^2) \\
&\quad - (\epsilon_{yy}\epsilon_{zz} + \epsilon_{yz}^2)(\mu_{xx}\mu_{zz} + \mu_{xz}^2), \\
\tilde{F}_1 &= -2(\epsilon_{xx}\epsilon_{yy} + \epsilon_{xy}^2)\mu_{xz}\mu_{yz} \\
&\quad - 2(\mu_{xx}\mu_{yy} + \mu_{xy}^2)\epsilon_{xz}\epsilon_{yz} \\
&\quad + 2(\epsilon_{xz}\mu_{yz} + \epsilon_{yz}\mu_{xz})\epsilon_{xy}\mu_{xy} \\
&\quad - 2\epsilon_{xx}\epsilon_{yz}\mu_{yy}\mu_{xz} - 2\epsilon_{yy}\epsilon_{xz}\mu_{xx}\mu_{yz}, \\
\tilde{F}_2 &= -2(\epsilon_{xx}\epsilon_{zz} + \epsilon_{xz}^2)\mu_{xy}\mu_{yz} \\
&\quad - 2(\mu_{xx}\mu_{zz} + \mu_{xz}^2)\epsilon_{xy}\epsilon_{yz} \\
&\quad - 2(\epsilon_{xy}\mu_{yz} + \epsilon_{yz}\mu_{xy})\epsilon_{xz}\mu_{xz} \\
&\quad + 2\epsilon_{xx}\epsilon_{yz}\mu_{xy}\mu_{zz} + 2\epsilon_{xy}\epsilon_{zz}\mu_{xx}\mu_{yz},
\end{aligned}$$

$$\begin{aligned}\tilde{F}_3 = & -2(\epsilon_{yy}\epsilon_{zz} + \epsilon_{yz}^2)\mu_{xy}\mu_{xz} \\ & - 2(\mu_{yy}\mu_{zz} + \mu_{yz}^2)\epsilon_{xy}\epsilon_{xz} \\ & + 2(\epsilon_{xy}\mu_{xz} + \epsilon_{xz}\mu_{xy})\epsilon_{yz}\mu_{yz} \\ & - 2\epsilon_{xy}\epsilon_{zz}\mu_{yy}\mu_{xz} - 2\epsilon_{xz}\epsilon_{yy}\mu_{xy}\mu_{zz};\end{aligned}$$

$$\det \tilde{\boldsymbol{\epsilon}} = \epsilon_{xx}\epsilon_{yy}\epsilon_{zz} + \epsilon_{xx}\epsilon_{yz} + \epsilon_{yy}\epsilon_{xz} + \epsilon_{zz}\epsilon_{xy}, \tag{19-20}$$

with a similar expression for $\det \tilde{\boldsymbol{\mu}}$.

If the medium is to behave symmetrically for propagation in the $\pm z$ direction, then $\tilde{F}_2 = \tilde{F}_3 = 0$, which is certainly the case if

$$\epsilon_{xz} = \epsilon_{yz} = \mu_{xz} = \mu_{yz} = 0. \tag{19-21}$$

Under conditions 19-19 and 19-21, $\boldsymbol{\epsilon}$ and $\boldsymbol{\mu}$ still represent most cases of practical interest, that is, gyroelectric, gyromagnetic, biaxial, and uniaxial media. The dispersion relation is now a quadratic in k_z^2, yielding two double roots for k_z. The two waves which can propagate in such media are, in general, both extraordinary waves.

19-4 SPECIAL ANISOTROPY

If, for the hth layer, the permittivity and permeability tensors are of the form

$$\boldsymbol{\epsilon} = \begin{pmatrix} \epsilon_1 & \epsilon_4 & 0 \\ \epsilon_5 & \epsilon_2 & 0 \\ 0 & 0 & \epsilon_3 \end{pmatrix}, \qquad \boldsymbol{\mu} = \begin{pmatrix} \mu_1 & \mu_4 & 0 \\ \mu_5 & \mu_2 & 0 \\ 0 & 0 & \mu_3 \end{pmatrix}, \tag{19-22}$$

where $\epsilon_i, \mu_i, i = 1, \ldots, 5$ are dimensionless and, in general, complex quantities, then the dispersion relation can be simplified somewhat. Setting

$$K_z = \frac{k_z}{k_0} \tag{19-23}$$

from Section 19-3 (the layer designation h has also been dropped for convenience), we find

$$K_z^4 + AK_z^2 + B = 0, \tag{19-24}$$

where

$$A = \left(\frac{v_\epsilon}{\epsilon_3} + \frac{v_\mu}{\mu_3}\right)\sin^2\gamma + \epsilon_4\mu_4 + \epsilon_5\mu_5 - (\epsilon_1\mu_2 + \epsilon_2\mu_1), \tag{19-25}$$

$$\begin{aligned}B = & \frac{\sin^4\gamma}{\epsilon_3\mu_3}\{\epsilon_1\mu_1\cos^4\beta + \epsilon_2\mu_2\sin^4\beta + [\epsilon_1\mu_2 + \epsilon_2\mu_1 \\ & - (\epsilon_4 + \epsilon_5)(\mu_4 + \mu_5)]\sin^2\beta\cos^2\beta + [v_\epsilon(\mu_4 + \mu_5) \\ & + v_\mu(\epsilon_4 + \epsilon_5)]\sin\beta\cos\beta\} - [(v_\epsilon/\mu_3)(\mu_1\mu_2 - \mu_4\mu_5) \\ & + (v_\mu/\epsilon_3)(\epsilon_1\epsilon_2 - \epsilon_4\epsilon_5)]\sin^2\gamma + (\epsilon_1\epsilon_2 - \epsilon_4\epsilon_5)(\mu_1\mu_2 - \mu_4\mu_5).\end{aligned} \tag{19-26}$$

In Eq. 19-26 we have set

$$\left.\begin{aligned} v_\epsilon &= \epsilon_1 \cos^2\beta + \epsilon_2 \sin^2\beta + (\epsilon_4 + \epsilon_5)\sin\beta\cos\beta, \\ v_\mu &= \cos^2\beta + \mu_2 \sin^2\beta + (\mu_4 + \mu_5)\sin\beta\cos\beta. \end{aligned}\right\} \tag{19-27}$$

The solutions for Eq. 19-24 are

$$\left.\begin{matrix} K_z^{(2)2} \\ K_z^{(1)2} \end{matrix}\right\} = -\frac{1}{2}(A \pm \sqrt{A^2 - 4B}). \tag{19-28}$$

If, in particular, the layers are all of uniaxial material, then Eq. 19-28 can be carried out explicitly to give

$$\begin{aligned} K_z^{(1)} &= \sqrt{\epsilon\mu - (\mu/\mu_3)\sin^2\gamma}, \\ K_z^{(2)} &= \sqrt{\epsilon\mu - (\epsilon/\epsilon_3)\sin^2\gamma} \end{aligned} \tag{19-29}$$

for each layer.

Before proceeding to a treatment of the general anisotropic layer problem, we examine the conditions under which a material may be described as biaxial.

19-5 NASC† THAT A MATERIAL BE BIAXIAL

Consider transformations of the tensors $\boldsymbol{\epsilon}$ and $\boldsymbol{\mu}$ for a rotation of coordinates under which a condition may be established for a material to be biaxial.

19-5-1 The Rotation Matrix

In going from a Cartesian coordinate system x, y, z to a rotated system x', y', z' we have

$$\begin{pmatrix} x' \\ y' \\ z' \end{pmatrix} = \mathfrak{A} \begin{pmatrix} x \\ y \\ z \end{pmatrix} \tag{19-30}$$

where the rotation matrix $\mathfrak{A}$ is an orthogonal matrix, that is,

$$\mathfrak{A}\mathfrak{A}^t = \mathfrak{A}^t\mathfrak{A} = \mathfrak{I}, \quad \text{or} \quad \mathfrak{A}^t = \mathfrak{A}^{-1}, \tag{19-31}$$

where t means transpose and $\mathfrak{I}$ is the identity matrix. Thus,

$$\begin{pmatrix} x \\ y \\ z \end{pmatrix} = \mathfrak{A}^{-1} \begin{pmatrix} x' \\ y' \\ z' \end{pmatrix} \tag{19-32}$$

$$= \mathfrak{A}^t \begin{pmatrix} x' \\ y' \\ z' \end{pmatrix}. \tag{19-33}$$

† NASC: Necessary and sufficient conditions.

19-5-2 The Most General Rotation—Euler's Angles

Consider a rigid rotation of the x- and y-axes about the z-axis by an angle ϕ. Then

$$\mathbf{x}' = \mathfrak{A}\mathbf{x},$$

where

$$\mathfrak{A} = \begin{pmatrix} \cos\phi & \sin\phi & 0 \\ -\sin\phi & \cos\phi & 0 \\ 0 & 0 & 1 \end{pmatrix}. \tag{19-34}$$

Follow this rotation with a rotation of the y'- and z'-axes about the x'-axis through an angle θ:

$$\mathbf{x}'' = \mathfrak{B}\mathbf{x}',$$

where

$$\mathfrak{B} = \begin{pmatrix} 1 & 0 & 0 \\ 0 & \cos\theta & \sin\theta \\ 0 & -\sin\theta & \cos\theta \end{pmatrix}. \tag{19-35}$$

Finally, rotate x''- and y''-axes about the z'-axis through an angle ψ:

$$\boldsymbol{\xi} = \mathfrak{C}\mathbf{x}'',$$

where

$$\mathfrak{C} = \begin{pmatrix} \cos\psi & \sin\psi & 0 \\ -\sin\psi & \cos\psi & 0 \\ 0 & 0 & 1 \end{pmatrix}. \tag{19-36}$$

The most general rotation possible is one characterized by $\mathfrak{A}$, followed by $\mathfrak{B}$, followed in turn by $\mathfrak{C}$; the angles ϕ, θ, ψ are called Euler angles. It follows that

$$\begin{pmatrix} \xi \\ \eta \\ \zeta \end{pmatrix} = \mathfrak{R}\begin{pmatrix} x \\ y \\ z \end{pmatrix}, \qquad \begin{pmatrix} x \\ y \\ z \end{pmatrix} = \mathfrak{R}'\begin{pmatrix} \xi \\ \eta \\ \zeta \end{pmatrix}, \tag{19-37}$$

where, from 19-34–19-36:

$$\mathfrak{R} = \mathfrak{C}\mathfrak{B}\mathfrak{A} = \begin{pmatrix} (\cos\psi\cos\phi - \cos\theta\sin\phi\sin\psi) & (+\cos\psi\sin\phi + \cos\theta\cos\phi\sin\psi) & (\sin\psi\sin\theta) \\ -(\sin\psi\cos\phi + \cos\theta\sin\phi\cos\psi) & (-\sin\psi\sin\phi + \cos\theta\cos\phi\cos\psi) & (\cos\psi\sin\theta) \\ \sin\theta\sin\phi & -\sin\theta\cos\phi & \cos\theta \end{pmatrix}. \tag{19-38}$$

(*Note*: We could as well have chosen the standard rotation axes of aerodynamics so that our rotation angles would become the standard pitch, roll and yaw; see Chapter 10.)

19-5-3 Transformation of a Tensor Under Rotation

Vectors transform as in Eq. 19-30, that is, if

$$\mathbf{E}(x, y, z) \rightarrow \mathbf{E}'(x', y', z'),$$

then

$$\mathbf{E}' = \mathfrak{A}\mathbf{E} \tag{19-39}$$

with $\mathfrak{A}$ as in Eq. 19-30.

Consider a transformation of the permittivity tensor $\boldsymbol{\epsilon}$ under a coordinate rotation characterized by a rotation matrix $\mathfrak{A}$ (similar considerations apply to the permeability tensor $\boldsymbol{\mu}$). Since

$$\mathbf{D} = \boldsymbol{\epsilon}\mathbf{E}, \qquad \mathbf{D}' = \boldsymbol{\epsilon}'\mathbf{E}', \tag{19-40}$$

then from

$$\mathfrak{A}\mathbf{D} = \mathfrak{A}\boldsymbol{\epsilon}\mathbf{E} = \mathfrak{A}\boldsymbol{\epsilon}\mathfrak{A}^{-1}\mathfrak{A}\mathbf{E} \tag{19-41}$$

and from

$$\mathbf{D}' = \mathfrak{A}\mathbf{D}, \qquad \mathbf{E}' = \mathfrak{A}\mathbf{E},$$

it follows that

$$\boldsymbol{\epsilon}' = \mathfrak{A}\boldsymbol{\epsilon}\mathfrak{A}^{-1} = \mathfrak{A}\boldsymbol{\epsilon}\mathfrak{A}'. \tag{19-42}$$

The transformation $\boldsymbol{\epsilon} \rightarrow \boldsymbol{\epsilon}'$ (and $\boldsymbol{\mu} \rightarrow \boldsymbol{\mu}'$) described by Eq. 19-42 is a similarity transformation, in which the trace (sum of diagonal entries) is preserved:

$$\operatorname{tr} \boldsymbol{\epsilon}' = \operatorname{tr} \boldsymbol{\epsilon}. \tag{19-43}$$

Example 1. Biaxial Material. Under a rotation of x- and y-axes about the z-axis through the angle ϕ, described by the matrix 19-34, the tensor

$$\boldsymbol{\epsilon} = \begin{pmatrix} \epsilon_1 & 0 & 0 \\ 0 & \epsilon_2 & 0 \\ 0 & 0 & \epsilon_3 \end{pmatrix}_{x,y,z} \tag{19-44}$$

becomes, according to 19-42:

$$\boldsymbol{\epsilon}' = \begin{pmatrix} \epsilon_1 \cos^2\phi + \epsilon_2 \sin^2\phi & (\epsilon_2 - \epsilon_1) \sin\phi \cos\phi & 0 \\ (\epsilon_2 - \epsilon_1) \sin\phi \cos\phi & \epsilon_1 \sin^2\phi + \epsilon_2 \cos^2\phi & 0 \\ 0 & 0 & \epsilon_3 \end{pmatrix}. \tag{19-45}$$

Note that Eq. 19-43 is satisfied. The rotation introduces off-diagonal entries $\epsilon'_{x'y'} = \epsilon'_{y'x'}$ proportional to $\epsilon_2 - \epsilon_1$ that become zero in the case of uniaxial material, as it must be on physical grounds.

Example 2. Biaxial and Gyrotropic Materials. Assume that

$$\epsilon = \begin{pmatrix} \epsilon_1 & \epsilon_4 & 0 \\ \epsilon_5 & \epsilon_2 & 0 \\ 0 & 0 & \epsilon_3 \end{pmatrix}_{x,y,z} . \tag{19-46}$$

Under rotation 19-34 we have, by 19-42

$$\begin{pmatrix} [\epsilon_1 \cos^2\phi + \epsilon_2 \sin^2\phi + (\epsilon_4 + \epsilon_5)\sin\phi\cos\phi] & [(\epsilon_2 - \epsilon_1)\sin\phi\cos\phi + \epsilon_4\cos^2\phi - \epsilon_5\sin^2\phi] & 0 \\ [(\epsilon_2 - \epsilon_1)\sin\phi\cos\phi + \epsilon_5\cos^2\phi - \epsilon_4\sin^2\phi] & [\epsilon_1\sin^2\phi + \epsilon_2\cos^2\phi - (\epsilon_4 + \epsilon_5)\sin\phi\cos\phi] & 0 \\ 0 & 0 & \epsilon_3 \end{pmatrix} \tag{19-47}$$

evaluated at x', y', z'. We see that the trace is preserved and that $\epsilon'_{x'y'} = \epsilon'_{y'x'}$ when $\epsilon_4 = \epsilon_5$. We may conclude that $\epsilon_4 = \epsilon_5$ is a necessary condition for the material to be biaxial.

19-5-4 Inverse Transformation for Biaxial Materials

Assume that

$$\boldsymbol{\epsilon}' = \begin{pmatrix} \epsilon'_1 & \epsilon'_4 & 0 \\ \epsilon'_4 & \epsilon'_2 & 0 \\ 0 & 0 & \epsilon'_3 \end{pmatrix}_{x',y',z'} \tag{19-48}$$

is known. Since the off-diagonal elements are equal, we hope to carry out the inverse transformation of the form of Eq. 19-44. To this end, we compare 19-48 with 19-45 and find that

$$\begin{aligned} (\epsilon_2 - \epsilon_1)\sin 2\phi &= \epsilon'_4/2, \\ (\epsilon_2 - \epsilon_1)\cos 2\phi &= \epsilon'_2 - \epsilon'_1, \\ \epsilon_1 + \epsilon_2 &= \epsilon'_1 + \epsilon'_2, \end{aligned} \tag{19-49}$$

which is a system of three equations in the unknowns ϕ, ϵ_1, and ϵ_2. From the first two equations

$$\tan 2\phi = \epsilon'_4/2(\epsilon'_2 - \epsilon'_1), \tag{19-50}$$

which determines ϕ uniquely. Note that we may restrict ϕ to one of the two intervals $-\pi/4 \le \phi \le \pi/4$ or $0 \le \phi \le \pi/2$; in either case the sign of $\epsilon_2 - \epsilon_1$ is known:

$$\operatorname{sgn}(\epsilon_2 - \epsilon_1) = \begin{cases} \operatorname{sgn}(\epsilon'_2 - \epsilon'_1), \text{ if } -\dfrac{\pi}{4} \le \phi \le \dfrac{\pi}{4}, \\ \operatorname{sgn} \epsilon'_4, \text{ if } 0 \le \phi \le \dfrac{\pi}{2}. \end{cases} \tag{19-51}$$

Also from the first two equations of 19-49 we get

$$\epsilon_2 - \epsilon_1 = \sqrt{(\epsilon_2' - \epsilon_1')^2 + (\epsilon_4'^2/4)} \cdot \operatorname{sgn}(\epsilon_2 - \epsilon_1). \tag{19-52}$$

From the third of Eqs. 19-49 and from 19-52 we find

$$\left.\begin{matrix}\epsilon_2\\ \epsilon_1\end{matrix}\right\} = \frac{1}{2}\left[\epsilon_1' + \epsilon_2' \pm \sqrt{(\epsilon_2' - \epsilon_1')^2 + \frac{1}{4}\epsilon_4'^2}\, \operatorname{sgn}(\epsilon_2 - \epsilon_1)\right]. \tag{19-53}$$

The unique solution for the inverse transformation is given by Eqs. 19-50, 19-51, and 19-53.

In conclusion, a necessary and sufficient condition for a material to be biaxial is that it be represented by a tensor of the form of 19-48. Then, a rotation reduces the tensor to its principal axes (Eq. 19-44) in a unique way. Note that if both $\boldsymbol{\epsilon}'$ and $\boldsymbol{\mu}'$ are of the form of 19-48, there is no assurance that ϕ is the same for both, that is, we may be able to reduce either $\boldsymbol{\epsilon}$ or $\boldsymbol{\mu}$ to principal axes, but not both.

19-6 GYROELECTROMAGNETIC LAYERS

We are now ready to proceed with the solution of the layer problem by first examining the field components in each layer. Then we set up a chain matrix relation between layers on the basis of the boundary conditions between layers. Finally, we obtain reflection and/or transmission coefficients for the layered structure.

19-6-1 Vector Wave Number and Field Components in Each Layer

In the hth layer the vector wave number $\mathbf{k}_h$ is given by

$$\mathbf{k}_h = (\mathbf{k}_h \cdot \hat{x}_0)\hat{x}_0 + k_{hz}\hat{z}, \tag{19-54}$$

where $\hat{x}_0$, $\hat{z}$ define the incidence plane and, in order to satisfy the boundary conditions at the interfaces between layers, $\mathbf{k}_h \cdot \hat{x}_0$ must be independent of h (phase condition). Thus, we may write

$$\mathbf{k}_h = k_x\hat{x} + k_y\hat{y} + k_z\hat{z}, \tag{19-55}$$

where k_x and k_y are the same for all incident, reflected, and transmitted fields across interfaces, whereas k_{hz} changes from one medium to another. Thus,

$$k_x = k_{0x} = k_0 \sin\gamma \cos\beta \tag{19-56}$$

$$k_y = k_{0y} = k_0 \sin\gamma \sin\beta \tag{19-57}$$

everywhere, and

$$k_{0z} = \pm k_0 \cos\gamma, \text{ for } \begin{matrix}\text{reflected}\\ \text{incident}\end{matrix} \text{ field in } z \geq 0. \tag{19-58}$$

In the hth layer all field components are linear combinations of terms of the type

$$\exp(-j\mathbf{k}_h \cdot \mathbf{r}) = \exp(-jk_x x - jk_y y)\exp(-jk_{hz}z), \tag{19-59}$$

each term corresponding to a different value of k_{hz}. We find values of k_{hz} from our foregoing discussion on the dispersion relation. For convenience in what follows we normalize the wave numbers to k_0 by setting (as we did in the discussion of the dispersion relation)

$$\mathbf{k}_h = k_0 \boldsymbol{\kappa}_h \tag{19-60}$$

so that, in particular, $k_{hz} = k_0 \kappa_{hz}$.

19-6-2 Field Components in Each Layer

If we neglect the factor $\exp[-j(k_x x + k_y y) + j\omega t]$, we find from Maxwell's equations that

$$E_{hz} = \sin\gamma[C_h^{(1)^{-1}}(a_{h1}\exp(-jk_0\kappa_{hz}^{(1)}z) + b_{h1}\exp(jk_0\kappa_{hz}^{(1)}z)) + (a_{h2}\exp(-jk_0\kappa_{hz}^{(2)}z) + b_{h2}\exp(jk_0\kappa_{hz}^{(2)}z))] \tag{19-61}$$

$$Z_0 H_{hz} = \sin\gamma[(a_{h1}\exp(-jk_0\kappa_{hz}^{(1)}z) - b_{h1}\exp(jk_0\kappa_{hz}^{(1)}z)) + C_h^{(2)}(a_{h2}\exp(-jk_0\kappa_{hz}^{(2)}z) - b_{h2}\exp(jk_0\kappa_{hz}^{(2)}z))] \tag{19-62}$$

$$E_{h{x \atop y}} = \sum_{i=1,2} P_{h{x \atop y}}^{(i)}(a_{hi}\exp(-jk_0\kappa_{hz}^{(i)}z) - b_{hi}\exp(jk_0\kappa_{hz}^{(i)}z)) \tag{19-63}$$

$$Z_0 H_{h{x \atop y}} = \sum_{i=1,2} Q_{h{x \atop y}}^{(i)}(a_{hi}\exp(-jk_0\kappa_{hz}^{(i)}z) + b_{hi}\exp(jk_0\kappa_{hz}^{(i)}z)) \tag{19-64}$$

where a_{hi} and b_{hi} are constants to be found by imposing the boundary conditions to the tangential (i.e., x- and y-) components above; (v is defined in Eq. 19-27).

$$P_{hx}^{(i)} = \frac{-1}{v_{h\epsilon}}[\epsilon_{h3}\kappa_{hz}^{(i)}\Lambda_h^{(i)}\cos\beta + \mu_{h3}(\epsilon_{h2}\sin\beta + \epsilon_{h4}\cos\beta)\Omega_h^{(i)}], \tag{19-65}$$

$$P_{hy}^{(i)} = \frac{-1}{v_{h\epsilon}}[\epsilon_{h3}\kappa_{hz}^{(i)}\Lambda_h^{(i)}\sin\beta - \mu_{h3}(\epsilon_{h5}\sin\beta + \epsilon_{h1}\cos\beta)\Omega_h^{(i)}], \tag{19-66}$$

$$Q_{hx}^{(i)} = \frac{-1}{v_{h\mu}}[\mu_{h3}\kappa_{hx}^{(i)}\Omega_h^{(i)}\cos\beta - \epsilon_{h3}(\mu_{h2}\sin\beta + \mu_{h4}\cos\beta)\Lambda_h^{(i)}], \tag{19-67}$$

$$Q_{hy}^{(i)} = \frac{-1}{v_{h\mu}}[\mu_{h3}\kappa_{hz}^{(i)}\Omega_h^{(i)}\sin\beta + \epsilon_{h3}(\mu_{h5}\sin\beta + \mu_{h1}\cos\beta)\Lambda_h^{(i)}]; \tag{19-68}$$

$$\Lambda_h^{(1)} = C_h^{(1)-1}, \qquad \Lambda_h^{(2)} = \Omega_h^{(1)} = 1, \qquad \Omega_h^{(2)} = C_h^{(2)}; \tag{19-69}$$

$$C_h^{(i)} = \epsilon_{h3}\kappa_{hz}$$
$$\times \frac{(\epsilon_{h1}\mu_{h4} - \epsilon_{h5}\mu_{h1}) \cos^2\beta + (\epsilon_{h4}\mu_{h2} - \epsilon_{h2}\mu_{h5}) \sin^2\beta + K_h \sin\beta\cos\beta}{v_{h\epsilon}v_{h\mu}\sin^2\gamma + \mu_{h3}\kappa_{hz}^{(i)2}v_{h\epsilon} + \mu_{h3}(\epsilon_{h4}\epsilon_{h5} - \epsilon_{h1}\epsilon_{h2})v_{h\mu}} \tag{19-70}$$

and

$$K_h = \epsilon_{h1}\mu_{h2} - \epsilon_{h2}\mu_{h1} + \epsilon_{h4}\mu_{h4} - \epsilon_{h5}\mu_{h5}. \tag{19-71}$$

In the particular case of a uniaxial nongyrotropic medium ($\epsilon_1 = \epsilon_2$, $\mu_1 = \mu_2$, $\epsilon_4 = \epsilon_5 = \mu_4 = \mu_5 = 0$),

$$C_h^{(1)-1} = 0, \qquad C_h^{(2)} = 0. \tag{19-72}$$

It should be noted that, in this case, $C_h^{(1)}$ assumes the indeterminate form 0/0; so a careful analysis is needed to obtain the limiting value above from the form of 19-70; numerically, the former should be inserted directly into our formulas, bypassing the form of 19-70.

19-7 CHAIN MATRIX ALGORITHM

The coefficients a_{hi} and b_{hi} are formed by imposing the continuity of tangential components of the electric and magnetic fields across the interfaces separating different layers. In many applications, we are interested only in the reflection and transmission coefficients for the entire layered structure, and the knowledge of the field coefficients inside each layer is then unnecessary. In such cases, we relate the coefficients a_{hi} and b_{hi} for the first layer to the coefficients a_{ni} and b_{ni} for the last layer without calculating coefficients for intermediate layers, by using a chain matrix technique.

We set

$$D_h = k_0 d_h, \tag{19-73}$$

$$A_{hi} = a_{hi}\exp(j\kappa_{hz}^{(i)}D_h),\ B_{hi} = b_{hi}\exp(-j\kappa_{hz}^{(i)}D_h), \tag{19-74}$$

$$\theta_{h+1,i} = \kappa_{h+1,z}^{(i)}(D_{h+1} - D_h); \tag{19-75}$$

then the boundary conditions at the interface $z = -d_h$ between layers h and $h+1$ yield

$$\boldsymbol{\alpha}_h \begin{pmatrix} A_{h1} \\ A_{h2} \\ B_{h1} \\ B_{h2} \end{pmatrix} = \boldsymbol{\beta}_h \begin{pmatrix} A_{h+1,1} \\ A_{h+1,2} \\ B_{h+1,1} \\ B_{h+1,2} \end{pmatrix} \tag{19-76}$$

with

$$\boldsymbol{\alpha}_h = \begin{pmatrix} P_{hx}^{(1)} & P_{hx}^{(2)} & -P_{hx}^{(1)} & -P_{hx}^{(2)} \\ P_{hy}^{(1)} & P_{hy}^{(2)} & -P_{hy}^{(1)} & -P_{hy}^{(2)} \\ Q_{hx}^{(1)} & Q_{hx}^{(2)} & Q_{hx}^{(1)} & Q_{hx}^{(2)} \\ Q_{hy}^{(1)} & Q_{hy}^{(2)} & Q_{hy}^{(1)} & Q_{hy}^{(2)} \end{pmatrix}; \tag{19-77}$$

$$\boldsymbol{\beta}_h = \begin{pmatrix} P_{h+1,x}^{(1)} \exp(-j\theta^{(1)}) & P_{h+1,x}^{(2)} \exp(-j\theta^{(2)}) & -P_{h+1,x}^{(1)} \exp(j\theta^{(1)}) & -P_{h+1,x}^{(2)} \exp(j\theta^{(2)}) \\ P_{h+1,y}^{(1)} \exp(-j\theta^{(1)}) & P_{h+1,y}^{(2)} \exp(-j\theta^{(2)}) & -P_{h+1,y}^{(1)} \exp(j\theta^{(1)}) & -P_{h+1,y}^{(2)} \exp(j\theta^{(2)}) \\ Q_{h+1,x}^{(1)} \exp(-j\theta^{(1)}) & Q_{h+1,x}^{(2)} \exp(-j\theta^{(2)}) & Q_{h+1,x}^{(1)} \exp(j\theta^{(1)}) & Q_{h+1,x}^{(2)} \exp(j\theta^{(2)}) \\ Q_{h+1,y}^{(1)} \exp(-j\theta^{(1)}) & Q_{h+1,y}^{(2)} \exp(-j\theta^{(2)}) & Q_{h+1,y}^{(1)} \exp(j\theta^{(1)}) & Q_{h+1,y}^{(2)} \exp(j\theta^{(2)}) \end{pmatrix} \tag{19-78}$$

where the dual subscript $h+1, i$ has been omitted from θ throughout Eq. 19-78.

If we define the matrix

$$\boldsymbol{\gamma}_h = \boldsymbol{\alpha}_h^{-1}\boldsymbol{\beta}_h, \tag{19-79}$$

then

$$\begin{pmatrix} A_{11} \\ A_{12} \\ B_{11} \\ B_{12} \end{pmatrix} = \boldsymbol{\Gamma} \begin{pmatrix} A_{n1} \\ A_{n2} \\ B_{n1} \\ B_{n2} \end{pmatrix} \tag{19-80}$$

where

$$\boldsymbol{\Gamma} = \{\Gamma_{jm}\} = \prod_{h=1}^{n-1} \boldsymbol{\gamma}_h, \qquad j, m = 1, 2, 3, 4. \tag{19-81}$$

Observe that $n \geq 2$; therefore, if we have one layer of material, we consider it to be composed of two layers with identical properties; the relative thicknesses of these two layers are unimportant.

The matrix $\boldsymbol{\Gamma}$ may be rewritten as

$$\boldsymbol{\Gamma} = \boldsymbol{\alpha}_1^{-1}\left(\prod_{h=2}^{n-1} \boldsymbol{\delta}_h\right)\boldsymbol{\beta}_{n-1}, \tag{19-82}$$

where the matrix produced is replaced by the identity matrix if $n = 2$, and the matrix

$$\boldsymbol{\delta} = \boldsymbol{\beta}_{h-1}\boldsymbol{\alpha}_h^{-1}, \qquad 2 \leq h \leq n-1, \tag{19-83}$$

contains the constitutive parameters of the hth layer. The elements of the matrices $\boldsymbol{\alpha}_h^{-1}$ and $\boldsymbol{\delta}_h$ are calculatable explicitly since $\boldsymbol{\alpha}_h$ can be inverted analytically. If we write, omitting the subscript h, since we shall be dealing with one representative layer,

$$\boldsymbol{\alpha}^{-1} = \{\tilde{\alpha}_{jm}; j, m = 1, 2, 3, 4\}, \tag{19-84}$$

then we may show that

$$\begin{aligned}
\tilde{\alpha}_{11} &= -\tilde{\alpha}_{31} = \tfrac{1}{2}PP_y^{(2)}; & \tilde{\alpha}_{41} &= -\tilde{\alpha}_{21} = \tfrac{1}{2}PP_y^{(1)};\\
\tilde{\alpha}_{32} &= -\tilde{\alpha}_{12} = \tfrac{1}{2}PP_x^{(2)}; & \tilde{\alpha}_{22} &= -\tilde{\alpha}_{42} = \tfrac{1}{2}PP_x^{(1)};\\
\tilde{\alpha}_{13} &= \tilde{\alpha}_{33} = \tfrac{1}{2}QQ_y^{(2)}; & \tilde{\alpha}_{23} &= \tilde{\alpha}_{43} = -\tfrac{1}{2}QQ_y^{(1)};\\
\tilde{\alpha}_{14} &= \tilde{\alpha}_{34} = -\tfrac{1}{2}QQ_x^{(2)}; & \tilde{\alpha}_{24} &= \tilde{\alpha}_{44} = \tfrac{1}{2}QQ_x^{(1)};
\end{aligned} \tag{19-85}$$

where

$$\begin{aligned}
P &= (P_x^{(1)}P_y^{(2)} - P_x^{(2)}P_y^{(1)})^{-1};\\
Q &= (Q_x^{(1)}Q_y^{(2)} - Q_x^{(2)}Q_y^{(1)})^{-1}.
\end{aligned} \tag{19-86}$$

The boundary conditions at the perfectly conducting surface $z = -d_n$ yield

$$A_{n1} = B_{n1}, \qquad A_{n2} = B_{n2}. \tag{19-87}$$

Using these and the boundary conditions at $z = 0$ yields a system of four equations in the four unknowns $E_{\parallel}$, $E_{\perp}$, A_{n1}, A_{n2}, where $\parallel$ and $\perp$ indicate parallel and perpendicular polarization components of $\mathbf{E}$:

$$\begin{aligned}
&\text{(a)}\quad E_{\parallel}\cos\beta\cos\gamma + E_{\perp}\sin\beta + A_{n1}(u_{1x} + u_{3x})\\
&\qquad\qquad + A_{n2}(u_{2x} + u_{4x}) = \sin\alpha\cos\beta\cos\gamma - \cos\alpha\sin\beta,\\
&(b)\quad E_{\parallel}\sin\beta\cos\gamma - E_{\perp}\cos\beta + A_{n1}(u_{1y} + u_{3y})\\
&\qquad\qquad + A_{n2}(u_{2y} + u_{4y}) = \sin\alpha\sin\beta\cos\gamma + \cos\alpha\cos\beta,\\
&\text{(c)}\quad -E_{\parallel}\sin\beta + E_{\perp}\cos\beta\cos\gamma + A_{n1}(v_{1x} + v_{3x})\\
&\qquad\qquad + A_{n2}(v_{2x} + v_{4x}) = \cos\alpha\cos\beta\cos\gamma + \sin\alpha\sin\beta,\\
&\text{(d)}\quad E_{\parallel}\cos\beta + E_{\perp}\sin\beta\cos\gamma + A_{n1}(v_{1y} + v_{3y})\\
&\qquad\qquad + A_{n2}(v_{2y} + v_{4y}) = \cos\alpha\sin\beta\cos\gamma - \sin\alpha\cos\beta,
\end{aligned} \tag{19-88}$$

where

$$\begin{aligned}
&\text{(a)}\quad u_{i{x \atop y}} = P_{1{x \atop y}}^{(1)}\xi_{1i}^{-} + P_{1{x \atop y}}^{(2)}\xi_{2i}^{-},\\
&\text{(b)}\quad v_{i{x \atop y}} = Q_{1{x \atop y}}^{(1)}\xi_{1i}^{+} + Q_{1{x \atop y}}^{(2)}\xi_{2i}^{+},\\
&\text{(c)}\quad \xi_{1i}^{\pm} = \Gamma_{1i}\exp(-j\kappa_{1z}^{(1)}D_1) \pm \Gamma_{3i}\exp(j\kappa_{1z}^{(1)}D_1),\\
&\text{(d)}\quad \xi_{2i}^{\pm} = \Gamma_{2i}\exp(-j\kappa_{1z}^{(2)}D_1)\\
&\qquad\qquad \pm \Gamma_{4i}\exp(j\kappa_{1z}^{(2)}D_1), \qquad i = 1, 2, 3, 4.
\end{aligned} \tag{19-89}$$

Once $E_{\parallel}$ and $E_{\perp}$ are found, the power reflection coefficient is

$$R = |E_{\parallel}|^2 + |E_{\perp}|^2. \tag{19-90}$$

19-8 PENETRABLE BACKING

If the half-space $z < -d_n$ is a homogeneous isotropic medium with permittivity ϵ_t and permeability μ_t, then the corresponding intrinsic impedance and refractive index are

$$Z_t = \sqrt{\frac{\mu_t}{\epsilon_t}}, \qquad N_t = \sqrt{\frac{\epsilon_t \mu_t}{\epsilon_0 \mu_0}}. \tag{19-91}$$

The vector wavenumber $\mathbf{k}_t$ for the transmitted ray is

$$\mathbf{k}_t = N_t k_0 \hat{k}_t, \tag{19-92}$$

$$\hat{k}_t = \hat{x} \cos\beta \sin\gamma_t + \hat{y} \sin\beta \sin\gamma_t - \hat{z} \cos\gamma_t, \tag{19-93}$$

$$\gamma_t = \arcsin\left(\frac{\sin\gamma}{N_t}\right) \tag{19-94}$$

(γ_t is the angle that $\hat{k}_t$ forms with the negative z-axis).

The reflected field can be written as before; but the transmitted field is

$$\begin{aligned} \mathbf{E}_t &= [\hat{x}(-E_{t\perp} \sin\beta + E_{t\parallel} \cos\beta \cos\gamma_t) \\ &\quad + \hat{y}(E_{t\perp} \cos\beta + E_{t\parallel} \sin\beta \cos\gamma_t) \\ &\quad + \hat{z} E_{t\parallel} \sin\gamma_t] \exp(jk_0 \Delta_t), \\ &= Z_t \mathbf{H}_t \times \hat{k}_t \end{aligned} \tag{19-95}$$

where

$$\Delta_t = -x \sin\gamma_t \cos\beta - y \sin\gamma_t \sin\beta + N_t(z + d_n) \cos\gamma_t. \tag{19-96}$$

When the boundary conditions at $z = 0$ and $z = -d_n$ are imposed, we get eight equations in the eight unknowns $E_{t\parallel}$, $E_{t\perp}$, $E_{\parallel}$, $E_{\perp}$, A_{n1}, A_{n2}, B_{n1}, and B_{n2}:

(a) $E_{\parallel} \cos\beta \cos\gamma + E_{\perp} \sin\beta + A_{n1} u_{1x} + A_{n2} u_{2x} + B_{n1} u_{3x} + B_{n2} u_{4x} = \sin\alpha \cos\beta \cos\gamma - \cos\alpha \sin\beta;$

(b) $E_{\parallel} \sin\beta \cos\gamma - E_{\perp} \cos\beta + A_{n1} u_{1y} + A_{n2} u_{2y} + B_{n1} u_{3y} + B_{n2} u_{4y} = \sin\alpha \sin\beta \cos\gamma + \cos\alpha \cos\beta;$

(c) $-E_{\parallel} \sin\beta + E_{\perp} \cos\beta \cos\gamma + A_{n1} v_{1x} + A_{n2} v_{2x} + B_{n1} v_{3x} + B_{n2} v_{4x} = \cos\alpha \cos\beta \cos\gamma + \sin\alpha \sin\beta;$

(d) $E_{\parallel} \cos\beta + E_{\perp} \sin\beta \cos\gamma + A_{n1} v_{1y} + A_{n2} v_{2y} + B_{n1} v_{3y} + B_{n2} v_{4y} = \cos\alpha \sin\beta \cos\gamma - \sin\alpha \cos\beta;$

(e) $E_{t\|} \cos \beta \cos \gamma_t - E_{t\perp} \sin \beta = \sum_{i=1,2} P_{nx}^{(i)}(A_{ni} - B_{ni});$

(f) $E_{t\|} \sin \beta \cos \gamma_t + E_{t\perp} \cos \beta = \sum_{i=1,2} P_{ny}^{(i)}(A_{ni} - B_{ni});$

(g) $E_{t\|} \sin \beta + E_{t\perp} \cos \beta \cos \gamma_t = Z_t Y_0 \sum_{i=1,2} Q_{nx}^{(i)}(A_{ni} + B_{ni});$

(h) $-E_{t\|} \cos \beta + E_{t\perp} \sin \beta \cos \gamma_t = Z_t Y_0 \sum_{i=1,2} Q_{ny}^{(i)}(A_{ni} + B_{ni}).$ (19-97)

Here the power transmission coefficient is

$$T = |E_{t\|}|^2 + |E_{t\perp}|^2. \tag{19-98}$$

Observe that the metal backing system may be considered as the limiting case of the system above when $A_{n1} = B_{n1}$ and $A_{n2} = B_{n2}$; for the metal case the last four equations become identities because $E_{t\|} = E_{t\perp} = 0$, $A_{n1} = B_{n1}$, $A_{n2} = B_{n2}$, and Z_t goes to zero as conductivity tends to infinity.

19-9 NUMERICAL RESULTS

A computer code has been developed and tested on a variety of layered structures. Consider the following example:

Two layers, each 1 cm thick; frequency 1 GHz;

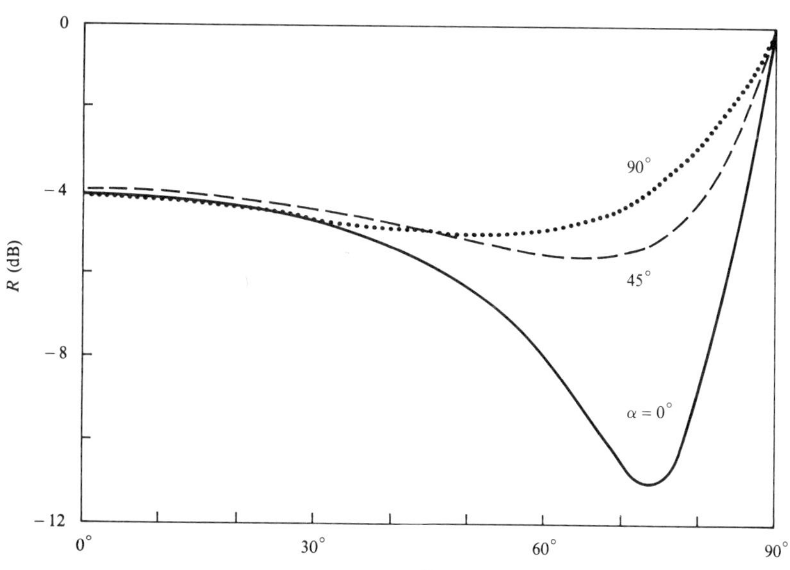

Figure 19-3 R versus θ for $\beta = 45°$. Reprinted with permission from Damaskos, Maffett, and Uslenghi (1985). "Reflection and Transmission for Gyroelectromagnetic Biaxial Layered Media," *Journal of the Optical Society of America*, **2**, pp. 454–461.

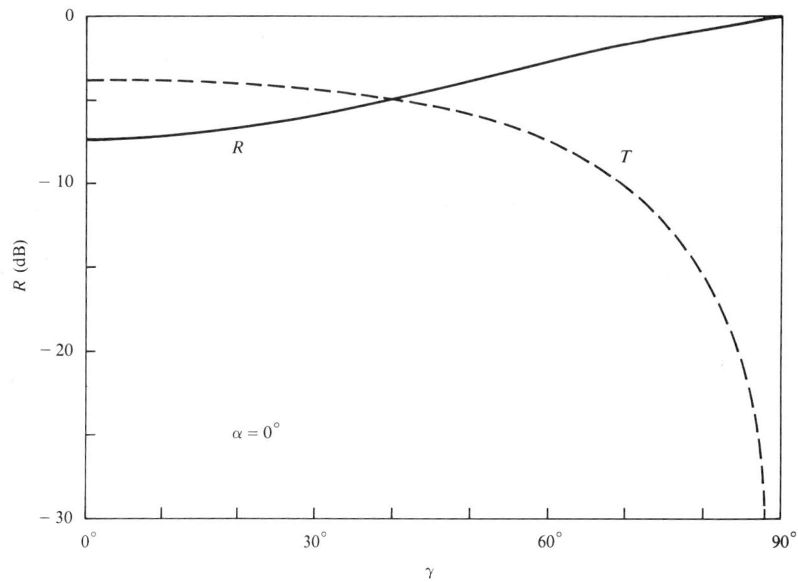

Figure 19-4 R and T versus γ for $\alpha = 0°$, $\beta = 45°$. Reprinted with permission from Damaskos, Maffett, and Uslenghi (1985). "Reflection and Transmission for Gyroelectromagnetic Biaxial Layered Media," *Journal of the Optical Society of America,* **2,** pp. 454–461.

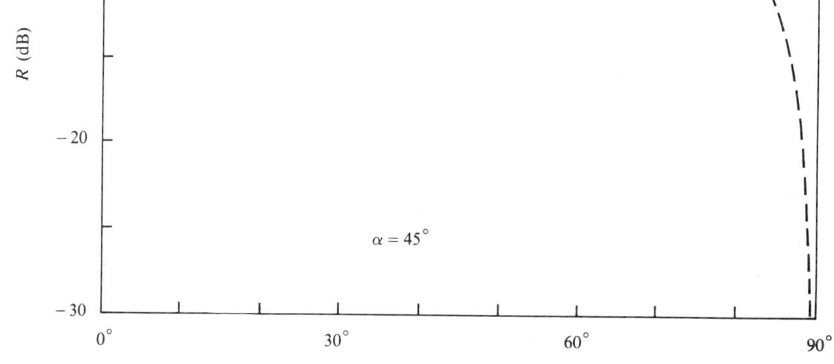

Figure 19-5 R and T versus γ for $\alpha = \beta = 45°$. Reprinted with permission from Damaskos, Maffett, and Uslenghi (1985). "Reflection and Transmission for Gyroelectromagnetic Biaxial Layered Media," *Journal of the Optical Society of America,* **2,** pp. 454–461.

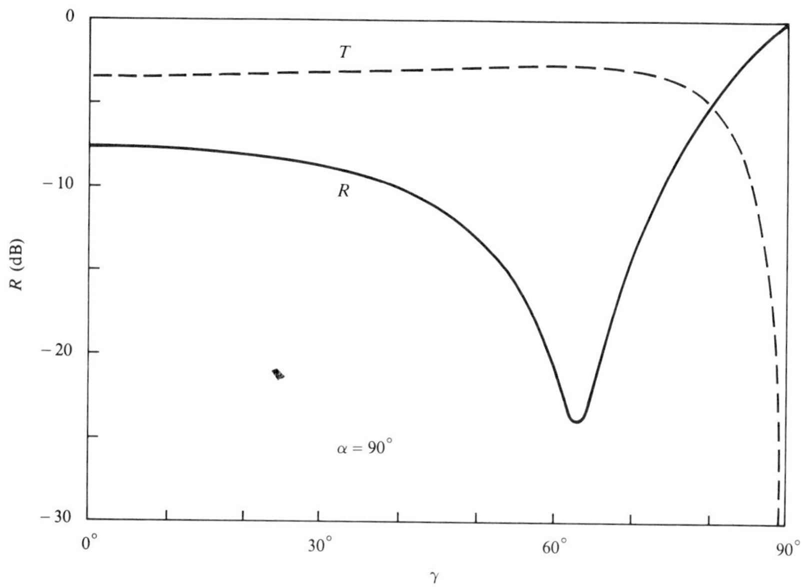

Figure 19-6 R and T versus γ for $\alpha = 90°$, $\beta = 45°$. Reprinted with permission from Damaskos, Maffett, and Uslenghi (1985). "Reflection and Transmission for Gyroelectromagnetic Biaxial Layered Media," *Journal of the Optical Society of America*, **2**, pp. 454–461.

Top layer: $\epsilon_1 = \epsilon_2 = 10$, $\epsilon_3 = 1$, $\epsilon_4 = \epsilon_5 = 0$; $\mu_1 = \mu_2 = 3 - j$, $\mu_3 = 1$, $\mu_4 = \mu_5 = j/2$ (electrically uniaxial, magnetically lossy, uniaxial, and gyrotropic);

Bottom layer: $\epsilon_1 = 5$, $\epsilon_2 = 10$, $\epsilon_3 = 1$, $\epsilon_4 = \epsilon_5 = 0$; $\mu_1 = \mu_2 = \mu_3 = 1$, $\mu_4 = \mu_5 = 0$ (lossless biaxial dielectric).

19-9-1 Metallic Backing

In Fig. 19-3 R is plotted versus γ for three values of α, with $\beta = 45°$.

19-9-2 Free-Space Backing

In Fig. 19-4 both R and T are plotted versus γ for $\alpha = 0°$, $\beta = 45°$. Figures 19-5 and 19-6 show effects of changing α to 45°, then to 90° (α is polarization angle).

In either case, the results vary by small amounts as β varies from 0–90°, but may vary by large amounts with α and γ; specifically, strong minima may occur in R.

20

THE INVERSE PROBLEM FOR BIAXIAL MATERIALS

In this chapter, we present the inverse problem for an important category of anisotropic materials: biaxial media, for which the relative permittivity and permeability tensors $\boldsymbol{\epsilon}$ and $\boldsymbol{\mu}$ are represented by diagonal matrices in a rectangular Cartesian reference system x, y, z. In Section 20-1 we set the boundary value problem to be solved for transmission and reflection coefficients from given constitutive parameters. Then, in Section 20-2, we fix the inverse problem: given reflection and transmission coefficients (from measured data, say), find the material constitutive parameters. Separate approaches are given for lossless nondispersive material (Section 20-3) and lossy dispersive material (Section 20-4). Finally, in Section 20-5, a measurement procedure is suggested.

A general treatment of fields in anisotropic guides was first given by Bresler (1959). Specific applications to gyromagnetic materials such as ferrites may be found in Kales (1953), Barzilai and Gerosa (1963), and Lax and Button (1962), among others, while an up-to-date treatment of gyrotropic guides is given by Hlawiczka (1981). The problem of a guide filled with uniaxial material was first treated by Kong and Cheng (1968); and recently, a dielectric uniaxial guide has been reexamined by Paul and Shevgaonkar (1981), whose work contains a good bibliography on the subject. A rectangular guide filled with a semiconductor in the presence of an external transverse magnetic field (Hall effect) was studied by Engineer and Nag (1965), who examined in detail the particular case when the diagonal elements of the complex permittivity tensor are equal. A guide filled with biaxial material was treated briefly by Goncharenko (1966) who, however, neglected an important category of possible modes.

While the electromagnetic theory of anisotropic materials is well established [see, for example, Kong, (1975)], work still needs to be done on experimental techniques for the determination of the constitutive parameters of these materials. The problem treated below is a typical inverse problem, in which we must ascertain first what measured data are sufficient (and, preferably, also necessary) for the unequivocal determination of the constitutive parameters; and second, what experimental setups are preferred to collect the required data. The problem is complicated by the fact that, in general, the material is both lossy and dispersive. For this case, a measurement setup involving a network analyzer and a microwave junction which allows for the separate independent excitation of a sample by either a TE_{10} or a TE_{20} mode is described in Section 20-5. In general, however, the measurements may be performed either in free space, such as an anechoic chamber, or inside a waveguide or resonant cavity.

The material of the following sections is adapted from the work of Damaskos et al.† Throughout we characterize biaxial media by the diagonal permittivity and permeability tensors

$$\boldsymbol{\epsilon} = \begin{pmatrix} \epsilon_1 & 0 & 0 \\ 0 & \epsilon_2 & 0 \\ 0 & 0 & \epsilon_3 \end{pmatrix}_{xyz} \quad \text{and} \quad \boldsymbol{\mu} = \begin{pmatrix} \mu_1 & 0 & 0 \\ 0 & \mu_2 & 0 \\ 0 & 0 & \mu_3 \end{pmatrix}_{xyz}, \tag{20-1}$$

respectively.

20-1 THE BOUNDARY VALUE PROBLEM

The six constitutive parameters ϵ_i and μ_i, $i = 1, 2, 3$, in Eq. 20-1 are dimensionless numbers and are, in general, complex and frequency-dependent. We seek their determination from measurements of reflection and transmission coefficients for a biaxial sample inserted in a rectangular waveguide.

Consider a metallic rectangular waveguide oriented along the z-axis, with horizontal walls of width a parallel to the x-axis and vertical walls of height $b \leq a$ parallel to the y-axis. We show below that TE_{m0} modes can exist, when the guide is filled with biaxial material whose principal axes coincide with the xyz-axes of the guide.

With time dependence $\exp(+j\omega t)$, Maxwell's equations are

$$\begin{aligned} &\text{A: } \nabla \times \mathbf{H} = j\omega\epsilon_0 \boldsymbol{\epsilon}\mathbf{E}, \\ &\text{F: } \nabla \times \mathbf{E} = -j\omega\mu_0 \boldsymbol{\mu}\mathbf{H}, \end{aligned} \tag{20-2}$$

with $\boldsymbol{\epsilon}$ and $\boldsymbol{\mu}$ given by Eq. 20-1. If $k_0 = \omega\sqrt{\epsilon_0\mu_0}$ is the wave number of free space, and if $\mathbf{E}$ and $\mathbf{H}$ for a given mode depend on z via the factor

† With permission. © 1984 IEEE. Reprinted from *IEEE Trans. on Microwave Theory and Techniques,* vol. MTT-32, (4), pp. 400–405, Apr. 1984.

$\exp(-\gamma z)$, then the transverse field components are

$$E_x = \frac{-1}{\gamma^2 + k_0^2\epsilon_1\mu_2}\left[\gamma \frac{\partial E_z}{\partial x} + j\omega\mu_0\mu_2 \frac{\partial H_z}{\partial y}\right] \tag{20-3}$$

$$E_y = \frac{-1}{\gamma^2 + k_0^2\mu_1\epsilon_2}\left[\gamma \frac{\partial E_z}{\partial y} - j\omega\mu_0\mu_1 \frac{\partial H_z}{\partial x}\right] \tag{20-4}$$

$$H_x = \frac{-1}{\gamma^2 + k_0^2\mu_1\epsilon_2}\left[\gamma \frac{\partial H_z}{\partial x} - j\omega\epsilon_0\epsilon_2 \frac{\partial E_z}{\partial y}\right] \tag{20-5}$$

$$H_y = \frac{-1}{\gamma^2 + k_0^2\epsilon_1\mu_2}\left[\gamma \frac{\partial H_z}{\partial y} + j\omega\epsilon_0\epsilon_1 \frac{\partial E_z}{\partial x}\right] \tag{20-6}$$

and the longitudinal components satisfy the coupled equations

$$\left(\frac{\epsilon_1}{\gamma^2 + k_0^2\epsilon_1\mu_2}\frac{\partial^2}{\partial x^2} + \frac{\epsilon_2}{\gamma^2 + k_0^2\mu_1\epsilon_2}\frac{\partial^2}{\partial y^2} + \epsilon_3\right)E_z = \frac{-\gamma}{j\omega\epsilon}\left(\frac{1}{\gamma^2 + k_0^2\epsilon_1\mu_2} - \frac{1}{\gamma^2 + k_0^2\mu_1\epsilon_2}\right)\frac{\partial^2 H_z}{\partial x\,\partial y} \tag{20-7}$$

$$\left(\frac{\mu_1}{\gamma^2 + k_0^2\mu_1\epsilon_2}\frac{\partial^2}{\partial x^2} + \frac{\mu_2}{\gamma^2 + k_0^2\epsilon_1\mu_2}\frac{\partial^2}{\partial y^2} + \mu_3\right)H_z = \frac{\gamma}{j\omega\mu}\left(\frac{1}{\gamma^2 + k_0^2\mu_1\epsilon_2} - \frac{1}{\gamma^2 + k_0^2\epsilon_1\mu_2}\right)\frac{\partial^2 E_z}{\partial x\,\partial y} \tag{20-8}$$

so that, in general, a hybrid mode is needed. However, decoupling of E_z from H_z occurs in two particular cases.

In the first case

$$\epsilon_1\mu_2 = \epsilon_2\mu_1 \tag{20-9}$$

and Eqs. 20-7 and 20-8 become

$$(\epsilon_1 D_x + \epsilon_2 D_y + \epsilon_3 h^2)E_z = 0 \tag{20-10}$$

$$(\mu_1 D_x + \mu_2 D_y + \mu_3 h^2)H_z = 0 \tag{20-11}$$

where

$$h^2 = \gamma^2 + k_0^2\epsilon_1\mu_2$$

and the subscripted D's denote differentiation with respect to that subscript; in particular,

$$D_x^2 = \partial^2/\partial x^2 \qquad \text{and} \qquad D_y^2 = \partial^2/\partial y^2. \tag{20-12}$$

The field is still a superposition of *TE* and *TM* modes, as Eq. 20-3 through Eq. 20-6 indicate; in general, condition 20-13 is not satisfied, and therefore we study this case no further.

In the second case, decoupling occurs if either

$$\partial/\partial x = 0 \tag{20-13}$$

or

$$\partial/\partial y = 0. \tag{20-14}$$

These two possibilities are essentially one and the same because we may go from 20-13 to 20-14 by a right-angle rotation of the coordinate system about the z-axis. We consider only 20-14 which, incidentally, is satisfied by the dominant TE_{10} mode in the guide filled with air.

Conditions 20-13 and 20-14 were neglected in the analysis by Goncharenko (1966); therefore, his statement that TM and TE modes can be supported separately only in uniaxial media ($\epsilon_1 = \epsilon_2$ and $\mu_1 = \mu_2$) is incorrect. In fact, uniaxial media are a particular case of condition 20-9.

Under condition 20-9, Eqs. 20-3–20-8 and the boundary conditions yield TM modes with identically zero field components, and TE_{m0} modes for which

$$\begin{aligned}
H_z &= A_m \cos\left(\frac{m\pi x}{a}\right) \exp(-\gamma_m z) \\
E_y &= -(j\omega\mu_0\mu_3 a/m\pi) A_m \sin\left(\frac{m\pi x}{a}\right) \exp(-\gamma_m z) \\
H_x &= (\mu_3\gamma_m a/\mu_1 m\pi) A_m \sin\left(\frac{m\pi x}{a}\right) \exp(-\gamma_m z) \\
E_x &= E_z = H_y = 0, \qquad m = 1, 2, 3, \ldots,
\end{aligned} \tag{20-15}$$

where

$$\gamma_m = jk_0\sqrt{(\mu_1/\mu_3)}\,\sqrt{\epsilon_2\mu_3 - \left(\frac{m\pi}{k_0 a}\right)^2} = j\beta_m. \tag{20-16}$$

Note that γ_m depends on ϵ_2, μ_1, and μ_3, and is independent of ϵ_1, ϵ_3, and μ_2. If $\epsilon_2\mu_3$ is real positive, then γ_m is pure imaginary at all operating frequencies above cutoff frequency

$$f_c = m(c_0/2a\sqrt{\epsilon_2\mu_3}) \tag{20-17}$$

where c_0 is the velocity of light in free space.

When the waveguide section $0 \le z \le L$ is filled with a sample of the biaxial material, and the incident TE_{m0} mode with

$$H_z^i = \cos\left(\frac{m\pi x}{a}\right) \exp(-j\beta_{0m} z) \tag{20-18}$$

$$\beta_{0m} = \sqrt{k_0^2 - (m\pi/a)^2} \tag{20-19}$$

exists in $z \le 0$, it produces a reflected TE_{m0} mode in $z \le 0$ with

$$H_z^r = R_m \cos\left(\frac{m\pi x}{a}\right) \exp(j\beta_{0m} z) \tag{20-20}$$

and a transmitted TE_{m0} mode in $z \geq L$ with

$$H_z^t = T_m \cos\left(\frac{m\pi x}{a}\right) \exp[-j\beta_{0m}(z - L)] \tag{20-21}$$

where we assume that the termination load at $z > L$ is matched. The TE_{m0} field inside the sample ($0 \leq z \leq L$) is the combination of two fields, such as 20-15, propagating in opposite directions. Imposition of the boundary conditions at $z = 0$ and $z = L$ yields the following expressions for the reflection coefficient R_m and the transmission coefficient T_m:

$$1/T_m = \cos(\beta_m L) + \frac{j}{2}\left(u + \frac{1}{u}\right) \sin(\beta_m L) \tag{20-22}$$

$$R_m/T_m = -\frac{j}{2}\left(u - \frac{1}{u}\right) \sin(\beta_m L) \tag{20-23}$$

where

$$u = \mu_1 \beta_m / \mu_3^2 \beta_{0m}. \tag{20-24}$$

20-2 THE INVERSE PROBLEM

The inverse problem consists in finding u and β_m from Eqs. 20-22 and 20-23, when L is given and amplitude and phase of R_m and T_m have been measured. Once u and β_m are known, the constitutive parameters are easily found.

In general, the parameters ϵ_2, μ_1, and μ_3 are complex, and therefore u and β_m are also complex. Adding Eqs. 20-22 and 20-23 and solving yields

$$u = \frac{j \sin(\beta_m L)}{(1/T_m) + (R_m/T_m) - \cos(\beta_m L)}. \tag{20-25}$$

Subtracting Eq. 20-23 from Eq. 20-22 and solving yields

$$u = \frac{(1/T_m) - (R_m/T_m) - \cos(\beta_m L)}{j \sin(\beta_m L)}. \tag{20-26}$$

Now, equating Eqs. 20-27 and 20-26, we find

$$\cos(\beta_m L) = (T_m^2 - R_m^2 + 1)/2T_m. \tag{20-27}$$

Separate real and imaginary parts in the right-hand side of Eq. 20-27 by setting

$$(T_m^2 - R_m^2 + 1)/2T_m = \alpha' + \alpha'' j \tag{20-28}$$

where α' and α'' are real numbers derived from measurements. Now set

$$\beta_m = \beta_m' - j\beta_m'' \tag{20-29}$$

where β' and β'' are real quantities such that

$$\beta'_m > 0, \qquad \beta''_m \geq 0. \tag{20-30}$$

The second part of Eq. 20-30 implies that the biaxial medium is lossy (in the case of an active medium, we would have $\beta''_m \leq 0$, and the discussion would proceed in a manner similar to the lossy case). Now Eq. 20-27 yields

$$\begin{aligned} \cos(\beta'_m L)\cosh(\beta''_m L) &= \alpha' \\ \sin(\beta'_m L)\sinh(\beta''_m L) &= \alpha'' \end{aligned} \tag{20-31}$$

and our inverse problem is reduced to finding β'_m and β''_m from the system of Eqs. 20-31, for given α' and α'' and under restrictions 20-30. Once β_m is known, u is given by either 20-25 or 20-26.

We see from Eq. 20-31 that $\cos(\beta'_m L)$ has the sign of α'; and, because of Eq. 20-30, $\sin(\beta'_m L)$ has the sign of α'':

$$\operatorname{sgn}[\cos(\beta'_m L)] = \operatorname{sgn} \alpha'; \qquad \operatorname{sgn}[\sin(\beta'_m L)] = \operatorname{sgn} \alpha''. \tag{20-32}$$

Elimination of the hyperbolic functions from Eq. 20-31 yields

$$\cos^2(\beta'_m L) = \xi = \frac{1}{2}[\alpha'^2 + \alpha''^2 + 1 - \sqrt{(\alpha'^2 + \alpha''^2 + 1)^2 - 4\alpha'^2}] \tag{20-33}$$

and therefore from Eq. 20-32

$$\begin{aligned} \cos(\beta'_m L) &= \sqrt{\xi}\,\operatorname{sgn}(\alpha'), \\ \sin(\beta'_m L) &= \sqrt{1-\xi}\,\operatorname{sgn}(\alpha''). \end{aligned} \tag{20-34}$$

From Eq. 20-34 we have

$$\beta'_m = \tilde{\beta}'_m + \frac{2\pi n}{L}, \qquad n = 0, 1, \ldots \tag{20-35}$$

where $\tilde{\beta}'_m$ is known, and $0 < \tilde{\beta}'_m L \leq 2\pi$. Elimination of the trigonometric functions from Eq. 20-31 and use of Eq. 20-33 yield

$$\sinh(\beta''_m L) = \sqrt{\alpha'^2 + \alpha''^2 - \xi} \tag{20-36}$$

from which β''_m is uniquely determined. Thus, β_m is determined, aside from the choice of the integer n in Eq. 20-35. If the length L of the sample is sufficiently small, then $n = 0$; however, in many practical cases the length L cannot be arbitrarily chosen; then, n is uniquely determined from nondispersive media by carrying out measurements at two different frequencies, as explained in the following section. If the material is dispersive and the sample is not sufficiently thin, a rough preliminary estimate of the values of the constitutive parameters still allows us to determine n for a given L.

20-3 LOSSLESS NONDISPERSIVE MATERIAL

If the material of the sample is lossless and nondispersive, the constitutive parameters may be determined by taking measurements at two different operating frequencies ω and $\omega^{(1)}$ under the same TE_{m0} mode (in practice, the dominant TE_{10} mode).

Let β_m and u be determined as indicated in the previous section for frequency ω, and let $\beta_m^{(1)}$ and $u^{(1)}$ be the corresponding values at frequency $\omega^{(1)}$. From Eq. 20-24,

$$\beta_m = (\mu_3^2/\mu_1)\beta_{0m}u, \qquad \beta_m^{(1)} = (\mu_3^2/\mu_1)\beta_{0m}^{(1)}u^{(1)} \tag{20-37}$$

whereas, from Eq. 20-35 and the lossless properties of the sample,

$$\beta_m = \tilde{\beta}_m + \frac{2\pi n}{L}, \qquad \beta_m^{(1)} = \tilde{\beta}_m^{(1)} + \frac{2\pi n}{L} \tag{20-38}$$

where n is the same integer in both of Eqs. 20-38, since $\tilde{\beta}_m \rightarrow \tilde{\beta}_m^{(1)}$ and $\beta_m \rightarrow \beta_m^{(1)}$ when $\omega \rightarrow \omega^{(1)}$. To determine n, we equate the ratio of the two Eqs. 20-37 to the ratio of the two Eqs. 20-38, obtaining

$$n = \frac{L}{2\pi}\,\frac{(\beta_{0m}u/\beta_{0m}^{(1)}u^{(1)})\tilde{\beta}_m^{(1)} - \tilde{\beta}_m}{1 - (\beta_{0m}u/\beta_{0m}^{(1)}u^{(1)})}. \tag{20-39}$$

By equating the ratio between Eq. 20-16 and the corresponding equation for $\beta_m^{(1)}$ to the ratio of the two equations 20-37, we find

$$\epsilon_2\mu_3 = \left(\frac{m\pi}{k_0 a}\right)^2 \frac{1 - (\beta_{0m}^2u^2/\beta_{0m}^{(1)2}u^{(1)2})}{1 - (\beta_{0m}^{(1)2}u^2k_0^{(1)2}/\beta_{0m}^{(1)2}u^{(1)2}k_0^2)} \tag{20-40}$$

and therefore the product $\epsilon_2\mu_3$ is known. Now the ratio μ_1/μ_3 is obtained from Eq. 20-16, and the ratio μ_3^2/μ_1 from either of Eqs. 20-37; hence, both μ_1 and μ_3 are known, and ϵ_2 is derived from the knowledge of the product $\epsilon_2\mu_3$.

In conclusion, if ϵ_2, μ_1, and μ_3 are real and independent of frequency, they can be found by measuring R_m and T_m at two different frequencies for the dominant TE_{10} mode. By changing the orientation of the sample in the waveguide, the other three constitutive parameters are similarly determined.

20-4 LOSSY DISPERSIVE MATERIAL

If the medium is dispersive, measured data at different frequencies cannot be mixed together, and the only way to obtain a second, independent measurement of R_m and T_m is to excite the guide with a different mode. Thus, we conduct two separate measurements at the same frequency ω, the first with the TE_{10} mode yielding R_1 and T_1, the second with the TE_{20} mode yielding R_2 and T_2. From Eq. 20-16:

$$\beta_1 = \sqrt{\mu_1/\mu_3}\,\sqrt{k_0^2\epsilon_2\mu_3 - (\pi/a)^2} \tag{20-41a}$$

$$\beta_2 = \sqrt{\mu_1/\mu_3}\,\sqrt{k_0^2\epsilon_2\mu_3 - 4(\pi/a)^2} \tag{20-41b}$$

and from Eq. 20-24:

$$\beta_1 = (\mu_3^2/\mu_1)\beta_{01}u_1 \tag{20-42a}$$

$$\beta_2 = (\mu_3^2/\mu_1)\beta_{02}u_2 \tag{20-42b}$$

where $u_{1,2}$ are the values of u measured with TE_{10}, TE_{20} excitation, respectively, and

$$\beta_{01} = \sqrt{k_0^2 - (\pi/a)^2}, \qquad \beta_{02} = \sqrt{k_0^2 - 4(\pi/a)^2}. \tag{20-43}$$

Equating the ratio of Eqs. 20-41 to the ratio of Eqs. 20-42 and solving for the product $\epsilon_2\mu_3$, we obtain

$$\epsilon_2\mu_3 = \left(\frac{\pi}{k_0 a}\right)^2 \frac{1 - 4(\beta_{01}\mu_1/\beta_{02}\mu_2)^2}{1 - (\beta_{01}\mu_1/\beta_{02}\mu_2)^2}. \tag{20-44}$$

Equating Eq. 20-41a to Eq. 20-42a and solving,

$$\mu_1^3/\mu_3^5 = \beta_{01}^2 u_1^2/[k_0^2\epsilon_2\mu_3 - (\pi/a)^2]. \tag{20-45}$$

If the sample is sufficiently thin, $\beta_1 L < 2\pi$. Let us assume that this is the case; then $\beta_1 = \tilde{\beta}_1$ and from Eq. 20-41:

$$\mu_1/\mu_3 = \tilde{\beta}_1^2/[k_0^2\epsilon_2\mu_3 - (\pi/a)^2]. \tag{20-46}$$

Now μ_1 and μ_3 are obtained at once from Eqs. 20-45 and 20-46 and ϵ_2 from Eq. 20-44. Similar measurements with a rotated sample produce ϵ_1, ϵ_3, and μ_2.

Of course, the method described in this section is also applicable to the simpler case of lossless nondispersive material considered in Section 20-3.

20-5 MEASUREMENT PROCEDURE

A waveguide that is convenient for exciting TE_{10} and TE_{20} modes has been suggested by Sucher and Fox (1963) and is shown in a form adaptable for use with modern network analyzers in Fig. 20-1. The TE_{10} mode is excited by the horizontal arms that are standard waveguides beyond the tapered transitions. The TE_{20} mode is excited by the branch arms that are also standard waveguides. Both modes can propagate in the central part of the fixture which is of double width. Beyond the transition region the horizontal arms are beyond cutoff for TE_{20} modes. With reasonable precision in fabrication of the fixture, TE_{10} modes will not be coupled into the branch arms. Hence, a circuit through the branch arms can be used to measure reflection and transmission coefficients due to the TE_{20} mode, while a circuit through the side arms can be used to measure reflection and transmission coefficients for the TE_{10} mode. Clearly, impedance matching will be required at the branch arms and at the tapered transitions.

Equations 20-27 and 20-41–20-46 require measurements of four complex

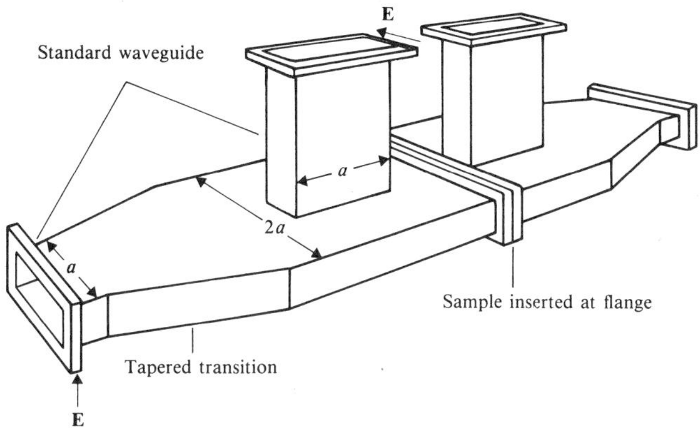

Figure 20-1 Sketch of TE_{10}–TE_{20} measurement fixture.

quantities to determine three unknown complex components of the constitutive parameters from each orientation of the sample. The additional measured quantity greatly simplifies the mathematical steps that are required to solve the inverse problem. In addition, if three orientations of the sample are used to determine the six unknowns, the measurements provide ample redundancies to evaluate immediately the quality of the measurement.

20-6 ADDITIONAL COMMENT

In the analysis above, the only point that requires special care is the determination of the integer n in Eq. 20-35. However, this value is easily found for nonmagnetic ($\mu_1 = \mu_2 = \mu_3 = 1$) materials; from Eq. 20-24, $\beta'_m = \beta_{0m}\,\mathrm{Re}(u)$, and thus n is found from Eq. 20-35.

Rather than measuring reflection and transmission coefficients in a guide with a matched load, we may inquire as to whether measuring the reflection coefficient $\tilde{R}_m$ for a sample backed against a short is sufficient to solve the inverse problem. The answer is negative, as we may see in the simple case of a lossless sample with a metal wall at $z = L$. Then the modulus of $\tilde{R}_m = 1$ and so only the phase of $\tilde{R}_m$ carries useful information. We find that

$$\frac{\mu_1 \beta_m L}{\mu_3^2 \tan(\beta_m L)} = \beta_{0m} L \tan\left[\frac{1}{2}\arg(\tilde{R}_m)\right]. \tag{20-47}$$

Even in the particular case $\mu_1 = \mu_3 = 1$, the transcendental Eq. 20-47 has an infinite number of solutions for β_m, each leading to a different value for ϵ_2 via Eq. 20-16.

To summarize: We have derived an analytical solution to the inverse problem of determining the constitutive parameters of a biaxial material, which may be lossy and dispersive, from measurements of reflection and transmission coefficients in a rectangular waveguide containing a sample of the material. We have also suggested an experimental setup with which to obtain the needed measurements.

REFERENCES

Abramowitz, M. and I. A. Stegun (eds.) (1964). *Handbook of Mathematical Functions,* National Bureau of Standards, Washington, D.C.

ALBIS PROGRAM (1966). *Ballistic Missile Radar Cross-Section Measurements,* The Johns Hopkins University Applied Physics Laboratory, Laurel, Maryland.

Alexopoulos, N. G. and P. L. E. Uslenghi (1981). "Reflection and transmission for materials with arbitrarily graded parameters," *J. Opt. Soc. Am.,* **71**(12), pp. 1508–1512.

Ali S. M. and S. F. Mahmoud (1979). "Electromagnetic fields of buried sources in stratified anisotropic media," *IEEE Trans. Ant. and Prop.,* **AP-27**(5), pp. 671–678.

Andreason, M. G. (1965). "Scattering from parallel metallic cylinders with arbitrary cross-sections," *IEEE Trans. Ant. and Prop.,* **AP-12,** pp. 746–754.

Bachman, C. G. (1982). *Radar Targets,* Lexington Books, Toronto.

Baker, B. B. and E. T. Copson (1953). *The Mathematical Theory of Huygens' Principle,* Third Edition, Oxford Univ. Press, London.

Barzilai, G. and G. Gerosa (1963). "A modal solution for a rectangular guide loaded with longitudinally magnetized ferrite." In *Electromagnetic Theory and Antennas,* E. C. Jordan (ed.), Pergamon, New York.

Bechtel, M. E. (1962). "Scattering coefficients for the backscattering of electromagnetic waves from perfectly conducting spheres," Report AP/R1S-1, Cornell Aeronautical Laboratory, Ithaca, New York.

Blacksmith, P., Jr., R. E. Hiatt and R. B. Mack (1965). "Introduction to Radar Cross Section Measurements," *Proc. IEEE,* **53**(8), pp. 901–920.

Booker, H. G. and P. C. Clemmow (1950). "The concept of an angular spectrum of plane waves, and its relation to that of polar diagram and aperture distribution" *J. Inst. Elec. Engrs.* (London), **97,** pp. 11–17.

Born, M. and Emil Wolf (1959). *Principles of Optics,* Pergamon, New York.

Bowman, J. J., T. B. A. Senior and P. L. E. Uslenghi (1969). *Electromagnetic and Acoustic Scattering by Simple Shapes,* North Holland, Amsterdam.

Bresler, A. D. and N. Marcuvitz (1956a). "Operator methods in electromagnetic field theory," Report R495-56-PIB425, 88 pp., Microwave Research Institute, Polytechnic Institute of Brooklyn, New York.

Bresler, A. D. and N. Marcuvitz (1956b). "Operator methods in electromagnetic field theory," Report R565-57-PIB493, 118 pp., Microwave Research Institute, Polytechnic Institute of Brooklyn, New York.

Bresler, A. D. (1959). "Vector formulations for the field equations in anisotropic waveguides," *IRE. Trans. Microwave Theory and Tech.,* **MTT-7,** p. 298.

Carslaw, H. S. (1930). *Introduction to the Theory of Fourier's Series and Integrals,* Dover, New York.

Chetaev, D. N. (1963). "On the field of a low-frequency electric dipole situated on the surface of a uniform anisotropic conducting halfspace," *Soviet Physics–Technical Physics,* **7**(11), pp. 991–995.

Chhikara, Raj S. and J. Leroy Folks (1974). "Estimation of the Inverse Gaussian Distribution Function," *J. Am. Stat. Assoc.,* **69**(345), pp. 250–254.

Churchill, Ruel V. and James Ward Brown (1984). *Complex Variables and Applications,* Third Edition, McGraw–Hill, New York.

Clemmow, P. C. (1966). *The Plane Wave Spectrum Representation of Electromagnetic Fields,* Pergamon, London.

Collin, R. E. (1960). *Field Theory of Guided Waves,* McGraw–Hill, New York.

Conover, W. J. (1980). *Practical Nonparametric Statistics,* Wiley, New York.

Corona, Paolo, Guiseppe Ferrara, and Claudio Gennarelli (1987). "Backscattering by loaded and unloaded dihedral corners," *IEEE Trans. Ant. and Prop.,* **AP-35**(10), pp. 1148–1153.

Crispin, J. W., Jr. and A. L. Maffett (1965a). "Radar cross-section estimation for simple shapes," *Proc. IEEE,* **53**(8), p. 833.

Crispin, J. W., Jr. and A. L. Maffett (1965b). "Radar cross-section estimation for complex shapes," ibid., p. 972.

Crispin, J. W., Jr. and K. M. Siegel (eds.) (1968). *Methods of Radar Cross Section Analysis,* Academic, New York.

Currie, Nicholas C. (ed.) (1984). *Techniques of Radar Reflectivity Measurement,* Artech House, Dedham.

Damaskos, N. J., A. L. Maffett and P. L. E. Uslenghi (1982). "Dispersion relation for general anisotropic media," *IEEE Trans. Ant. and Prop.,* **AP-30**(5), pp. 991–993.

Damaskos, N. J., R. B. Mack, A. L. Maffett, W. Parmon, and P. L. E. Uslenghi (1984). "The inverse problem for biaxial materials," *IEEE Trans. Microwave Theory and Techn.,* **MTT-32**(4), pp. 400–405.

Damaskos, N. J., A. L. Maffett, and P. L. E. Uslenghi (1985). "Reflection and transmission for gyroelectromagnetic biaxial layered media," *J. Opt. Soc. Am.,* **A2,** pp. 454–461.

Daniele, V. G. and R. S. Zich (1973). "Radiation by arbitrary sources in anisotropic stratified media," *Radio Sci.,* **8**(1), pp. 63–70.

Dixon, Wilfrid J. and Frank J. Massey, Jr. (1969). *Introduction to Statistical Analysis,* Third Edition, McGraw–Hill, New York.

Elderton, W. P. (1938). *Frequency Curves and Correlation,* Cambridge University Press, London.

Engineer, M. H. and B. R. Nag (1965). "Propagation of electromagnetic waves in rectangular guides filled with a semiconductor in the presence of a transverse magnetic field," *IEEE Trans. Microwave Theory and Techniques,* **MTT-13,** pp. 641–646.

Erdelyi, A., W. Magnus, F. Oberhettinger, and F. G. Tricomi (eds.) (1955). *Higher Transcendental Functions,* Vols. 1, 2, 3, McGraw–Hill, New York.

Fenlon, F. H. (1969). "Calculation of the acoustic radiation field at the surface of a finite cylinder by the method of weighted residuals," *Proc. IEEE,* **57,** pp. 291–306.

Fisher, R. A. (1925). "Applications of Student's distribution," *Metron,* **5**(3), p. 90.

Fock, V. A. (1957). *Diffraction, Refraction, and Reflection of Radio Waves,* Thirteen Papers, N. A. Logan and P. Blacksmith, Jr. (eds.), AFCRC-TN-57-102, ASTIA Doc. No. AD 117 276.

Follin, J. W., Jr. and A. L. Maffett (1981). "RCS scintillations and their statistical description," Proc., Second Annual Tactical Air Surveillance and Control Conference, Rome Air Development Center, Griffiss AFB, New York.

Follin, J. W., Jr., F. C. Paddison, and A. L. Maffett (1984). "RCS scintillation and glint statistics," *Electromagnetics,* **4**(2–3), pp. 139–164.

Fulks, Watson (1969). *Advanced Calculus,* Wiley, New York.

Goncharenko, A. M. (1966). "Electromagnetic properties of an anisotropically-loaded rectangular waveguide," *Soviet Phys. Tech. Phys.,* **10,** pp. 1121–1122.

Guinard, N. W. and J. C. Daly (1970). "An experimental study of a sea clutter model," *Proc. IEEE,* **58**(4), pp. 543–550.

Graglia, R. D. and P. L. E. Uslenghi (1987). "Anisotropic layered absorbers on cylindrical structures," *Electromagnetics,* **7**(2), pp. 117–127.

Hardy, G. H., J. E. Littlewood, and G. Polya (1959). *Inequalities,* Cambridge Univ. Press, London.

Harrington, R. F. (1968). *Field Computation by Moment Methods,* Macmillan, New York.

Hlawiczka, P. (1981). *Gyrotropic Waveguides,* Academic, New York.

Howard, Dean D. (1959). "Radar target angular scintillation in tracking and guidance systems based on echo signal phase front distortion," *Proc. National Electronics Conference,* **15,** pp. 840–849.

James, G. L. (1976). *Geometrical Theory of Diffraction for Electromagnetic Waves,* Peter Peregrinus, Ltd., Herts.

Johnson, Richard N. (1986). "Radar-absorbing material: a passive role in an active scenario," *The International Countermeasures Handbook,* Eleventh Edition, EW Communications, Inc., New York.

Jones, D. S. (1964). *The Theory of Electromagnetism,* Pergamon, New York.

Joy, Edward, B., W. Marshall Leach, Jr., George P. Rodrigue, and Demetrius T. Paris (1978). "Applications of probe-compensated near-field measurements," *IEEE Trans. Ant. and Prop.,* **AP-26**(3), pp. 379–389.

Kales, M. L. (1953). "Modes in wave guides containing ferrites," *J. Appl. Phys.*, **24,** pp. 604–608.

Kell, Robert E. (1965). "On the derivation of bistatic RCS from monostatic measurements," *Proc. IEEE,* **53**(8), pp. 983–988.

Keller, J. B. (1962). "Geometrical theory of diffraction," *J. Opt. Soc. Am.*, **52,** p. 116.

Keller, J. B. and B. R. Levy (1959). "Decay exponents and diffraction coefficients for surface waves on surfaces of nonconstant curvature," *IRE. Trans. Ant. and Prop.*, **AP-7,** pp. S68–S70. See also Levy and Keller (1959) below.

Kerker, Milton (1975). "Invisible bodies," *J. Opt. Soc. Am.*, **65**(4), pp. 376–379.

Kerr, Donald, E. (ed.) (1951). *Propagation of Short Radio Waves,* McGraw–Hill, New York.

Kleinman, R. E. (1965). "The Rayleigh region," *Proc. IEEE,* **53**(8), pp. 848–855.

Kleinman, R. E. and T. B. A. Senior (1972). "Rayleigh scattering cross sections," *Radio Sci.*, **7,** pp. 937–942.

Knott, E. F. and T. B. A. Senior (1974a). "Comparison of three high frequency diffraction techniques," *Proc. IEEE,* **62**(11), pp. 1468–1474.

Knott, E. F. and T. B. A. Senior (1974b). "Nonspecular radar cross section study," Technical Report AFAL-TR-73-422, Wright–Patterson AFB, Ohio.

Knott, E. F. and T. B. A. Senior (1974c). A program incorporating diffraction for the computation of radar cross sections, University of Michigan Radiation Laboratory report 011758-2-T, Ann Arbor, Michigan.

Knott, E. F. (1985a). "The relationship between Mitzner's ILDC and Michaeli's equivalent currents," *IEEE Trans. Ant. and Prop.*, **AP-33**(1), pp. 112–114.

Knott, E. F. (1985b). "A progression of high-frequency RCS prediction techniques," *Proc. IEEE,* **73**(2), pp. 252–264.

Knott, E. F., J. F. Shaeffer, and M. T. Tuley (1986). *Radar Cross Section,* Artech, Norwood.

Kong, J. A. and D. K. Cheng (1968). "On guided waves in moving anisotropic media," *IEEE Trans. Microwave Theory and Tech.*, **MTT-16,** pp. 99–103.

Kong, J. A. (1972). "Electromagnetic fields due to dipole antennas over stratified anisotropic media," *Geophys.*, **37**(6), pp. 985–996.

Kong, J. A. (1975). *Theory of Electromagnetic Waves,* Wiley, New York.

Kouyoumjian, R. G. and L. Peters, Jr. (1965). "Range requirements in radar cross section measurements," *Proc. IEEE,* **53**(8), pp. 920–928.

Krueger, C. H., Jr. (1967). "A computer program for determining the reflection and transmission properties of multilayer plane impedance boundaries," Report AFAL-TR-67-191, 27 pp., Wright–Patterson AFB, Ohio.

Kulp, R. L. (1984). "Detection probabilities for β-distributed scattering cross sections," *Electromagnetics,* **4**(2–3), pp. 165–183.

Lax, B. and K. J. Button (1962). *Microwave Ferrites and Ferrimagnetics,* McGraw–Hill, New York.

Levy, B. R. and J. B. Keller (1959). "Diffraction by a smooth object," *Commun. Pure and Appl. Math.*, **12,** pp. 159–209.

Liepa, Valdis V., Eugene F. Knott, and Thomas B. A. Senior (1974). "Scattering from two-dimensional bodies with absorber sheets," Technical report AFAL-TR-74-119 (University of Michigan Radiation Laboratory), Wright–Patterson AFB, Ohio.

Lighthill, M. J. (1959). *Fourier Analysis and Generalized Functions,* Cambridge Univ. Press, London.

Lyzenga, D. R., A. L. Maffett, and R. A. Shuchman (1983). "The contribution of wedge scattering to the radar cross section of the ocean surface," *IEEE Trans. Geosci. and Remote Sensing,* **GE-21**(4), pp. 502–505.

Mack, R. B. (1981). "Basic design principles of electromagnetic scattering measurement facilities," Rome Air Development Center Report RADC-TR-81-40, Griffiss AFB, New York.

Magnus, W. and F. Oberhettinger (1954). *Formulas and Theorems for the Functions of Mathematical Physics,* Chelsea, New York.

Marcum, J. I. (1960). "A statistical theory of target detection by a pulsed radar," *IRE. Trans. Info. Theory,* **IT-6,** pp. 59–144.

Marcum, J. I. and P. Swerling (1960). "Studies of target detection by pulse radar," *IRE. Trans. Info. Theory,* **IT-6**(2), pp. 145–267.

Martynova, T. A. (1973). "Propagation of electromagnetic waves in a waveguide with local layered gyrotropic filling," *Radio Eng. and Electr. Phys.,* **18**(9), pp. 1320–1323.

Maue, A. W. (1949). "Zur Formulierung eines Allgemeinen Beugungsproblems durch eine Integralgleichung," *Zeitschrift für Physik,* **126,** pp. 601–618.

Mautz, J. R. and R. F. Harrington (1969). "Radiation and scattering from bodies of revolution," *Appl. Sci. Res.,* **20,** pp. 405–434.

Meade, J. E. (1955). "Target considerations," in *Guidance,* by A. S. Locke et al., Van Nostrand, Princeton, pp. 420–442.

Medgyesi-Mitschang, Louis N. and John M. Putnam (1987). "Electromagnetic scattering from electrically large coated flat and curved strips: Entire domain Galerkin formulation," *IEEE Trans. Ant. and Prop.,* **AP-35**(7), pp. 790–801.

Mentzer, J. R. (1955). *Scattering and Diffraction of Radio Waves,* Pergamon, London.

Michaeli, Arie (1984). "Equivalent edge currents for arbitrary aspects of observation," *IEEE Trans. Ant. and Prop.,* **AP-32,** pp. 252–258. (Correction: (1985) **AP-33,** p. 227.).

Michaeli, Arie (1986a). "Elimination of infinities in equivalent edge currents," Part I, *IEEE Trans. Ant. and Prop.,* **AP-34,** pp. 912–918.

Michaeli, Arie (1986b). "Elimination of infinities in equivalent edge currents," Part II, *IEEE Trans. Ant. and Prop.,* **AP-34**.

Michaeli, Arie (1987). "Equivalent currents for second-order diffraction by the edges of perfectly conducting polygonal surfaces," *IEEE Trans. Ant. and Prop.,* **AP-35**(2), pp. 183–190.

Mieras, H., J. N. Buchman, R. M. Barnes, C. L. Bennett, G. M. Vachula and Wolfgand M. Boerner (1983). "Polarization null characteristics of simple targets," Report RADC-TR-82-335, Rome Air Development Center, Griffiss AFB, New York.

Millar, R. F. (1956). "An approximate theory of diffraction of an electromagnetic wave by an aperture in a plane screen," *Proc. Inst. Elec. Eng.*, **103** (part C), pp. 177–185.

Millar, R. F. (1957). "The diffraction of an electromagnetic wave by a circular aperture," *Proc. Inst. Elec. Eng.*, **104** (part C), pp. 87–95.

Mittra, R. (ed.) (1973). *Computer Techniques for Electromagnetics,* Pergamon, London (see chapter by A. J. Poggio and E. K. Miller).

Mittra, R. (ed.) (1975). *Numerical and Asymptotic Techniques in Electromagnetics,* Springer–Verlag, Berlin (see chapters by Harrington and by Kouyoumjian).

Mitzner, K. L. (1974). "Incremental length diffraction coefficients," Northrup Corp., Aircraft Division Technical Report AFAL-TR-73-296, Wright–Patterson AFB, Ohio.

Morse, P. M. and H. Feshbach (1953). *Methods of Theoretical Physics,* Vols. I and II, McGraw–Hill, New York.

Negi, J. G. and P. D. Saraf (1973). "Effects of anisotropy on quasistatic fields of a vertical electric dipole buried in a layered anisotropic earth," *Radio Sci.*, **8**(2), pp. 155–165.

Novak, L. N. and F. W. Vote (1979). "Millimeter airborne radar target detection and selection techniques," *IEEE NAECON-79 Conf. Record,* **2**. pp. 807–817.

Oshiro, F. K. (1965). "Source distribution techniques for the solution of general electromagnetic scattering problems," *First GISAT Symposium,* Mitre Corp., **1,** pp. 83–107.

Oshiro, F. K. and K. M. Mitzner (1967). "Digital computer solution of three-dimensional scattering problems," *Symposium Digest, Internat. Ant. and Prop. Symp.*, Ann Arbor, Michigan, pp. 257–263.

Paddison, F. C., J. W. Follin, Jr., F. Mitchell, and A. L. Maffett (1985a). "Total scattering from a long cylinder," *Proc. First Ann. Rev. Numerical Electromagnetic Code Applic.* Lawrence Livermore Laboratory, Livermore, CA.

Paddison, F. C., J. W. Follin, Jr., F. Mitchell, and A. L. Maffett (1985b). "Large bistatic angle radar cross section of a right circular cylinder," *Electromagnetics,* **5,** pp. 63–77.

Paddison, F. C., C. A. Shipley, A. L. Maffett, and M. H. Dawson (1978). "Radar cross section of ships," *IEEE Trans. Aerospace and Electronic Sys.*, **AES-14**(1), pp. 27–34.

Panofsky, Wolfgang K. H. and Melba Phillips (1956). *Classical Electricity and Magnetism,* Addison–Wesley, Reading, MA.

Paul, D. K. and R. K. Shevgaonkar (1981). "Multimode propagation in anisotropic optical waveguides," *Radio Sci.*, **16,** pp. 525–533.

Peake, W. H. and D. E. Barrick (1967). "Scattering from surfaces with different roughness scales: Analysis and interpretation," Report No. 1388-26, The Ohio State University ElectroScience Laboratory, Columbus.

Pearson, K. (1902). "On the systematic fitting of curves to observations and measurements, *Biometrika,* **1,** p. 265.

Peters, L., Jr. (1958). "End-fire echo area of long thin bodies," *IRE. Trans. Ant. and Prop.*, **AP-6,** p. 133.

Pollon, G. E. (1970). "Statistical parameters for scattering from randomly oriented arrays, cylinders, and plates," *IEEE Trans. Ant. and Prop.*, **AP-18**(1), pp. 68–75.

Ramo, S., J. R. Whinnery, and T. Van Duzer (1965). *Fields and Waves in Communication Electronics,* Wiley, New York.

Rheinstein, John (1965). "Scattering of short pulses of electromagnetic waves," *Proc. IEEE,* **53**(8), pp. 1069–1070.

Rice, S. O. (1945). "Mathematical analysis of random noise," *Bell Sys. Tech. J.*, **24,** p. 100. (See also: Wax, Nelson (1954). *Selected Papers on Noise and Stochastic Processes,* Dover, New York).

Richmond, Jack H. (1966). "A wire-grid model for scattering by conducting bodies," *IEEE Trans. Ant. and Prop.*, **AP-14,** pp. 782–786.

Richmond, Jack H., Leon Peters, and Roger A. Hill (1987). "Surface waves on a lossy planar ferrite slab," *IEEE Trans. Ant. and Prop.*, **AP-35**(7), pp. 802–808.

Ridenour, Louis N. (ed.) (1947). *Radar System Engineering,* McGraw–Hill, New York.

Ryan, C. E. and L. Peters, Jr. (1970). "Evaluation of edge-diffracted fields including equivalent currents for the caustic regions," *IEEE Trans. Ant. and Prop.*, **AP-7,** pp. 292–299. Correction: **AP–18,** p. 275.

Schwartz, L. (1961). *Methodes mathematiques pour les Sciences Physiques,* Hermann, 115 boulevard Saint-Germain, Paris VI.

Semenenko, A. I. and F. S. Mironov (1976). "Matrix method in the optics of anisotropic layered media," *Opt. Spectrosc.*, **41**(3), pp. 262–265.

Senior, T. B. A. (1962). "A note on impedance boundary conditions," *Can. J. Phys.*, **40,** pp. 663–665.

Senior, T. B. A. (1965). "A survey of analytical techniques for cross-section estimation," *Proc. IEEE* **53**(8), pp. 822–833.

Senior, T. B. A. (1976). "Low-frequency scattering by a dielectric body," *Radio Sci.*, **11,** pp. 477–482.

Siegel, K. M., H. A. Alperin, R. R. Bonkowski, J. W. Crispin, Jr., A. L. Maffett, I. V. Schensted, and C. E. Schensted (1955). "Bistatic radar cross sections of surfaces of revolution," *J. Appl. Phys.*, **26**(3), pp. 297–305.

Siegel, K. M. (1958). "Far field scattering from bodies of revolution," *Appl. Sci. Res.*, **7**(4), pp. 293–328.

Siegel, K. M. (1963). "Low frequency radar cross section computations," *Proc. IEEE,* **51,** pp. 232–233.

Shen, L. C. and J. A. Kong (1983). *Applied Electromagnetics,* Wadsworth (Brooks/Cole), Monterey, CA.

Shotland, E. (1967). "Statistical analysis of radar target scintillation," Report TG 955, The Johns Hopkins University Applied Physics Laboratory, Laurel, Maryland.

Stratton, J. A. (1941). *Electromagnetic Theory,* McGraw–Hill, New York.

Sucher, M. and J. Fox (eds.) (1963). *The Handbook of Microwave Measurements,* Vol. 1, Polytechnic, New York, p. 327.

Swerling, P. (1960). "Probability of detection for fluctuating targets," *IRE. Trans. Info. Theory,* **IT-6,** pp. 268–308.

Swerling, P. (1970). "Recent developments in target models for radar detection analysis," *AGARD Conf. Proc.*, (66).

Tai, C. T. (1971). *Dyadic Green's Functions in Electromagnetic Theory,* Intext, San Francisco.

Tang, C. M. (1979). "Electromagnetic fields due to dipole antennas embedded in stratified anisotropic media," *IEEE Trans. Ant. and Prop.*, **AP-27**(5), pp. 665–670.

Tikhonov, A. W. (1959). "The propagation of a continuous electromagnetic wave in a laminarly anisotropic medium," *Soviet Phys.-Doklady,* **4**(3), pp. 566–570.

Tweedie, M. C. K. (1957). "Statistical properties of inverse gaussian distributions, I and II," *Ann. Math. Stat.*, **28,** pp. 362–377 and pp. 696–705.

Überall, H. (1964). "Radar scattering from coated perfect conductors: Application to the semi-infinite cone and use of exact eikonal," *Radio Sci. J. Res.*, **68D**(6).

Ufimtsev, Peter Ya. (1962). "The method of fringe waves in the theory of diffraction," *Sovyetskoye,* Moscow. Translation by U.S. Air Force Foreign Technology Division, Wright–Patterson AFB, OH with release for public distribution 1971.

Ufimtsev, Peter Ya. (1975). "Comments on 'Comparison of three high-frequency diffraction techniques," *Proc. IEEE,* **63,** pp. 1734–1737.

Ulaby, Fawwaz T., Richard K. Moore, and Adrian K. Fung (1982). *Microwave Remote Sensing,* Vol. II, Addison–Wesley, Reading.

Uslenghi, P. L. E. (ed.) (1978). *Electromagnetic Scattering,* Academic Press, New York.

van der Pol, B. and H. Bremmer (1950). *Operational Calculus,* Cambridge Univ. Press, London.

van Zyl, Jakob J., Howard A. Zebker, and Charles Elachi (1987). "Imaging radar polarization signatures: Theory and observation," *Radio Sci.*, **22**(4), pp. 529–543.

Visnovsky, S. and R. Krishnan (1981). "Complex Faraday effect in multilayer structures," *J. Opt. Soc. Am.*, **71**(3), pp. 315–322.

Wait, J. R. (1966). "Electromagnetic fields of a dipole over an anisotropic half-space," *Can. J. Phys.*, **44,** pp. 2387–2401.

Weston, V. H. (1963). "Theory of absorbers in scattering," *IEEE Trans. Ant. and Prop.*, **AP-11,** pp. 578–584.

Whalen, A. D. (1971). *Detection of Signals in Noise,* Academic, New York.

Whittaker, E. T. and G. N. Watson (1958). *A Course of Modern Analysis,* Cambridge Univ. Press, London.

Wright, James W. (1984). "Glint—A survey," *Electromagnetics,* **4**(2–3), pp. 205–228.

Young, J. D. and E. K. Walton (1984). "The Ohio State University compact cross-section measurement range," *IEEE Trans. Ant. and Prop.*, **AP-32**(11).

Zebker, Howard A., Jakob J. van Zyl, and Daniel N. Held (1987). "Imaging radar polarimetry from wave synthesis," *J. Geophys. Res.*, **92**(B1), pp. 683–701.

AUTHOR INDEX

SUBJECT INDEX